Exploration and Engineering

NEW SERIES IN NASA HISTORY

Exploration and Engineering

The Jet Propulsion Laboratory and the Quest for Mars

ERIK M. CONWAY

Johns Hopkins University Press
Baltimore

© 2015 Johns Hopkins University Press
All rights reserved. Published 2015
Printed in the United States of America on acid-free paper

Johns Hopkins Paperback edition, 2016
2 4 6 8 9 7 5 3 1

Johns Hopkins University Press
2715 North Charles Street
Baltimore, Maryland 21218-4363
www.press.jhu.edu

The Library of Congress has cataloged the hardcover edition of this book as follows:

Conway, Erik M., 1965–
Exploration and engineering : the Jet Propulsion Laboratory and the quest for Mars / Erik M. Conway.
pages cm
Includes bibliographical references and index.
ISBN 978-1-4214-1604-5 (hardcover : alk. paper) — ISBN 978-1-4214-1605-2 (electronic) — ISBN 1-4214-1604-2 (hardcover : alk. paper) — ISBN 1-4214-1605-0 (electronic) 1. Space flight to Mars. 2. Mars (Planet)—Exploration. 3. Jet Propulsion Laboratory (U.S.) I. Title.
TL799.M3C66 2015
523.43072'3—dc23
2014018368

A catalog record for this book is available from the British Library.

ISBN-13: 978-1-4214-2122-3
ISBN-10: 1-4214-2122-4

Special discounts are available for bulk purchases of this book. For more information, please contact Special Sales at 410-516-6936 or specialsales@press.jhu.edu.

Johns Hopkins University Press uses environmentally friendly book materials, including recycled text paper that is composed of at least 30 percent post-consumer waste, whenever possible.

CONTENTS

ACKNOWLEDGMENTS

I began this work about a year after moving to the Jet Propulsion Laboratory from the NASA Langley Research Center, where I had been a contract historian. JPL had not had a staff historian since the mid-1970s, and during 2004, Blaine Baggett, director of the Office of Communication and Education at JPL, and Charles Elachi, JPL director, decided to hire me. I have them to thank for supporting this work. I also need to thank Margo Young and Robert Powers, former and current heads of JPL's library and archives section, for their assistance, as well as Julie Cooper, JPL's archivist; Mary Beshid, Susan Hendrickson, and Kay Schardein, its records managers; and project librarians Julie Reiz, Brent Shockley, and Suzanne Sinclair. Video librarian Sherri Rowe-Lopez provided rapid access to JPL's huge collection of audio-visual material. And thanks also to Mickey Honchell, JPL's interlibrary loan librarian, for handling myriad odd requests with alacrity.

This narrative has been woven from a variety of technical and scientific documents, from media accounts, and from extensive interviews with JPL engineers, managers, and scientists. The interviewees are too numerous to thank individually, but several of them read and critiqued all or part of the book. Thanks to Blaine Baggett, John Callas, Glenn Cunningham, Barry Goldstein, Matthew Golombek, Fuk Li, Robert M. Manning, Daniel J. McCleese, Brian Muirhead, George Pace, and Guy Webster for the extensive and valuable comments. All errors that remain are, of course, my own.

This is a JPL-centric history of Mars exploration, as the Laboratory has been the lead center (though the term has fallen from use) for Mars exploration since the late 1980s. But the Mars Observer, Mars Global Surveyor, Mars Climate Orbiter, Mars Polar Lander, Mars Reconnaissance Orbiter, and Mars Phoenix missions were carried out primarily by what are called "systems contractors" in the aerospace business. For all but Mars Observer, the systems contractor was Lockheed

Martin's Astronautics group in Colorado. LMA, as it is known colloquially, allowed me to interview several of its senior engineers and managers for this history, and thanks go to Edward Euler, Steven Jolly, Claude "Bud" McAnally, Parker Stafford, and Loren Zumwalt, for their time and interest in this project. I also gained great insight from an interview with Pioneer Aerospace's parachute expert, Allen Witkowsky, whom I met accidently one evening at Lucky Baldwin's in Pasadena.

At the NASA History Office in Washington, chief historians Steven Dick and Bill Barry, and their deputy, Steve Garber, supported this work through many twists and turns. Jane Odom, the NASA chief archivist, and her staff archivists Colin Fries, John Hargenrader, and Elizabeth Suckow, provided access to materials from early in this period. Former NASA chief historian and long-time friend Roger Launius supported this endeavor in myriad ways, as well.

My acquisitions editor at Johns Hopkins University Press, Bob Brugger, patiently awaited this manuscript, which took a couple years longer and several more chapters than originally planned, and handled its production with aplomb; and I want to thank the Press staff for preparing a beautiful volume.

Finally, several individuals at JPL gave me access to personal files that filled various-size holes in my research. JPL's current director, Charles Elachi, granted me access to some files from his tenure as head of the old Space and Earth Science Projects Directorate (SESPD); chief scientist Daniel McCleese provided records earlier in the 2000s; Samuel Thurman offered a large digital collection of his records from the Mars Climate Orbiter and Polar Lander projects; Robert M. Manning and Tommaso Rivellini lent records of the Mars Pathfinder, Mars Exploration Rover, and Mars Smart Lander projects as well as the "Bubble Team" studies of 2000; Mark Adler shared his original Mars Mobile Pathfinder proposal; and Donna Shirley, retired Mars program director, gave access to her records of those years. All of these documents are archived in an electronic repository at JPL known as the Historian's Mars Exploration Collection. This history would be far poorer without them.

While this book is the product of many years' research at JPL and was read and critiqued by many people there, it represents my opinions and interpretations alone, not those of JPL, the California Institute of Technology, or the National Aeronautics and Space Administration. It was released for public audiences under URS234014.

Exploration and Engineering

Introduction

Mars has a peculiar hold on the minds of Americans. A vast range of American science fiction posits martian civilizations. Scientific literature of the early twentieth century assumed martian civilizations, too, although as evidence accumulated during the century, native martians were reduced successively from ancient civilizations to fungus-like plantlife to microbes in the soil to, finally, maybe nanomicrobes under the surface.[1] Policy literature, echoing modern Mars fiction, suggests Mars as a future abode for humans.[2] Fascination with the Red Planet extends back at least into the nineteenth century and is clearly linked to the longstanding scientific claims that Mars must have life. It also seems strongly associated with American frontier mythology. Mars, like the American West, would be not only a future home but the salvation of civilization.[3]

Driven by science, fantasy, and the Cold War, the United States and former Soviet Union sent many robotic visitors to Mars. The earliest successful voyages, by the American robotic Mariners 4, 6 and 7, carried cameras and atmospheric remote sensing instruments.[4] These showed a very different Mars than had been expected by scientists of the 1950s. Mars had a heavily cratered, Moonlike surface with no apparent water, no visible signs of life, and barely a wisp of an atmosphere. This thin atmosphere was almost purely carbon dioxide, with only small amounts of nitrogen, argon, and water vapor. The small community of planetary astronomers had expected Mars to have an atmospheric composition close to Earth's. Finding that this wasn't true forced them to accept that Earth's atmosphere was a product of living processes—and, in a larger view, that Earth had been radically altered by biotic processing.[5] But Mars appeared sterile, and not much at all like Earth, despite being counted among the "Earthlike" planets of the solar system.

The first planetary orbiter, the Jet Propulsion Laboratory's Mariner 9, completed a photographic reconnaissance of the entire surface of Mars in 1971. It arrived in the middle of a planetary-scale dust storm and was reprogrammed to wait out the storm. The resulting photographs revealed a terrain of enormous diversity. In addition to the cratering, Mars had volcanoes far larger than anything on Earth and a canyon system whose length was larger than North America. The planet was also lopsided, with its southern hemisphere having a much higher average elevation than the northern hemisphere. And if crater counts were a reliable guide to age, the southern hemisphere's surface was much older than the northern. Mars was a very strange place.

Mars also had no visible water. But it did have terrain features that looked a lot like water-carved features on Earth—river valleys and the like. And it had water vapor in its atmosphere, although not much. Water vapor molecules would disintegrate over time from ultraviolet radiation bombardment (as Mars has no atmospheric oxygen, it also has no ozone layer), so there had to be water or water ice somewhere within reach of the atmosphere to replenish even that small amount. But nobody could find this ice; there was some evidence that the martian polar "ice" caps were mostly water ice, with a thin top layer of carbon dioxide ice ("dry ice"), but it wasn't widely accepted. Arguments raged over the whereabouts of Mars's water within the small community of Mars specialists.

The first scientifically successful Mars landers were the American Vikings. These orbiter-lander pairs arrived in 1976 to great fanfare. But instead of trying to resolve the central geological and climatological question of water, these probes were aimed at finding life.[6] The two landers dug up samples and subjected them to a set of complex chemical experiments, hoping to find the chemistry of life. While these results remain somewhat controversial to this day, the majority opinion of the involved scientists was that the landers did not find traces of life.[7]

Instead of beginning a great new era of Mars exploration, the Viking missions nearly signified an end. The failure to find life cooled political interest in the planet. And the decision by the National Aeronautics and Space Administration (NASA) to place the majority of its resources into development of the Space Shuttle program focused the agency on low Earth orbit. Planetary exploration experienced a funding crisis from 1977 through 1982, as funds were diverted to finance shuttle program overruns; in 1981, the NASA administrator threatened to eliminate planetary missions entirely and close NASA's principal center for planetary exploration, the Jet Propulsion Laboratory. The Reagan administration also sought deep cuts to NASA's science budget overall.[8]

As the United States was relinquishing its leadership in planetary science, the Soviet Union was forging ahead. It continued its long series of Venus missions and, shortly after the Viking missions, had initiated an ambitious Mars mission of its own. Called Fobos, it consisted of a pair of large orbiters carrying landers destined for the moon Phobos. These were to be launched in 1988 and would host instruments from western European and Scandinavian nations in addition to those from the USSR.[9]

New competitors in planetary exploration also arose. Between 1960 and 1980, planetary exploration had been the exclusive province of the United States and Soviet Union. Canada, Japan, and several European nations had flown Earth orbiting missions during that period but had not ventured to deep space. That was changing. A small, international flotilla of spacecraft was being prepared to greet Halley's comet in March 1986, conspicuously without U.S. participation. The European Space Agency was also building a solar mission called Ulysses that represented the surviving portion of the joint NASA/ESA International Solar Polar Mission and would mark Europe's first (albeit brief) exploration of Jupiter. Going forward, NASA would have to cope with competition from some of its past scientific partners.

Mars and Exploration on the Cheap

This was the context in which Mars scientists had to operate in during the 1970s and early 1980s. Scientists outside NASA began advocating for new missions, often very ambitious ones, while some inside began questioning the approach their agency took to planetary missions. They began to argue that lower-cost missions were both possible and desirable. In 1978, the National Academy of Sciences's Committee on Planetary and Lunar Exploration advocated for using commercially procured spacecraft designed as Earth orbiters for planetary missions to reduce costs. This was predicated on a perception that JPL-built spacecraft—the Mariner missions to Venus, Mars, and Mercury; the Viking orbiters; and the Voyager missions to the outer planets—were too expensive and that that the Laboratory's painfully acquired engineering rigor was unnecessary.[10] The reasoning was simply that JPL, as a university-operated nonprofit, was less focused on cost than a for-profit corporation would be, and thus commercialization would reduce costs. Within NASA's Office of Space Science and Applications, this idea was implemented in the Planetary Observer program, intended to be a series of low-cost planetary missions using a common, commercially procured, spacecraft "bus" (see chapter 1).

For JPL, the NASA drive for commercially built, low-cost planetary missions during the 1980s and 1990s was problematic. It threatened the survival of JPL's spacecraft engineering culture. The Caltech-operated facility had been transferred to NASA from the U.S. Army after successfully launching the first American spacecraft, Explorer 1, in 1958. The Laboratory was able to build nearly every part of a spacecraft itself; its Viking and Galileo orbiters were "in-house" projects, meaning they were designed, built, and tested by JPL engineers in Pasadena. JPL had been kept alive, and even growing, during the late 1970s nadir of planetary science by increasing its role in military research and development. But planetary exploration was the Laboratory's particular expertise. Its Mariner and Viking spacecraft had explored Mercury, Venus, and Mars, and its Voyager spacecraft had visited Jupiter and Saturn, with one scheduled to visit Saturn and Neptune late in the 1980s. Its Galileo orbiter was largely finished in 1984 for a 1986 launch, but after the destruction of the space shuttle *Challenger,* Galileo was put in storage.[11] The Laboratory's directors saw JPL as the vanguard of humanity's exploration of the cosmos. Military work was a sidelight and a source for technology development funding, not the core of the institution's identity. While JPL had once been a weapons laboratory, its people did not wish it to become one again just to survive.

JPL also managed spacecraft development by private firms via "systems contracts," a practice that dated back to the 1950s, when the Lab had developed ballistic missiles for the U.S. Army. For JPL, the systems contracts was a blessing and a curse. Each project management office at JPL employed about 20 people directly—trivial in an institution of more than 7,000 employees by the mid-1980s. So they were not large sources of employment. Nor did they contribute much to the Lab's technological knowledge, as these spacecraft would be integrated and tested elsewhere. Instead, the systems contracts kept the lab from getting too insular. JPL's way of designing and building spacecraft was not the only way, and the project teams were exposed to different methodologies. And, of course, contract management is itself a skill that the in-house projects needed. At JPL, like the rest of the aerospace industry, project teams had to decide whether to make each piece of a spacecraft in their own facilities or to buy it from a subcontractor, so even a "JPL-built spacecraft" had many commercially procured components and subsystems and, with those, contracts to manage.

The problem JPL leaders faced in the late 1980s was simply that there was little enthusiasm in NASA for allowing JPL to continue building spacecraft in house. The Committee on Planetary and Lunar Exploration's advocacy of the use of commercial Earth orbiters—the "Planetary Observer" idea—would eventually put JPL out of the spacecraft construction business altogether.

No one at JPL was terribly enthusiastic at becoming nothing more than a contract management organization (a "job shop") for NASA headquarters. Talented engineers generally want to design and build things, not watch other people do it. Further, most experienced engineering managers believed that one needed to have technical experience of one's own to be an effective overseer of technical contracts. So becoming nothing but a contract manager would inevitably lead to failure. How would JPL sustain its technical culture without new projects of its own? No one knew.

The Planetary Observer program itself would fail, but the idea of low-cost planetary missions conducted by private industry did not vanish. Instead, the 1990s saw a new cost-cutting effort known as "faster, better, cheaper."[12] Like the concept behind Planetary Observer, this idea revolved around the use of private contracting, though it was broader. Faster-better-cheaper advocates sought simplified scientific goals, reduced documentation, curtailed systems engineering, and less testing. This approach appeared to work for a few years before a series of failures, including two highly publicized Mars missions, caused NASA to abandon the experiment. The 2000s then witnessed the reimposition of engineering rigor, resulting in a return to higher costs.

This emphasis on cost played out on many levels in Mars exploration during the 1990s and 2000s. At the level of high programmatic strategy, it affected what mission concepts were and were not acceptable. For scientists, it offered opportunity and risk: one component of cost-reduction strategy was the adoption of "principal-investigator led missions," done to reduce mission complexity versus the tried-and-often-failed strategy of missions designed to satisfy whole communities of practitioners. (In later years, community missions would be called "flagship missions.") For project managers, the emphasis on cost biased decisions of all kinds, from requirements development through design to testing—especially testing, as it is somewhat scalable. One could test to only normal, or "nominal," conditions; test to encompass some defined range of conditions; or test to determine a design's "margins"—in other words, to figure out what will break it. Cost and risk are considered dependent variables in the aerospace literature, and it's this flexibility of testing that makes the linkage.

NASA, Planetary Science, and Human Exploration

During the period examined in this book, NASA witnessed several redirections of its overall goals, as well as of its management paradigms. During the 1980s, NASA pursued a permanent space station called Freedom, which was never built; in the first few years of the 1990s, it began a short-lived effort to send astronauts to Mars. For the rest of the decade, the agency's emphasis was on completing the International Space Station. Late in the decade, NASA administrator Dan Goldin began preparing for space station completion by seeking a new presidential initiative to send astronauts to Mars, but in 2004, the decision made by the White House was not for Mars expeditions. The new policy, the Vision for Space Exploration, focused on sending astronauts to the Moon, instead. The Vision for Space Exploration survived barely four years before being replaced, this time by an initiative to commercialize astronaut travel to the International Space Station coupled with development of a new "heavy lift" launch vehicle, with no explicit destination.

Each of these high-level changes affected Mars exploration. The decision to "return to the Moon" under the Vision for Space Exploration, for example, resulted in the creation of a new lunar science program, in part funded by reversing rapid growth in the Mars exploration budget. Mars specialists, in turn, suddenly saw declining resources when their plans had projected growth. This threw them into a morass of replanning that was still not complete as this book went to press.

This interaction between NASA's science program and its human exploration program is an important theme of this book. Outsiders often see the space agency as a monolith, and NASA officials often try to foster that appearance. (Readers unfamiliar with the internal organization of NASA may benefit from reading the appendix before diving into the main narrative. While this book is *not* about organizational change, a basic understanding of NASA's structure and evolution will be helpful.) But the agency is structured into relatively independent directorates. Space science is the purview of what's currently known as its Science Mission Directorate and, for most of the agency's past, as the Office of Space Science ("Code S," to hardcore insiders). Human spaceflight belongs to a directorate currently known as the Human Exploration and Operations Directorate, though it has a rather complex organizational past. These two directorates are coequal organizationally (though not financially). The science directorate relies on committees of scientists to generate its future programs and plans—usually, though not exclusively, via the National Academy of Sciences. The human exploration directorate receives its mission goals from the White House—in other words, though a process much more influenced by national politics. That's not to argue that science is apolitical; science has its own internal politics. Rather, it's rare that the White House dictates specific scientific missions, though it has happened. But the White House always approves (and disapproves) the goals of the human spaceflight program.

A consequence of this organization is that the science directorate is relatively autonomous. It's largely able to choose its own missions, with the White House Office of Management and Budget possessing veto authority, which it uses on occasion. But the science directorate is only rarely compelled to do what human mission planners want. It happened thrice during the period of this book, though. As we'll see, those moments of intersection between scientists' desires and those of human exploration advocates generated chaos. The two communities were not well aligned in their goals.

All Mars exploration to date has been carried out by robotic explorers, machines built on Earth and fired off to Mars in lieu of sending people there.[13] So far, no nation-state has been willing to spend the hundreds of billions of dollars or euros or rubles necessary to send people. We, and the support systems we need to survive the hostile space and martian environments, have been too expensive for our political systems to contemplate shipping off to the Red Planet with any seriousness. Instead, as we largely do for deep-ocean research, we have sent robotic agents in our stead.

In his book *Digital Apollo*, historian David Mindell examined the human role in carrying out the Apollo missions to the Moon.[14] He focused on the human-machine interface between the astronauts-as-pilots, and the vehicles they most directly flew, the Lunar Excursion Modules. When Apollo was established, NASA engineers began designing a fully automated spacecraft in which astronauts would be merely passengers, not pilots; in Mindell's narrative, the astronauts, who were nearly all test pilots, wielded their considerable public clout to gain influence over the design process. Like commercial airline pilots, they sought to retain control of the most crucial portion of flight, the actual Moon landing. That meant constraining the engineers' desire to automate, and instead crafting compromises that were engraved, or designed, into the hardware and software of flight. The design compromises permitted astronauts to be pilots, and each of the Apollo Moon landings was, in the end, "hands on," flown by the pilots. This wasn't strictly necessary, as Mindell also makes clear. JPL had sent a series of Surveyor robotic landers to the Moon; five of the seven Surveyor spacecraft landed successfully, too. One spun out of control due to a thruster failure; in the other, the solid-fuel retrorocket apparently exploded. Neither could be salvaged by the operations team in Pasadena.

Mindell raises the Surveyor program only as a counterpoint, to ensure he's not misread as arguing the astronauts were necessary to successful landings. He does not dig into the Surveyor program in detail, and the manned program remained his focus. But he does raise a challenge to other historians to illuminate the human role in robotic space exploration.[15]

As this book's narrative shows, the human role in creating and operating exploring machines is complex and multifaceted. Throughout, scientists, engineers, and managers at many levels struggled to develop programs for carrying out scientific exploration on Mars; to design, build, and test robotic agents for science; to innovate new automation techniques; to respond to demands from the human spaceflight part of NASA; and to do it all under sometimes-extreme financial pressure. They didn't always succeed.

The Plan of the Book

The Viking missions to Mars were followed by a 20-year hiatus in Mars exploration, partly intentional, and partly not. After 1996, though, NASA and JPL sent a nearly continuous stream of vehicles to Mars every 26 months, with the 2009 launch opportunity the first since 1994 to close without a launch. JPL either built or managed each one of this series of spacecraft, and its role in advocating, managing, designing, testing, and operating them is my primary subject. The book is therefore a view from a NASA center, not from headquarters, and it is not primarily about the evolution of Mars science, though because JPL engineers scientific spacecraft, the desires and involvement of scientists is woven throughout.

The book begins with the advocacy among scientists for new Mars missions after Viking and ends with the Mars Phoenix mission, which landed in the martian Arctic in 2008 and wraps up the book's faster-better-cheaper story. It treats successful and failed missions as symmetrically as possible, given the availability of documents and people for interviews, and it investigates what engineers took from one project to the next to illuminate the development of engineering knowledge. One will find never-flown missions discussed in this work, too, as they are often essential to understanding the origins of flown missions. I also interweave programmatic concerns at JPL and, as records allow, at NASA headquarters, in order to examine the interplay of politics, funding, science, and engineering.

CHAPTER ONE

Planetary Observers, Mars Observer

The Mars science community is spread across the country, within universities, at NASA centers, and in the U.S. Geological Survey. They are a subdivision of a larger community of planetary scientists that has always had larger dreams than NASA has had funds. The resulting cacophony of demands for missions and money produces confusion over priorities—and, of course, over who and what should be funded. So NASA gains scientific credibility, and, it hopes, prioritization, for its planetary missions from a National Academy of Sciences committee known as the Committee on Planetary and Lunar Exploration (COMPLEX).

In its first post-Viking mission report, published in 1978, this committee called for an expansive program of martian science, with a faint echo of Kennedy's speech to land a man on the Moon. The committee members called the detailed exploration of Mars's many environments "a worthy goal in its own right and [one that] should be accomplished within the next decade." These explorations, they recommended, should be carried out via remote sensing and in situ measurements and should chemically characterize the surface, define the general circulation of the atmosphere, examine whether or not Mars had a magnetic field, explore Mars's internal structure, and return samples to Earth so that they could be accurately dated. They sought the development of "surface mobility" (long-range roving laboratories) or multiple landers with experiment arms to examine different terrains, including the polar regions.[1]

But the committee explicitly rejected a continuing emphasis on the search for martian life. Drawing on the dominant view that Viking had not found significant indications of life on Mars, the panel recommended that the next series of missions focus on characterization of the planet's surface chemistry, mineralogy, and geology. This was necessary to establish the environmental context of whatever life might have evolved on Mars, and knowledge of the martian chemical environment would also aid in the interpretation of future life-seeking instruments' results.[2] Understanding this context would help resolve the ambiguities that had undermined the Viking experiments, while also satisfying planetary geologists' desires.

But the panel's ambitions were the polar opposite of political realities of late-1970s Washington. In 1980, after several fiscal years had passed with only the Galileo mission to Jupiter gaining approval, John Naugle, the associate administrator for space science, created the new Solar System Exploration Committee, chartered to "translate the scientific strategy developed by COMPLEX into a realistic, technically sound sequence of missions consistent with that strategy and with resources expected to be available for solar system exploration."[3] Geoffrey Briggs, a Mars scientist who had been an experimenter on the Mariner 9 and Viking missions and who served as executive secretary for the new committee during 1981–1983, recalls that the group's emphasis was on creating a program that was reasonable. "We felt we were under a lot of pressure to come back with a program that nobody could argue was not well thought out and attractive from every point of view, not just the science arguments as to why certain missions were important, but also the way we structured and managed the program and kept the efforts to make the program as affordable as possible."[4]

Many years later, Briggs recalled that the issue of "resources" was key. The declining number of planetary missions was having the pernicious effect of raising the cost of proposed new missions while also making it more difficult to retain the talented spacecraft engineers that future missions would depend on. The increasing cost meant NASA could afford even fewer missions, reinforcing the downward trend in mission approvals. Briggs thought reforming the way NASA managed its programs could result in significant cost savings. NASA typically rationed the number of "new starts" it asked for from Congress each year, leading to internecine struggles among its program managers in Washington for one of the coveted slots. Once a manager gained a new start for a project, it was largely in his or her interest to protect that project, regardless of its performance. There was little incentive for effective cost management; as costs soared on the approved project, the ability of other program officials to gain new starts of their own declined, but it did not affect the approved project. Nor did it really affect that program's manager, who, because of the rationing scheme, had had no hope of gaining another new start for several more years anyway.

The only exception to this rule in NASA was the Explorer program, which consisted of small, relatively simple, Earth-orbiting science missions. The Explorer program manger had a fixed annual budget and did not have to seek new start clearance from Congress or the White House for each mission. About every three years, the Explorer program issued an "announcement of opportunity" that invited mission proposals from scientists, and a formalized selection process took place to choose the winner. The Explorer program manager then had to manage all of his approved projects within the fixed program budget. This structure provided some incentive for effective cost management, as it was in the manager's interest to maintain a reasonably stable sequence of missions. (The scientists he served could be trusted to squawk loudly to Congress if the time between announcements grew too long.) Geoff Briggs wanted to make the Explorer program the model for a revamped planetary science program.

Briggs also asked the solar system committee to explore technological means for reducing mission costs. He invited A. Thomas Young, who had been Viking's mission manager, to chair a panel to study these possibilities. Young's team identified three factors that tended to drive mission costs: the "heritage" of a spacecraft, the scope or complexity of the mission, and "requirements creep," that is, the amount of change from the initial project concept allowed after its inception.

The heritage argument was a simple one. The greater commonality a given spacecraft had with its ancestors, the less inherent risk there was to the project. A spacecraft with high heritage would not need a large investment in systems engineering and new technology development, reducing the mission's cost and risk.[5] Of course, it also meant that the demand for high heritage served to restrict the potential missions to those that could be accomplished with extant technologies.

Mission complexity related primarily to instrumentation. One tendency the solar system committee had identified was the "last ship sailing" syndrome. Since the planetary mission rate had fallen so low, NASA tried to squeeze many instruments onto a single spacecraft. If there was going to be only one Mars mission a decade, every Mars scientist wanted to be part of it, and the spacecraft mass and electronic complexity soared as they were accommodated. In turn, the mission became unaffordable.[6]

The high-heritage argument for future low-cost missions led the solar system committee to split future missions into two large categories: ones that seemed amenable to currently available technologies and ones that needed new technologies. In the first category were missions involving planetary orbiters. The committee argued, based on analyses carried out by JPL and Ames Research Center and site visits to several manufacturers of satellites, that commercial Earth orbiters could be adapted to inner solar system missions, potentially lowering costs.

They recommended NASA adopt this approach in the near term, calling these conceptual missions Planetary Observers and embedding them within a "Core program" of low-cost exploration. Earth orbiters were not useable beyond Mars, however, due to differing power, thermal, and telecommunications conditions. So for outer planet missions, the committee recommended development of a "new modular spacecraft free of unnecessary complexity and suitable to adaptation with maximum inheritance."[7] This conceptual vehicle had been in the works at JPL, under the name Mariner Mark II. The committee's analysis of the Planetary Observer program suggested that it could be fit within a $60 million per year budget.[8]

Geoff Briggs recalls that the Core program was intended to look like a serious science effort that had been tightly focused on affordability. It was also boring. Just as it did not aim at new technological capabilities, it did not seek to entice the portion of the space advocacy community that sought grand displays of technological prowess. Three years later, the committee described a second category of missions in part two of its study: the Augmented program.

In the Augmented program, the committee placed all the planetary missions that required significant technological breakthroughs to accomplish: future landed missions, surface rovers, and sample return missions to Mars and comets. These missions could not be fit within NASA's current fiscal situation, which was problematic, as the lack of new technology would eventually leave the planetary science program moribund. Further, NASA had always pursued multiple goals. One of these was new technology itself; another was "enhancement of national pride and prestige."[9] The reliance on existing technology that was the centerpiece of the Planetary Observer concept would not serve these purposes. The Augmented program was designed to appeal to members of the Reagan administration who were known to be technophiles and fans of grand displays of technological mastery. In essence, it was a form of advocacy for technologically advanced missions.

At least during the 1980s, though, only the Core program gained much traction in Washington. One of the missions defined in it was the Mars Geoscience/Climatology Orbiter. The mission was to conduct global mapping of Mars's mineral composition and magnetic field from a low altitude polar orbit, to generate a global topographic map using a radar altimeter, and to examine the role of water in the martian climate.[10] The mission was actually a blend of two different conceptual Mars missions. One had been an orbiter aimed at determining Mars's surface mineralogy and topography. The other was an aeronomy and climate orbiter, designed to study the planet's atmosphere. Daniel J. McCleese, an atmospheric scientist who had come to JPL to study water vapor in the Mars atmosphere and wanted to be one of the experimenters on the aeronomy and climate mission, recalls that neither mission had sufficient support on the committee to be adopted on its own. The committee merged the two to increase support for the orbiter mission overall. The two-instrument aeronomy mission had been designed to be small, tightly focused, and inexpensive to operate (like a previous JPL mission, the Solar Mesosphere Explorer, it would have been operated by students at the University of Colorado).[11] Merging the two made the mission more expensive and complex. The resulting notional spacecraft had six instruments to perform the broader mission.

The White House Office of Management and Budget (OMB) approved the Mars Geoscience/Climatology Orbiter mission in 1983, but it did not accept Briggs's programmatic concept. It would not permit the Planetary Observer program to be managed the way the Explorer program was. Each mission would require new start approval from the budget office. Since a basic idea behind the Planetary Observer concept had been to save money through sequential purchases of the same spacecraft, this goal was already in jeopardy. But it was still possible that the next mission intended for Planetary Observer, Lunar Observer, might be approved later; the project manager JPL appointed to the Mars Geoscience/Climatology Orbiter mission, William I. Purdy, kept one eye on that possibility.

Mars Observer

Purdy had been involved with systems-contract projects at JPL before, including the Mariner Venus-Mercury mission, the first mission to visit Mercury, and Seasat-A, the first oceanographic satellite. Most recently, Purdy had been manager of a classified Defense Advanced Research Projects Agency effort at JPL called Talon Gold that had also involved extensive contracting.[12] His Mars Geoscience/Climatology Orbiter project office at JPL consisted of about 20 full-time staff.

The first task of Purdy's project office was to prepare a request for proposals (RFP in government parlance) for the spacecraft. This was scheduled to be released to industry on June 4, 1984.[13] The RFP had to tell prospective bidders what the spacecraft needed to supply to the scientific payload in terms of power, data transfer, and pointing accuracy. To accomplish this, NASA formed a science working group to define the scientific requirements for the mission and to prepare a "strawman" payload. NASA did not intend to choose the actual payload until after the spacecraft contractor had been selected. This was a reversal of the traditional NASA approach. In the past, scientific objectives had been defined first, then instruments had been chosen, and finally the spacecraft was designed to support the chosen instruments. Use of a preexisting commercial design meant tailoring the payload to the spacecraft, reversing the previous process.[14] Briggs and other promoters of the switch to commercial spacecraft hoped this would prevent the payload from driving cost escalation in the spacecraft, a long-standing problem in the space industry. But the potential bidders needed to have some idea what the spacecraft had to do, so the strawman payload essentially defined a range of potential resource requirements—power, communications capability, on-orbit stability—that prospective bidders would use to define their spacecraft concepts.

Mike Carr, from the United States Geological Survey's astrogeology division in Menlo Park, chaired the science working group. Carr had been a leader of Viking's imaging science team as well as a member of Mariner 9's imaging team, and over the years since those missions, he had worked up a theory that Mars had a lot of water ice lurking underground. Like Earth's Moon, he thought, the martian crust was probably heavily fractured from meteor bombardment, and very early in the planet's history, it had sunk into the crustal fractures and disappeared from view. But because the ice would be heated from below by the planet's core, and because it would be under enormous pressure from the rock above it, outbreaks (and large ones) could still have happened. He thought the water-related features still visible on Mars were probably a product of those outbreaks. That water would be on the surface, or near it, and could be found by the right sorts of measurements.[15] His working group met seven times between July 1983 and February 1985, using the science objectives laid out by Briggs' Solar System Exploration Committee as guidelines to develop the Mars geoscience orbiter's scientific approach.[16]

While the science working group was debating the potential payload, Purdy's project office had to cope with an additional challenge: how to write an RFP that could encompass two possible, but different, booster concepts. The new mission, quickly renamed Mars Observer, was scheduled to be launched by a space shuttle in 1990. The shuttle could only reach low Earth orbit, and so Mars Observer needed an additional rocket stage to send it on its way to Mars. One possible method was to use what NASA was calling a "transfer orbit stage," a separate, solid-fueled upper stage that would accelerate the vehicle toward Mars and then release it. But this rocket hadn't been developed yet. Another was essentially to build Mars Observer around a solid-fuel rocket (an "integral rocket," in the jargon). At least one prospective bidder, Hughes, built its commercial communications satellites this way. JPL did not want the RFP to exclude the integral approach by requiring use of the transfer orbit stage, while NASA wanted to use Mars Observer as a means of developing the transfer stage so that future missions could use it. To make things worse, NASA could not decide how to structure a competition for the stage, either.

Finally, in what became known as the "Ten Point Agreement," NASA and JPL leaders agreed that bidders would be allowed to propose both transfer-stage-based and integral rocket designs, and that the NASA Marshall SpaceFlight Center would write a separate request for proposals for the transfer orbit stage. Should a transfer-stage-based proposal be accepted, Marshall would handle the development contract for it. The stage would also be paid for separately and provided to Mars Observer for free. The time consumed in negotiating this agreement delayed the release of the spacecraft RFP for a year, so it did not go out until June 1985.[17] This meant that the spacecraft had not been selected when NASA headquarters released the announcement of opportunity for the scientific payload in April 1985. The announcement of opportunity, in Briggs's conception of the Observer program, was supposed to be released after the spacecraft had been chosen so the spacecraft's abilities would put a hard limit on the instruments. The delay in deciding whether or not to use the conceptual transfer stage caused another violation of the basic Observer program cost-containment idea.

NASA's Office of Space Science received more than 90 proposals in response to the announcement of opportunity, and selected the final payload through a complex and confusing process. A series of NASA review panels evaluated the instrument proposals, placing 14 into the top category. Then Purdy's engineers, and a separate science working group chaired by Caltech geologist Arden Albee, examined 60 possible eight-instrument combinations as well as several seven-instrument combinations. None of the eight-instrument combinations was supportable by any of the spacecraft Purdy expected to have proposed in response to the RFP; they either weighed too much, cost too much, or required too much power. The seven-instrument payloads, however, could all be accommodated.

On April 8, 1986, Purdy received the surprising news that NASA had chosen one of the eight-instrument, oversubscribed payloads. Many years later, he remarked that he shouldn't have been surprised, given all the other violations of the program's basic tenets.[18] Due to the data rate limitations of the available commercial spacecraft, the announcement of opportunity had not provided any allowance for a camera. But the selected payload did include a camera. The Mars Observer camera, proposed by an Arizona State University geologist named Michael Malin, had been an innovative design. It would be fixed to the bottom of the spacecraft, so it didn't require an expensive scan platform; it didn't need a sophisticated cooling system; and it used charge-coupled detectors, a relatively recent development that eliminated large, heavy television-style detector tubes. It also promised much higher spatial resolution than any previous planetary imager, an enticing prospect for researchers. Its principal drawback as far as the project was concerned was that it also effectively doubled the amount of data that the spacecraft had to be able to send back to Earth, even with the clever, self-contained data compression that Malin had devised for it.

Most of the other instruments chosen were expected. There were two "facility" instruments, the gamma ray spectrometer (GRS), to be built by Lockheed under contract to Goddard Space Flight Center, for sensing near-surface ice and certain minerals, and a visual and infrared mapping spectrometer (VIMS) to be built by JPL. VIMS, whose proposal had been spearheaded by Larry Soderblom of the United States Geological Surveyor, was designed to sense a wide variety of surface minerals by infrared reflectance. NASA also chose William Boynton of the University of Arizona as the leader for the GRS science team. One unexpected choice, though, was a proposal by a young postdoctoral researcher at Arizona State, Philip Christensen, for a thermal emission spectrometer (TES). It was supposed to look for carbonates in the martian surface. Carbonates on Earth are deposited exclusively in liquid water; finding them on Mars would be strong evidence of the existence of surface water in the martian past.[19]

The instruments aimed at studying the martian climate were also unsurprising. JPL's Dan McCleese had proposed an atmospheric temperature profile instrument (a "sounder") based on technology for measuring the temperature

Mars Observer initial payload selection

Instrument	Investigators (institution)
Gamma ray spectrometer (GRS)	William Boynton (University of Arizona)
Magnetometer	Mario Acuna (Goddard Space Flight Center)
Mars Observer camera	Michael Malin (Arizona State University)
Pressure modulator infrared radiometer	Daniel J. McCleese (JPL)
Radar altimeter*	David A. Smith (Goddard Space Flight Center)
Radio science	G. L. Tyler (Stanford University)
Thermal emission spectrometer (TES)	Phil Christensen (Arizona State University)
Visual and infrared mapping spectrometer (VIMS)**	Larry Soderblom (US Geological Survey / JPL)

Source: Adapted from Charles Polk, "Mars Observer Project History," December 1990, EDS D-8095, p. 27.

*Renamed the Mars Observer laser altimeter in June 1988.

**Deleted from the payload in June 1988.

of Earth's stratosphere. McCleese's instrument also would produce water vapor profiles. Another experiment, an ultrastable oscillator to be added to the spacecraft's radio so that it could also be used for atmospheric measurements, would allow independent temperature and pressure measurements to be made, checking the performance of McCleese's instrument. NASA's science directorate had also chosen two radar altimeters, with the caveat that only one would actually fly and the two science teams would be merged. The review panels had not been able to determine which was the best technical proposal based only on the proposals, so during the six-month accommodation phase, Purdy's engineers would make this assessment. Ultimately, in August 1986, Goddard Space Flight Center's altimeter, a derivative of one flown on Seasat A, was selected and geophysicist David A. Smith was named the principal investigator.

While NASA was choosing instruments, JPL was engaged in selecting the spacecraft contractor. In February, the NASA administrator decided to restrict the competition to only those designs that relied on a separate upper stage. This produced a challenge to the selection process by one of the bidders, which delayed the final awarding of the contract.[20] On March 24, JPL selected RCA's Astro Electronics division in East Windsor, New Jersey, for the spacecraft contract; simultaneously, NASA awarded the upper stage contract to the only bidder, Orbital Sciences Corporation, for its transfer orbit stage.[21] RCA had proposed a variant of its Satcom-K communications satellite that used electronics from the Defense Department's weather satellites. It was essentially a cube, and RCA had proposed putting most of the instruments on a nadir-pointing plate (so that they would always point down toward the martian surface). But both the magnetometer and the gamma ray spectrometer had to be deployed away from the spacecraft body on booms. A single, five-panel solar array and the high-gain antenna also had to deploy on booms. In addition to looking rather ungainly, the resulting spacecraft had to manage a complicated set of deployments once it was launched.

In celebration of the completion of this complex process of selection, Bill Purdy had big, yellow lapel buttons made up for his team that shouted "I survived the 10-Point Agreement!" He still had his 20 years later.

JPL's leadership chose Arden Albee as the project scientist. Albee, a California Institute of Technology geologist who had built much of his reputation on analysis of the Moon rocks returned to Earth by the Apollo astronauts, explained to his new principal investigators in April 1986 that during the next six months, they would be tasked with finalizing their instrument specifications and costs.[22] At the same time, George Pace, a lanky, laconic engineer who was Purdy's spacecraft manager and a veteran of the big Surveyor Moon-landing program of the

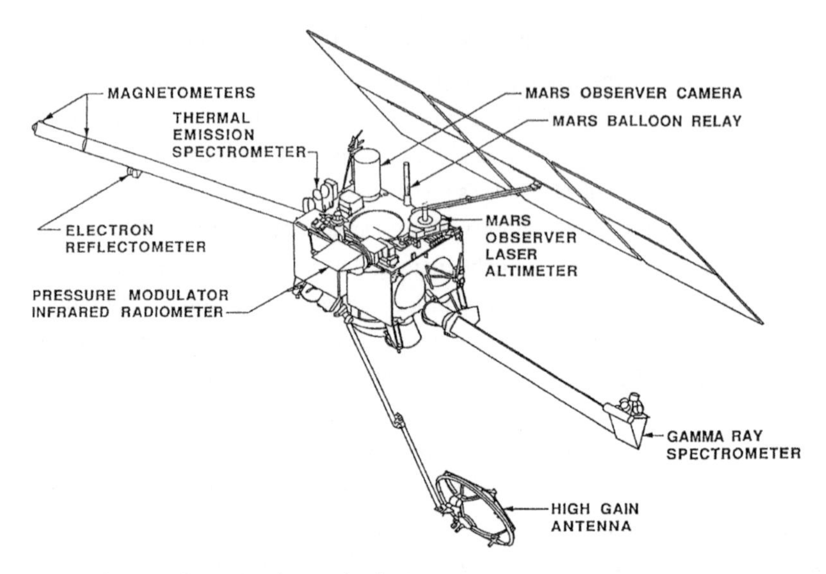

Drawing of Mars Observer. The "nadir deck" containing the instruments that would have pointed toward Mars faces up on this diagram.

1960s, would be negotiating the final spacecraft contract.[23] That negotiation included alterations to the proposed spacecraft to accommodate the oversubscribed payload. At the end of this accommodation period, NASA would hold a science confirmation review that would determine the final payload. If Pace and his counterpart at RCA could not find a way to increase the spacecraft's telecommunications capabilities so that it could support all the instruments, an additional instrument might have to be dropped.

One of the principles of the Observer program had been that instruments chosen to fly should already be relatively mature technologically. That was supposed to ensure low costs and reliable cost estimates. But this wasn't the case for many of Mars Observer's instruments. Instead, throughout the life of the project, instrument development costs rose. Smith's altimeter, for example, nearly doubled in its projected cost during the six-month accommodation phase despite being derived from an already-flown design. Its mass also increased by a quarter.

Purdy's science manager remembered many years later that the Observer concept's demand for cutting edge science without technology development had put scientists interested in participating in the mission in a bind. NASA reviewers might not see a truly mature, flight-proven instrument as providing worthwhile new results, so proposing one might result in nonselection. A new instrument concept would violate the program objectives, but might be seen positively by the reviewers because of its proposed capabilities. So there had been a strong bias among the proposers and the reviewers for new techniques, despite the official program guidelines.[24]

George Pace got RCA to make sufficient improvements to the spacecraft design to support the oversubscribed payload. A larger high-gain antenna solved the data transfer rate problem. The payload and larger antenna had also made the spacecraft overweight, and clever changes to the propulsion system and mission flight profile restored some mass "margin," the difference between what the launch vehicle could lift and what the spacecraft was predicted to weigh. Margin management was key to any aerospace project. This early in a project, engineers generally want the spacecraft's predicted mass to be around 30 percent less than the launcher's capability because most spacecraft gain weight in development as the true mass of various subsystems become better known. But Mars Observer began its life with less than a 10 percent mass margin, creating a large risk of being overweight on launch day.

JPL and RCA signed a fixed-price contract for the Mars Observer spacecraft on September 30, 1986, specifying a cost of $55.3 million and a 15 percent award fee based upon on-orbit performance. The contract also contained options for up to three additional spacecraft buses, with option availability dates, differing in price, from October 1986 through October 1991. This clause enabled the overall Observer concept to continue, with a putative Lunar Observer as the next potential mission.[25] Shortly after the contract was signed, General Electric purchased RCA, and the East Windsor, Connecticut, plant Mars Observer was to be built in became known as GE Astro Space.

Getting the Boot: Mars Observer Leaves the Shuttle

If 1986 had been a rough year for Purdy's Mars Observer planning, 1987 would turn out to be worse. On January 28, 1986, the space shuttle *Challenger* had exploded shortly after liftoff, killing its crew of seven and destabilizing all of NASA's programs. Mars Observer, which was supposed to fly on the shuttle in 1990, suddenly faced uncertainty over whether it had a launch vehicle or not. When NASA had set out to develop the shuttle, it had also decided to stop using expendable launch vehicles, and the Defense Department (with much foot-dragging) was following suit. A major policy change would have to occur to get Mars Observer off shuttle, and those take time to bring about. Mars Observer also no longer had a stable launch date, even if it stayed on the shuttle manifest, because NASA had no idea how long it would take to return the shuttles to service. It took time for this fact to emerge, however, so for 1986 and early 1987, the project operated under the assumption that it still had a 1990 launch date on a shuttle.

The first solid indication that Mars Observer might not fly in 1990 came in August 1986, the month before Purdy had signed the contract with RCA. After a conversation with Briggs at headquarters, Purdy explained the situation to the project science group. The post-Challenger shuttle manifest had only three planetary launch slots between 1989 and 1990, but four launches were planned: Galileo, Ulysses, Magellan, and Mars Observer. Of these, Mars Observer was the lowest priority to NASA. This meant either Mars Observer slipped to 1992 or it went on an expendable rocket in 1990. The only rocket of appropriate power for Mars Observer was a leftover Titan III, and Purdy asked its manufacturer to brief his staff on its capability for the 1990 launch window.[26]

The launch vehicle briefers argued that their Titan provided one big advantage over a shuttle launch. It could throw more mass toward Mars. This was an important gain to the project, as it had low mass margins. Switching to the Titan would enable Purdy to regain mass margins without having to undertake an expensive mass-reduction effort on the spacecraft and instruments, a clear benefit. The launcher representatives also contended that it was also still possible, at this point, to make the 1990 launch date, although only because the Air Force had a "spare" Titan III on order that it was willing to offer to NASA.[27]

By this time, however, the 1990 launch was looking unlikely for another reason: money. In June 1986, NASA told JPL that the Mars Observer project would get significantly less funding during fiscal years 1987 and 1988 than it had planned on. For 1987, the reduction eliminated all of the project's financial reserve (which already had significant liens against it from the overcost instruments) and for fiscal year 1988 it cut the planned $69 million budget to $56 million.[28] These numbers would make it difficult to still make the 1990 launch window. But it was still possible that NASA would change the funding profile to restore the cuts later in the year.

On April 14, 1987, however, NASA told Purdy that Mars Observer would be launched on a Titan III, but not until 1992. This two-year delay allowed NASA to cut Mars Observer's budget still further. The project's budget for fiscal year 1988 was slashed nearly in half, from $56 million to $29.3 million.[29] Purdy redirected most of the funds to the instruments, to electronic parts procurement, and to the development of the payload data system. One of the few new spacecraft technologies being developed for Mars Observer, this represented the first "packetized" data system designed for a planetary spacecraft. The ARPANET, designed during the 1970s, had developed the ability to encode origin and destination information on individual "packets" of data, which allowed them to be automatically sorted and reassembled. Mars Observer's science plan depended on transmitting data directly to computer workstations at the scientists' home institutions, so they didn't have to all gather at JPL for the two (Earth) year long mission—the way Viking and Mariner 9 investigators had been gathered in the past, though only for a few months. Instead, the science operations concept entailed figuring out how to continually move the data to the scientists' home institutions; it also meant finding ways to coordinate the scientists' demands for spacecraft maneuvers without having them at JPL, where the spacecraft operators would be. The new data system was a key component of this new operations concept, but it was severely behind schedule, leading to its priority.

For Purdy, the decision to boot Mars Observer into 1992 was disastrous. "All bets were off," as he put it years later.[30] It meant all of the contracts had to be renegotiated, and the project's costs would soar. The project's initial projection of the slip's cost was $125.7 million.[31] This did not include the cost of buying the Titan III, which would be around $150 million.[32] This effective doubling of the original project cost, however, rapidly got even worse. NASA officials began to realize that Mars was politically popular. So it changed the design basis of the mission.

Mars Enthusiasm and Mars Observer

Outside NASA, advocacy and lobbying for Mars missions had been going on for years. Astronomer Carl Sagan, geologist Bruce Murray, and engineer Louis Friedman had formed the Planetary Society in 1979, as Murray put it in 1989, to "prove that there was great support among the general public for continuing the exploration of the planets that had begun so auspiciously during the Apollo era of the late 1960s and early 1970s."[33] Located in Pasadena, the Planetary Society had launched a publicity campaign to save Mars Observer at the first hint of the launch slip in August 1986, hoping to convince NASA, or Congress, to keep Mars Observer on schedule. While they ultimately failed to change Mars Observer's fate, they succeeded in raising its public profile (and its profile in Congress) significantly.[34]

Independently, other scientists had formed a "Mars Underground" to promote both Mars science and Mars settlement. The group had organized two big conferences in 1981 and 1985 entitled, appropriately enough, the Case for Mars. Papers presented at these conferences spanned the full range of Mars interests: engineers presented potential mission profiles, scientists considered what might be learned by explorers (both robotic and human), activists positioned Mars as a future ecotopia—or merely as more resources to be exploited.[35] Mars Observer got swept up in the new enthusiasm for Mars, as NASA officials began to see it as an opportunity.

At the second of these conferences, former NASA administrator Thomas O. Paine gave an impassioned speech promoting Mars colonization. Mars *was* the future frontier. With more gravity than the Moon, an atmosphere containing a useful gas (if not much of it), a useable surface area equal to all of Earth's continents, and similar mineral composition, and almost certainly lots of water, Mars was a more likely future human residence than the airless, waterless Moon. In 1985, President Reagan asked Paine to chair a commission on the future of space exploration. After a year of public meetings and report-writing, Paine published *Pioneering the Space Frontier.* Drawing on one of the American conservative movement's favorite tropes, Frederick Jackson Turner's 1893 argument for the importance of the frontier in American history, Paine and his commissioners called on the United States to establish lunar and asteroid mining colonies. These "Free Societies on New Worlds" would integrate the nearly infinite mineral wealth of space into the American economic sphere. But Mars, they concluded, for the next 50 years could only be the target of exploration. It was too distant for commerce, and permanently inhabited research bases could wait a few decades. This, to the commissioners, also rendered its exploration suitable for international cooperation instead of Cold War competition. So long as Mars was not part of America's economic sphere, cooperation was politically acceptable. Nonetheless, robotic exploration should expand immediately, so that suitable locations for robotic and later human missions could be identified.[36]

These intertwined threads of space commerce and space colonization began to draw in Mars Observer at the same time Purdy's team was trying to reconstruct the project's schedules and budgets around the new 1992 launch date. Purdy had initiated study of the second proposed mission in the Observer program, Lunar Geophysical Orbiter, in April 1986. Late in March 1987, shortly before the contract called for a significant increase in cost of the second set of Observer hardware, NASA had authorized buying the second spacecraft, accepting the project's argument that having spare hardware would speed the testing program. The option cost was $22 million. Twenty years later, George Pace remembered that as being almost insanely low. Purdy hoped that the lunar mission would move forward, but even if it didn't, the availability of a complete set of spare parts during Mars Observer's own assembly, test, and launch operations phase (known as ATLO at JPL) reduced the risk that the team wouldn't make the 30-day launch window in 1992.[37] The second spacecraft purchase made sense regardless of what happened to the lunar mission.

Just a few weeks after ordering the spares, and at almost the same time as the launch slip, Purdy was directed to terminate the Lunar Geophysical Orbiter mission studies and redirect those funds to support of a new task: a Mars rover sample return study.[38] The rapidly growing external advocacy for an expansive space program, and for Mars specifically, gave officials at NASA headquarters some hope that a more comprehensive Mars program could be built around the technologically ambitious goal of returning Mars rocks to Earth over the next few years.

Mars Observer was directly affected by the sudden interest in Mars. If a sample return mission were to go forward, its mission planners would need information that Mars Observer could provide. Mars Observer would provide a great deal of vital information about landing sites, making sample return dependent on Mars Observer's own success and capabilities. But Mars Observer's payload had been designated a "Class B" mission, in which electronic redundancy was minimized to save money. The instruments largely were susceptible to what are known as "single string" failures, in which a single component failure can disable the instrument. This drew concern from the sample return study team and from a review board formed to examine Purdy's new plans for the 1992 launch. These groups recommended redesigning the instruments to reduce single-string failure potential, altering the spacecraft's design so that it could point more accurately at potential landing spots and expanding the data return capability of the Mars Observer camera by adding greater on-board data compression. Further, NASA changed its position on instrument spares, requiring them for the sake of risk reduction.[39] These changes would increase the probability that Mars Observer would return the data that the sample return mission planners needed.

But the changes implemented to the mission and to its instruments caused costs to rise even above the $127 million increase that the Mars Observer team had anticipated from the two-year launch delay. Purdy held a project retreat January 20–22, 1988, to finalize his cost estimate; at this point, conducting the mission as planned with "minimal spares" looked to be $450 million. NASA was now demanding full spares (thus permitting an almost immediate reflight in the event of failure), and this costed out at $480 million. Neither figure included the cost of the Titan III. NASA's Office of Space Science, however, wanted the project held to $395 million. So Purdy had to find ways to eliminate $85 million in costs. The only way was to drop one or more instruments.

First NASA asked Arden Albee's project science working group to make a prioritized list of instruments that could be dropped and a description of the relative impacts of dropping some. His group refused, believing that even participating in the exercise "might be interpreted as acquiescence."[40] NASA then approached the Committee on Planetary and Lunar Exploration for a prioritization. It also refused to provide one. Mars Observer would, the committee opined, be the only mission to Mars in the next decade. So it could no longer be evaluated on the original Observer basis of frequent, cost-constrained missions.[41] Instead, they insisted that Mars Observer be carried out as planned.

Purdy reflected later that at this point, the decision on what to cut was no longer terribly difficult. Only deletion of one of the two most expensive instruments, JPL's own VIMS and Goddard Space Flight Center's altimeter, would provide savings even close to what NASA was demanding.[42] These totaled about $66 million. In the last week of June 1988, Purdy issued a stop-work order to the JPL team building VIMS.[43] At about the same time, radar altimeter scientist Dave Smith, recognizing that the radar altimeter was not going to be affordable, had informally proposed switching to a laser altimeter. This would be built in-house at Goddard Space Flight Center, using mostly civil-service labor, and so would cost less than half as much as the radar-based instrument, no more than $10 million. This new instrument became the Mars Observer laser altimeter (MOLA).

Finally, in what came to be called the "Handshake Agreement" between JPL director Lew Allen and NASA associate administrator Lennard Fisk, the Mars Observer project's new budget was set at $419 million.[44] The agreement did not resolve the issue of launch vehicle, however, and the project was required to maintain "dual compatibility," the ability to launch either on a Titan III or on a space shuttle, in 1992. This, of course, cost money, too. Purdy was not allowed to drop shuttle compatibility until September 1988.[45]

While all this was going on, one more change occurred to Mars Observer that had ramifications well beyond the project. The Soviet Union was planning to launch a Mars orbiter in 1994 that would carry a pair of French balloons to Mars, arriving during Mars Observer's primary mission. The balloons would be released to float in the martian atmosphere and transmit data to the Soviet orbiter. But the balloons would not be able to record data for replay later, so their data could not be captured except when the orbiting spacecraft was in radio range of them. Having both the Soviet orbiter and Mars Observer receiving data in different orbits would increase the data return. Mars Observer could collect the data via a balloon relay added to the mission and then send it back to Earth with the rest of its information.[46]

Michael Malin, the Mars Observer camera principal investigator, was willing to add an interface for the balloon relay electronics to his camera (the data would be stored as a digital image file and sent back to Earth that way), so the spacecraft's telecommunications system wouldn't have to be redesigned again. As part of the Handshake Agreement, NASA directed that the experiment be added. The agency also agreed to pay for it outside Mars Observer's budget, so it did not raise the project's costs further.[47]

The Mars balloon relay experiment was part of a "Mars Together" effort, in which the Cold War competitors would cooperate in exploring Mars. First somewhat jokingly suggested at a 30th anniversary of Sputnik conference in Moscow, the Planetary Society had backed the idea, then Soviet leader Mikhail Gorbachev had spoken favorably of it in May 1988. Mars Together even became part of the American presidential campaign that year.[48] The advocacy for Mars going on outside NASA combined with a new drive for U.S.–Soviet cooperation had fundamentally altered the political framework in which Mars missions would take place.

The one significant check on Mars enthusiasm occurred with the loss of the ambitious Soviet Fobos missions. Two orbiters equipped with small landers destined for the moon Phobos were launched toward Mars in July 1988, the same time NASA and JPL were finalizing the Handshake Agreement. Fobos 1 was lost when controllers uplinked an incorrect command, while Fobos 2 reached Mars and went into orbit properly in January 1989. It was then lost after returning imagery, and just before it was to release the moon landers.[49] This was a portent of things to come for the Soviets.

Getting Mars Observer to the Launch Pad, 1989–1993

Between 1989 and Mars Observer's September 1992 launch date, the project became legendary within NASA for its cost. NASA's decision to push the launch back two years had already imposed a large cost increase. But during the next few years, the remaining assumptions behind the Observer concept also fell before engineering and institutional realities. It turned out not to be quite so easy to turn an Earth orbiter into a Mars orbiter. And the instruments proved to be difficult as well.

The Planetary Observer concept had been based on the use of "heritage" to reduce cost. JPL had chosen to base Mars Observer on an Earth orbiter that had a substantial flight record. But heritage turned out not to be the benefit it was supposed to be. The first substantial blow came during the contract renegotiation resulting from the launch slip. The project's mission assurance engineer discovered that the Defense Meteorological Satellite Program weather satellite series Mars Observer was derived from had suffered many avionics failures during its eight years of operations. The military weather satellites had experienced an average of three thermal failures each during their first 45 days of flight. Their civilian near-twins had had 17 failures aboard seven spacecraft.[50] This had been acceptable because the United States maintained two civilian and two military weather satellites in orbit, while only needing one of each for effective forecasting. The United States also kept completed satellites in storage on the ground. These stored spares could be launched relatively quickly to replace a failed spacecraft. The high level of electronic risk was thus mitigated through multiple duplication—the electronics on each spacecraft were redundant, and the spacecraft were themselves redundant.

This level of redundancy would not exist for Mars Observer. The spacecraft was to be built with redundant electronic systems, just as its ancestral Earth orbiters were. But the flight spares that the project had bought would not be assembled into a complete, tested spacecraft. They would only permit a reflight of the mission in the next launch window to Mars in 26 months, and at a substantial additional cost for assembly, testing, and launch. And, of course, NASA and JPL would suffer the embarrassment of an avoidable failure of what had already become a fairly expensive mission. So with NASA's concurrence, JPL imposed a much more expansive, and expensive, test program on GE Astro Space than was the company's practice.

George Pace, Mars Observer's spacecraft manager, recalls another challenge to the project imposed by differences between Earth orbiters and Mars-bound spacecraft: boom deployments. On an Earth orbiter, all engine firings occur before the spacecraft activates its telecommunications system and begins deploying its solar array, instrument, and antenna booms. The booms do not need to be strong enough to withstand engine firings when deployed. But Mars-bound spacecraft make several course adjustments during cruise, and when they reach Mars undergo a long firing to slow enough to enter Mars orbit. Either Mars Observer's booms would have to be strong enough to withstand multiple firings or they would have to remain undeployed until after the spacecraft reached orbit. The solar array and high-gain antenna booms had to be deployed, however, at least partly. The mass penalties involved were so high that the project developed a set of compromises that made the booms lighter but more complex. The high-gain antenna boom was designed with a joint partway up its length that allowed the antenna to see Earth over the spacecraft but without being fully extended. And instead of deploying fully during cruise, the solar array kept four of its six panels against the spacecraft body, deploying only two until it reached Mars orbit.[51]

The instrument booms posed a different set of challenges. Mars Observer was not going to perform any scientific studies during its cruise phase, so initially the project's engineers had thought they wouldn't need to deploy the magnetometer and gamma ray spectrometer booms until the spacecraft reached Mars orbit. But in fact both the gamma ray spectrometer and magnetometer sensor heads needed to be moved away from the spacecraft during the cruise phase so that their scientists could measure the spacecraft's own influence on their measurements. Without knowing the spacecraft's emissions, the instrument's science team wouldn't be able to make sense of their data when the vehicle got to Mars. So the GRS boom had to deploy part way early in the mission. The magnetometer principal investigator presented the project engineers a slightly different challenge. His instrument had two sensor heads, not one, and one of them needed to be halfway up the boom when it was deployed. A typical boom deployed out of a canister like a big jack-in-the-box, with the sensor attached to the free end. The second head couldn't be stored in the canister with the compressed boom, so the boom designers had to figure out how to make the boom "pick up" the middle sensor head as the boom came out. The complex boom deployments, of course, cost more money to design than simple ones would have, and they weighed more.

As the project evolved, other changes to the spacecraft design had to be made as well. The time delay for communications between Earth and Mars forced redesign of some of the telecommunications system. Failure of a new transponder development at JPL also forced redesign of some of the system, and the project had to buy a copy of the transponder used by the Magellan mission to Venus instead.[52]

Mars Observer also had to be more autonomous than an Earth orbiter. Earth orbiters pass over their ground stations on every orbit. Because of the need to communicate with other spacecraft, JPL's Deep Space Network would only be available to Mars Observer for an hour or two per day, so the spacecraft had to be able to care for itself for at least a day after a failure, and be able to get its antennae into position to contact Earth again for the next communication window. This is called a "safe mode," built into every planetary spacecraft. "Fault protection" software puts the spacecraft into its safe mode in the event of an onboard failure. GE's approach to fault protection was the inverse of JPL's. JPL-built spacecraft had essentially autonomous subsystems, each having its own data-processing capability and software. Mars Observer, like its weather satellite forebears, had a single data-processing subsystem that handled all of the spacecraft's functions—attitude control, telecommunications, power management, and fault protection. The resulting integrated software architecture bore little resemblance to JPL practice. This fundamental difference in approach made it harder—and more expensive—for the GE and JPL engineers to come to terms over an acceptably robust fault-protection design.

The difficulties in adapting an Earth orbiter to Mars were paralleled by instrument development difficulties. The most challenging instrument proved to be Dan McCleese's atmospheric temperature profile sounder, which was being built at JPL. The design was derived from an instrument aboard the U.S. polar orbiting weather satellites but would be the first instrument built at JPL to operate in the thermal infrared spectrum. It had the ability to see Mars looking straight down (called "nadir viewing") and to see Mars's horizon, or "limb." For calibration purposes, it also had the ability to view space and was cooled to 80K (–193°C) by a large radiator. The instrument overran its budget repeatedly; in 1991, the project office pushed the instrument workforce off the job for the last couple of months of the fiscal year to keep within budget. NASA ultimately had McCleese descope the instrument, reducing its testing requirements and severely cutting funding for development of its data-processing system.[53]

The gamma ray spectrometer also had development problems, but these were less technical than managerial. The institutional arrangements for the production of the GRS were complex. Goddard Space Flight Center in Maryland was handling the procurement under a contract from JPL; in turn, Goddard had subcontracted the instrument to Lockheed Martin in Denver. The instrument's science team leader, William Boynton of the University of Arizona, was initially discouraged from direct supervision of the instrument's progress by the project office.[54] This "hands-off" form of contracting was done in the hope of keeping confusion to a minimum over who was actually empowered to request changes. Boynton recalled, "They were very concerned that if I were to talk to Lockheed-Martin directly, I might suggest improvements to the instrument, and Lockheed-Martin would go ahead and build those and send the bill to Goddard, and NASA would end up paying for more than they bargained for. It's quite understandable. So they kept me very much at arm's length, and they would not let me talk to them except under very well regulated, scrutinized conditions."[55]

But as the instrument overran anyway, Mars Observer's instrument manager changed his mind and had Boynton visit the contractor every few weeks. Boynton's view of the contractor's problems was simply that their project manager wasn't good at fostering communications within his team, leading to duplication and wasted effort. He, Goddard, and JPL eventually got the company to replace its project manager, resolving this issue.

Mario Acuna's magnetometer also suffered from the hands-off management that JPL enforced. Acuna's magnetometers had flown on many different spacecraft to the Moon, Venus, Jupiter, and Saturn, so he had plenty of prior experience. He thought Mars Observer itself would impair his instrument because it was magnetically dirty. It had stray magnetic fields produced by the spacecraft's power system and electronics that would contaminate his data, and possibly prevent detecting a martian magnetic field if it were extremely weak. He thought a different solar array design could effectively neutralize the problem. But there was nothing he could do about it because he was not allowed to work with the component vendors to fix the problem.[56] The decision to redesign the magnetometer boom so that it could be partially deployed prior to Mars orbit entry was an effort to mitigate this problem. It would enable Acuna to determine the stray magnetic fields of the spacecraft while in interplanetary space.[57]

Malin's Mars Observer camera, being built in the basement of Caltech's South Mudd laboratory, presented still other developmental problems. He had proposed using a variety of new and untested technologies to achieve very high resolution and data return. He had included a large amount of random access memory in the design to store the image data, intending to use a batch of chips designed for use in talking dolls—obviously never space qualified! Here JPL's mission assurance branch intervened to force him to provide additional redundancy. Ultimately, instead of using expensive space-qualified memory, he argued them into a deal that let him provide two complete sets of memory banks for the camera, with management software that could "map around" a failed chip so the rest could keep functioning.

The development difficulties left the Mars Observer team increasingly behind schedule. The cost increases created by the launch delay and the instruments had frustrated Bill Purdy, who took a job with Thompson-Ramo-Wooldridge (TRW) at the end of 1988. His successor, David D. Evans, also had extensive experience in the systems contract management side of JPL's operations. Evans liked to use a chart with colored dots like those on a stoplight on it to indicate the health of various aspects of the project (somewhat confusingly, they were often called "fever charts"); in September 1990, the payload and spacecraft development efforts went red. Despite descopings, reformulations of the test schedule, and the addition of more personnel, they remained red until December 1991.[58] That month was the first that GE Astro Space was able to keep to project schedule. In part, a new vice president made Mars Observer a higher priority. But it also reflected normal practice for the company. George Pace recalled that the company's way of working was rather alien for the JPL staff sent to the factory to oversee the effort. Because GE normally prepared several spacecraft a year (JPL built one during the entire 1980s), their assembly and test employees would work them sequentially in a more industrial fashion. Pace remembers, "That's just the way they worked. They had all these other missions, and they were launching eleven and twelve spacecraft a year, eleven or twelve. I mean, that's unbelievable from what they were cranking out of that little plant in East Windsor." Late in 1991, workers started to descend on Mars Observer and, with all of the instruments and major assemblies delivered, began making progress against the schedule.[59]

In March 1992, only a few months before launch, though, the project staff determined that there was a potential problem with the propulsion system. Mars Observer's propulsion system was derived from Earth orbiters, which used a set of nitrogen tanks to pressurize the fuel tanks through a regulator valve. The regulator kept a constant pressure in the fuel tanks as the fuel was burned; the fuel tanks were also pressurized on the ground right before launch, but without the valve the tank pressure would decline during each firing of the engine. The problem was that the regulator tended to leak, and JPL's propulsion specialists had concluded that the leakage might overpressurize the tanks during the spacecraft's one-year voyage to Mars. For an Earth orbiter, the slow leakage didn't matter; the propulsion system only had to work for the one-day-long trip to orbit, and then was shut down forever. Mars Observer, of course, had to fire its engine several times on the way to Mars for course corrections, and then do a long firing to go into orbit. So suddenly the leakage mattered.[60]

Because this disturbing problem was not uncovered until three months before the spacecraft had to be shipped to Kennedy Space Center, it left the project with few options. The team investigated the possibility of adding more valves to the system to protect against the leakage, but they figured it would take three to four weeks to install and test new valves. They didn't have that much schedule margin left. Instead, they decided to change the way the propulsion system was operated during the trip to Mars. They had intended originally to activate the pressurization system right after launch, allowing the regulator to maintain a constant fuel tank pressure throughout the cruise. Instead, they decided to leave the regulator shut and blocked off by existing valves during the cruise and only open them the day before the spacecraft had to do its orbit insertion burn. The course corrections could be done using just the pressure already in the fuel tanks. They called this a "blowdown" operation mode.

Mars Observer left the factory June 15, 1992, in a NASA-owned oversized truck designed for moving spacecraft around the country. The beleaguered project had one more incident once Mars Observer reached Kennedy Space Center; the carefully cleaned spacecraft had a load of dust dumped on it when the launch pad's payload air conditioning system was turned on. It had to be removed from the rocket, taken back to a clean room at Kennedy Space Center, cleaned, and put back out on the Titan III. That took five days.[61]

Mars Observer's Titan III sent it off to Mars on September 25, 1992. There was a brief moment of terror when the new transfer orbit stage appeared to have failed. The ground teams weren't receiving the data stream it was supposed to be sending back about its performance. This turned out to be because the rocket's transmitter had not turned on. The launch had actually gone perfectly, and the operations team in Pasadena heard from Mars Observer itself right on schedule. Ten long years in the making, Mars Observer was finally on its way.

Conclusion

Mars Observer's development team took out of the Mars Observer experience that the contract relationship between JPL, RCA, and its successor GE Astro Space; and between JPL and the various instrument subcontractors, had not worked well. The fixed-price contracts had not really controlled costs, and the enforced hands-off management approach had prevented JPL technical experts and the project scientists from engaging with the engineering early on, when issues might have been resolved less expensively. The late discovery of the propulsion valve problem was particularly vexing to George Pace because engineers from JPL's propulsion section had been involved in the project from the very beginning. Pace attributes the late discovery to a lack of a feeling of ownership of the mission. While JPL's name was attached to the mission, Mars Observer was not seen as a JPL spacecraft within JPL—it was GE's spacecraft, and GE's problem if it did not work properly. So the JPL engineers hadn't been really committed to finding whatever problems might exist in GE's vehicle.

Mars Observer cost $484.3 million to develop, ultimately, in 1993 dollars. The Viking Orbiters had cost $217 million in 1976; inflating both to 2007 dollars, the Viking Orbiters had cost $687.9 million and Mars Observer $652.4 million.[62] There had, of course, been two Viking Orbiters, so Mars Observer, intended to cost less, had actually cost more than each of the Vikings. The launch slip had produced the majority of the cost increase, while of the project's management elements, the payload had the largest cost increase, reflecting the impact of instrument immaturity at selection and NASA's decision to require higher reliability.

Mars Observer costs (millions of U.S.$)

Project element	Estimate March 1987	Cost at launch +30 days October 1992	Increase (%)
Project management	8,450	18,182	46
Project science	780	1,251	62
Mission	22,694	46,650	49
Payload	90,473	184,471	49
Spacecraft	89,214	231,281	37
Contingency	45,357	2,459	n/a
Total	256,986	484,284	53

Sources: JPLA 265, folders 53, 135.

Note: March 1987 estimate reflects cost prior to the decision to delay the launch by two years. Neither figure includes cost of launch vehicle or mission operations.

Mars Observer, then, had not lowered the cost of planetary exploration at all. Over its lifetime, all of the basic cost-control measures built into the original program had been violated. NASA's science officials had been unwilling to select a payload in keeping with the spacecraft's capabilities; the two-year delay had forced renegotiation of the original contracts and enabled most of the cost explosion; the lack of follow-on missions and the political enthusiasm for Mars at the end of the 1980s simply encouraged further cost growth by empowering demands for greater reliability; an Earth orbiter proved a poor choice of "heritage" for a Mars orbiter. In short, by the time Mars Observer itself reached space, it had already failed at its programmatic purpose of lowering the cost of planetary exploration.

CHAPTER TWO

Politics and Engineering on the Martian Frontier

During Mars Observer's years of development, the political stage for Mars missions began to change. The most direct reason for this was the slow dawning of recognition in the Reagan administration that space was popular. Despite the firestorm of controversy surrounding his Strategic Defense Initiative, and the tepid congressional reaction to Space Station Freedom, President Reagan continued using space-related imagery in his speeches. Throughout his presidency, following the rhetoric of space advocacy groups, he presented space as a new frontier for American pioneers. On the landing of the space shuttle *Columbia* in 1982, for example, he called on this imagery with the comment "the fourth landing of the *Columbia* is the historical equivalent to the driving of the golden spike which completed the first transcontinental railroad."[1]

In 1986, after release of the National Commission on Space's report *Pioneering the Space Frontier* and the *Challenger* explosion, NASA administrator James Fletcher asked astronaut Sally Ride to chair another commission to make recommendations on NASA's future course. Her report recommended, among several options, a long-term program to send humans to Mars. The first step of this option was expanded robotic exploration of Mars during the 1990s, culminating in a pair of Mars rover and sample return missions to be launched in 1996.[2] That, in company with the Solar System Exploration Committee's Augmented Program report, set in motion a joint study of a sample return mission by JPL and the Johnson Space Center. But the dream of sample return quickly disappeared, replaced by a very different mission that, initially, wouldn't involve JPL. In the rapidly changing political climate of Mars exploration, however, that changed too. By 1992, JPL had an opportunity to do something really new at Mars.

Sample Return Studies, 1986–1989

The 1978 Committee on Planetary and Lunar Exploration had called for a Mars sample return mission following the global mineralogical reconnaissance that was part of Mars Observer's task (see chapter 1).[3] But Mars scientists generally didn't want to pick up whatever rock happened to sitting next to the lander wherever it came to rest and ship that back home. They wanted to do what geologists on Earth do, go looking for specific rocks. That's how Earth-bound geologists construct their interpretations of Earth's history, and that's how planetary scientists wanted to go about constructing Mars's history. That meant they wanted rovers, and more: the rovers needed to carry instrumentation that would enable choosing rocks intelligently.

Don Rea, JPL's assistant director for technology and space program development and a planetary scientist who specialized in atmospheres, was appointed to head the Lab's sample return study. Glenn Cunningham, a systems engineer who had worked on the Mariner 6 and 7 missions to Mars and the Voyager outer planet missions, became one of his two deputies.[4] Their task was to produce a plan to return a variety of soil and rock samples from the martian surface sometime in the 1990s. The only limitations set by NASA headquarters were to accomplish the scientific goals within the constraints of available launch vehicles. They were not given a cost limitation to meet. The timeframe was established by the possibility of Soviet competition; in April 1988, Rea told the study team that the target was a 1998 launch for sample return. "This is thought to be essential if the US program is not to be surpassed by the Soviet missions to Mars in the mid and late 1990's," according to the meeting's notes.[5]

The rover design task was given to Donna Shirley.[6] Hired in 1966 as an aerodynamicist, she originally had worked on atmospheric entry shapes for Mars landers and then went to the Mariner-Venus-Mercury mission as a project engineer. She recruited a pair of robotics specialists to work on the study part-time to provide analyses of rover navigation tasks. In August 1987, Shirley's engineers gave Carr's scientists their first taste of how difficult the rover challenge was going to be. The first cut at understanding a Mars rover's "day" was disheartening to the scientists. The rover team had run two conceptual rovers, one using a relatively manual surface navigation method and one a semiautonomous method, through a 17 day mission. The scenario had included the telecommunications delay back to Earth, as well as the time the Earth-bound science and rover planning, driving, and collecting teams would need to perform their duties. It also assumed the rovers could operate at night and that each would have to collect ten samples.[7]

The more manually controlled rover would be able to travel only about 800 meters in 17 days. It could move no more than 30 meters before needing new instructions from Earth and possibly would move much less. It would only actually be moving for five minutes out of every two hours; the rest of the time it would be stationary while awaiting new instructions. This situation also meant that the human operators had to be able to prepare and transmit new instructions for the rover in a bit under three hours and maintain that cycle throughout the mission. That gave the science team only 30 minutes to decide where the rover should go and what it should do in each cycle, a difficult thing to accomplish in and of itself![8]

The semiautonomous rover performed somewhat better. In 17 days, it might cover 10 kilometers. It also would require a less intensive Earth operation, with its operators performing their tasks in eight hour cycles. After receiving its instructions for the day, the rover would be on its own. The advantages were that it could collect its samples from a somewhat greater diversity of sample sites and that it reduced the burden on the ground operators and on the Deep Space Network, as it would not require near-continuous communications. Its disadvantage was that designing it would be a "major technology challenge." The rover needed many times more computing power than had ever flown in space.[9]

Yet even the semiautonomous rover results were unsatisfying to Mike Carr's scientists. They didn't like the fact that the rover would be stationary for long periods of time, or that it would travel at best dozens of kilometers. The preceding, late-1970s effort to plan a rover mission had assumed (if unrealistically) that it would collect samples across hundreds of kilometers, allowing a detailed examination of martian geology. Shirley's engineering effort suggested that this was impossible, at least in the near term.

At the same meeting, Shirley made an effort to get the scientists to think more clearly about what they really needed the rover to achieve. They had been thinking along the lines of what her team called a "Godzilla" rover: a huge, robust rover carrying a large payload. She wanted them to consider other possibilities; as things stood at the end of 1987, the Godzilla rover concept not only wouldn't really serve the scientists' desires, it was looking to be so large it would be too heavy for a shuttle launch and would need to go on the Titan IV / Centaur, the most powerful (and expensive) expendable rocket. And some estimates made it marginal even for the Titan. In short, she needed them to start thinking about a less ambitious rover.[10]

The chastened science working group responded with reduced ambitions the next year. At the science working group's next meeting in January 1988, Mike Carr had the scientists work out goals for three different mission variants. Rover capability was the distinguishing feature of the three. In the simplest version, the rover had to travel no more than 100 meters from the lander, would carry few instruments, and would have its data-processing and analysis needs handled by the lander, not on board. It would collect samples identified on the basis of the lander's instruments and would perform no analysis of its own. The rover would be, in short, merely a "sample gathering device," not a doer of science itself.[11]

This minimal mission would be the least expensive, and it might be possible with a single Titan/Centaur launch. The ability to place the rover and return vehicle in a single package was particularly attractive, as it meant the rover would not have to find its way to an ascent vehicle. Because landing accuracies on Mars had uncertainties of tens of kilometers, even a very accurate pair of landings could leave the two vehicles too far apart for the rover's capabilities. This element of mission risk could only be reduced by a significant research and development program to enable greater landing accuracy, which would raise the cost of a two-launch mission substantially.

The science working group's other two missions were both two-launch efforts, reflecting the limitations of available launch vehicles. In one version, the rover would be capable of traveling 10–20 km. Because it would be collecting samples far from the ascent vehicle, it would have to carry its own analytic instruments to characterize the samples prior to collection so that its Earth-bound scientists could evaluate the sample's potential value. In the longer-range variant, the rover would be capable of traversing 100 km over two Mars years. It would also be a true "robot geologist," with a wide variety of analytic capabilities of its own and "science alarms" that would stop the rover if it traversed scientifically interesting locations.[12]

An important outcome of this set of sample return studies was the definition of what needed to be known about a landing site to certify it as safe for a lander or rover. At JPL, a "scale workshop" had defined the maximum size obstacle a reasonably sized lander could tolerate and used that information to help determine future imaging needs for Mars. A landing site's safety could be determined by orbital imaging, and the question this group answered was how high a resolution did the imagery have to possess to be acceptable? The Viking orbiters had been able to discern objects of about 38 meters in size, minimum, and his team had declared that unacceptable. Due to constraints imposed by the payload shrouds of the available launch vehicles, the leg length of a lander couldn't be more than about a meter. That also became the tolerable size of potential obstacles, and the imaging goal. In practice, detection of 1-meter obstacles from orbit meant a camera resolution of something like 25 cm/pixel.[13] The Mars Observer camera resolution, by contrast, was expected to be about 1.4 m/pixel.[14] Certifying a landing site as truly "safe" meant sending a much bigger camera to Mars.

By the beginning of 1989, the sample return effort had defined the "local" rover mission variant as its baseline mission, with a few modifications. An imaging orbiter would fly first, in 1996, carrying the big camera necessary to perform the landing site selection and safety imaging; the rover, a communications orbiter, and ascent and Earth-return vehicles would all go in 1998.[15] As things stood in early 1989, though, the funding needs for even this limited version of sample return seemed far out of reach.

But the political situation in Washington was about to change.

Sample Return and the Space Exploration Initiative

In July 1987, NASA administrator James Fletcher had established the Office of Exploration at NASA headquarters to begin planning and promoting future human missions to the Moon and Mars. But a humans-to-Mars focus grew out of interactions between a newly revived National Space Council, led by Vice President Dan Quayle and Fletcher's successor as NASA administrator, Admiral Richard Truly. Quayle wanted to establish a major new space initiative to enhance his own political career; as chairman of the Space Council, it was also within his authority to initiate a new policy process. But he wanted President George H. W. Bush to be able to announce the new initiative on the twentieth anniversary of the first Apollo Moon landing, July 20, 1989, which gave him a very short time to prepare it. Truly, in turn, had asked the Johnson Space Center (JSC) director to assemble a working group to develop a new Moon-Mars initiative.

This ad hoc working group, as it was known, met in secret June 4, 1989, in a remote building on the JSC campus to assemble a proposal. Truly was briefed on the plan on June 13.[16] At this stage, it incorporated a Moon base, robotic exploration of Mars, and finally a human Mars expedition. It would require raising the NASA budget 10 percent each year until it reached $30 billion, twice the agency's 1989 appropriation. After a bit of refinement of the plan, the administration began briefing space advocates, scientists, congressional staffers, and finally aerospace executives. One of the aerospace executives promptly leaked it to the *Washington Times* and *Aviation Week and Space Technology*; out in the open, the plan was instantly criticized for its huge costs. President Bush made the announcement anyway, called the Space Exploration Initiative, on the steps of the National Air and Space Museum.[17]

On July 20, 1989, Bush said, "In 1961 it took a crisis—the space race—to speed things up. Today we don't have a crisis; we have an opportunity. To seize this opportunity, I'm not proposing a 10-year plan like Apollo; I'm proposing a long-range, continuing commitment. First, for the coming decade, for the 1990s: Space Station Freedom, our critical next step in all our space endeavors. And next, for the new century: Back to the Moon; back to the future. And this time, back to stay. And then a journey into tomorrow, a journey to another planet: a manned mission to Mars."[18]

Bush's announcement was immediately followed by what has become known as the 90-Day Study.[19] This effort pulled in the Mars Rover Sample Return Science Working Group, which suddenly found itself charged with defining all of the robotic missions to Mars that would have to precede a human landing.[20]

At JPL, Glenn Cunningham was reassigned to coordinate the precursor mission studies. His team got moving on this August 1. Cunningham's directions were to develop a robotic program that would include launches every 26-month opportunity to Mars, provide the scientific and relevant technological information needed by human missions, be implemented exclusively by the United States, and not exceed $1.5 billion/year.[21] Johnson Space Center personnel would define the needs of the human mission planners.

The human Moon-Mars initiative devised down at JSC had started out with return to the Moon and construction of a permanent outpost there, and the first precursor mission was defined around that goal. It was the Lunar Observer mission that had already been defined as the second mission in the Planetary Observer program. It would launch in October 1995 and carry out a mineralogy, topography, and gravity-mapping mission. Very similar in its instrumentation to Mars Observer, Lunar Observer had one unique addition. It would carry a "sub-satellite," a second satellite that it would deploy into a lunar orbit for gravity studies.[22]

Just as the Lunar Observer concept was lightly customized to fill the mission's new role as a human mission precursor, the work the group had already done drove most of the recommendations for Mars precursors. The basic technological needs of the sample return mission seemed well aligned with the needs of future human landings: better knowledge of the martian surface; improved entry, guidance, and navigation techniques; and automation. Similarly, since the human missions being planned depended on manufacturing fuel, water, and other necessities on the martian surface, the human mission planners needed definitive answers on the composition of the surface. Sample return would provide them.

One new mission, however, did wind up being added. As part of its earlier planning effort, NASA's Office of Exploration had commissioned a study at its Ames Research Center of what future Mars missions might be needed; they had responded with the Global Network mission. This mission would emplace 20 penetrators, or "hard landers," at various spots on Mars's surface, including some near the polar ice caps, to carry out meteorological and seismological studies. This meteorology network was seen as necessary to provide "ground truth" for Mars atmosphere models.[23] The Viking lander data sets, from two equatorial locations, were not sufficient to properly constrain global atmosphere models. Yet these models were the basis for designing atmosphere entry systems for spacecraft, since variation in upper atmosphere density would affect the entry vehicle.

The Mars robotic precursor plan that evolved from this effort had a second Mars Observer flight in October 1996, in the event that Mars Observer itself failed, and then in December 1998 the Global Network mission would be launched. The network mission would be followed by a sample return mission in 2001, using the local rover concept for sample collection. In 2003, large, high-resolution imaging and communications orbiters would fly to Mars. Finally, beginning in 2005, longer-range rover and sample return missions would be conducted in alternating opportunities through 2011.[24] The later rover missions would be targeted at the chosen human landing sites to provide detailed site surveys so that astronauts would know exactly what to expect. They would also be the means for certifying the safety of the first human landing sites.[25]

The robotic explorers would be followed to Mars by automated cargo vehicles that would emplace supplies, habitats for the astronauts, instruments, and other needs at the rover-certified sites. A new, heavy lift launch vehicle would be necessary for this phase of the Mars program, as well as for lofting the components of the astronauts' vehicle into Earth orbit. The piloted vehicle would be assembled at Space Station Freedom and would use nuclear thermal propulsion to expedite the trip to Mars. These missions would last 500 or 1000 days, with 400 of those days in transit to and from the Red Planet.[26] The 90-Day Study plan was, accordingly, expensive, with a set of options that ranged from $471 billion to $541 billion, beginning in fiscal year 1991 and extending through 2025.[27]

Vice President Quayle's National Space Council was horrified at the 90-Day Study's price tag. Quayle had Truly remove the offending cost estimates from the public version of the study, and he had Space Council staff start looking for less expensive options. The council also embarked on a public relations strategy aimed at discrediting the report, knowing that the costs would leak out anyway.[28] It was obvious already that Congress would not finance a program of this magnitude; salvaging Quayle's agenda meant finding a way to make it cheaper—or, at least, to make it appear cheaper.

The 90-Day Study was released in December 1989, as the president's fiscal year 1991 budget was being prepared for submission to Congress. In an early sign of trouble, the White House Office of Management and Budget reduced NASA's Space Exploration Initiative–related request substantially, although it still allowed an increase in the agency's budget. In Congress, however, most agencies were facing cuts, not increases, and NASA was no different. A White House "space summit" on May 1, 1990, failed to convince congressional leaders that the new initiative was worth funding. Two more revelations in 1990, that the Hubble Space Telescope's primary mirror had been manufactured incorrectly and that two of the four space shuttles had developed dangerous hydrogen leaks, damaged NASA's technical credibility and its political chances.[29]

It did not help the administration's case that the Space Exploration Initiative, a quintessential Cold War–era giant space endeavor, happened to be working its way through the political process at the same time as the Soviet Union was collapsing. Soviet leader Mikhail Gorbachev had embarked on a series of political and economic reforms in 1985 that had been intended to strengthen the nation's destitute economy, but they had the opposite effect. Political freedoms Gorbachev introduced undermined the Communist Party and its hold on the USSR as well as the USSR's hold on the states of Eastern Europe. Cheering crowds dismantled one of the Cold War's greatest symbols, the Berlin Wall, in November 1989 before a shocked American television audience. East and West Germany merged during 1990, and the Soviet Union itself dissolved in December 1991. The new Russian Federation, as the Soviet Union's core state was renamed, did not even retain control of the USSR's old launch facilities at Baikonur, which were in a new state: Kazakhstan.

The collapse of the Soviet Union deprived NASA of its founding function, the creation of Cold War techno-spectacles.[30] While the agency also performed outstanding scientific research, that wasn't the agency's driving purpose. Congress had been willing to fund space exploration for its political uses in the Cold War; with the USSR rapidly imploding, there was no longer any reason to grant large funding increases while most other portions of the federal budget were being cut.

In this atmosphere, the House Appropriations Committee chose to remove all Space Exploration Initiative–related money from the fiscal year 1991 budget, while still allowing an increase in NASA's overall budget. It specifically stripped out Lunar Observer and new launch vehicle development funds and deleted funding for new human missions and vehicles. The Senate agreed. The White House continued to promote the initiative, and although NASA officials continued to act as if the initiative could be revived, it was effectively over.[31]

Geoff Briggs had converted the Mars Rover Sample Return Science Working Group into the Mars Science Working Group in mid-1989. The reconstituted group met in August 1990, where a major topic of discussion was the ramifications of the cuts for Mars science. The budget would no longer support a sample return of any description. Given the constraints, Briggs had asked Ames Research Center to put together an alternative to sample return that might fit the budget. G. Scott Hubbard, a physicist hired by Ames in 1987 to revitalize that center's space technology group, landed that task. Ames had developed a reputation for doing small, low-cost space science missions during the 1970s, but like JPL, it had seen little space-related work in the 1980s.

Hubbard approached the problem by constraining the mission to the Delta II rocket, a medium-size launch vehicle that had never been used for planetary missions. This led to a mission design consisting of a network of small, inexpensive, identical landers to be launched over three Mars opportunities.[32] These would be "hard landers," unlike the Vikings, using some sort of airbag to cushion the probes' final descent. They would also enter the martian atmosphere directly from their interplanetary trajectory, not from Mars orbit. This would save the weight of an orbit-entry propulsion system and fuel, at the cost of a much higher entry velocity. Initially, Hubbard's mission concept was called Evolutionary Mars Network because it would evolve as it was built up over several launch periods. NASA headquarters didn't like that name, so it became the Mars Environmental Survey (MESUR). At a proposed $750 million to $1 billion, MESUR was not "cheap," but it was a tenth the sample return mission's cost.

While Hubbard's group at Ames had been interested in developing the MESUR network mission for atmospheric science, another group of scientists saw value in it for geology and exobiology. Led by Cornell University astronomer turned geologist Steven Squyres, this group had taken on the task of pulling together geochemistry and exobiology objectives for it.[33] Both groups of objectives seemed to be compatible at this point in the planning, as meteorologists and geologists each wanted data from a variety of locations on Mars, including the polar regions. For geologists, the poles were where most of Mars's water was expected to be. Meteorologists, of course, needed polar data for model validation purposes. As long as the small landers could carry the instruments each group desired, the entire Mars community could support this network concept. MESUR, or something very much like it, was what Geoff Briggs needed to keep the planetary science program going after the demise of the Space Exploration Initiative. Sample return, despite its universally regarded high scientific priority, was simply unaffordable.

MESUR Gets a Rover

JPL's scientists and engineers faced a problem: the Mars Environmental Survey mission was an Ames mission. It left many of JPL's scientists and engineers irrelevant, and potentially unemployed, because it would swallow most of NASA's planetary budget for at least a dozen years. If MESUR continued as it was defined in 1990, there would be no rover missions to Mars within most people's careers. Yet the scientific desire to reach out and grab rocks just beyond a lander's reach continued to exist, a legacy of the Viking veterans' frustration with not being able to pick up rocks they could see. And, of course, JPL's engineers had strong desires to foster robotics technology, too. So an almost immediate effort started to get a rover added to MESUR. But it couldn't be the big rover of the sample return planning. A rover had to be tiny if it were to go on the network's small landers.

In his memoir relating the evolution of the Mars Pathfinder rover, roboticist Andrew Mishkin explains that Donna Shirley brought in a second opinion on how a surface rover might operate to challenge the Godzilla rover mentality at JPL. Built on the idea of insect-like responses to stimuli, this small-scale effort produced a microrover named Tooth, and then a somewhat larger, but still small compared to the sample return rover, called Rocky 3.[34]

At the same time JPL's robotics engineers were working on their vehicles, the MESUR mission was running into engineering troubles. Several problems had become clear to the science working group members during 1990 and the first half of 1991. To build up MESUR's network across three launch windows meant that the first set of landers had to be long-lived. They would have to survive at least eight years for the full network to provide two years of global atmospheric data. This was a difficult technical challenge. Another problem turned out to be the network configuration. Put simply, the geologists and meteorologists needed the landers to be in different locations. Thirdly, getting all the data back to Earth was turning out to be difficult, too. The landers were also going to need a communications orbiter. The orbiter hadn't been in Ames's original plan and wasn't part of the network's cost estimate. Adding the orbiter to the network plan put the mission cost well above a billion dollars.

Finally, the science working group's geologists still wanted MESUR to perform geochemistry. But the design hadn't included any way to collect a rock to experiment on. The key instruments Steve Squyres and Mike Carr wanted to deploy were seismometers, to detect marsquakes and investigate the planet's internal structure, and a mineralogical instrument called an alpha-proton X-ray spectrometer (APXS). The APXS, a version of which had been used successfully on the Surveyor Moon landers in the 1960s, had to be put in direct contact with its target to give a good result. It needed a deployment mechanism. The seismometer needed to be moved off the lander, too, after landing so that it was free from the lander's inevitable vibrations. This was a matter of Viking history. Only one of the two Viking landers' seismometers had actually worked, and its results were uninterpretable. As Matt Golombek, the JPL geologist who had been serving as secretary for the Mars Rover Sample Return Science Working Group, put it years later, the observations "correlated with strong wind."[35] The mission scientists couldn't tell the difference between a marsquake and a passing breeze.

So to satisfy a geologist, the network landers needed a way of getting two instruments onto the surface. Ames engineers presented a variety of possible methods. Spring-loaded devices could simply toss them out. A Viking-like arm or boom could pick them up and drop or place them (the meteorologists also wanted an instrument arm). But neither approach was terribly convincing. The instruments were relatively delicate, so throwing them made little sense. One couldn't accurately place them this way in any case. The Viking veterans didn't like the arm idea, as the Viking arms hadn't been able to pick up the rocks they had wanted. The network, in short, was starting to look unappetizing to its own scientists. Put another way, scientists' desires were getting in the way of a simple, cheap mission.

During the spring of 1991, Golombek had been attending a (mostly) weekly lunch with Charles Elachi, the director JPL's space science division. They had been discussing the possible scientific uses of a Rocky-like microrover. One of these possible uses was as an instrument deployer for MESUR. In this environment, a summer 1991 Rocky 3 sample demonstration proved to be incredibly well-timed. Rocky trundled across the Arroyo Seco, picked up a predetermined scoop of dirt, retraced its path and dumped the dirt into a box.

Rocky 3's success enabled Elachi to get more money for a more flightlike rover for still another demonstration. This $2.5 million project became the Mars Science Microrover, more commonly known as Rocky 4. Matt Golombek was initially in charge of the effort but was replaced by Arthur "Lonne" Lane. Lane was a veteran JPL project manager experienced in carrying out short-fused efforts. In the face of considerable skepticism, Golombek led the task of convincing his fellow scientists that a tiny rover was worth having.

All of this Mars-oriented effort occurred against a backdrop of shrinking NASA budgets and personnel changes at headquarters. Geoff Briggs stepped down as head of NASA's planetary science directorate and moved back to Ames Research Center in 1990, and Wesley T. Huntress Jr., succeeded him. Huntress had been a researcher in cosmochemistry at JPL beginning in the late 1960s. In 1985, he became the pre-project study scientist for the mission that became Cassini-Huygens; in 1988, he had moved to NASA headquarters to help assemble the Earth Observing System that was being proposed. Huntress, like Briggs, had been frustrated by the slow pace of planetary missions and their high costs. In particular, he was frustrated by JPL's engineering culture, and the way it stifled efforts toward low-cost missions. "JPL did not know how to do it. They just simply did not. Their whole system here, the way JPL does missions, it's anathema when it comes to low cost. You just can't do them here," Huntress commented many years later.[36]

Once Huntress was established in his new job as director of planetary programs, the ongoing reductions in the science budget led him to do a deal with the director of NASA's Office of Exploration for a $75 million lunar mission called Clementine to prove that planetary missions didn't have to be expensive. The Office of Exploration, which had been created to support the Space Exploration Initiative, wanted to show some early accomplishments for the not-quite-dead initiative, and Clementine was a way to do it without violating the congressional ban.[37]

Huntress also planned the new Discovery program in 1991 to carry out Explorer class planetary science missions. It would fulfill the role that the Planetary Observer program had been designed for, and failed in, low-cost science. He initiated project concept studies with both JPL and the Applied Physics Laboratory (APL), a unit of the Johns Hopkins University and a longtime defense space contractor. APL's reputation was for low-cost, quick-development missions; most recently it had done the Delta 180 antiballistic missile demonstration for the Strategic Defense Initiative Office in 18 months.[38]

The Discovery program concept would also break a logjam in space science created by the scientific planning committees themselves. As had happened with sample return and was happening again with MESUR, as the science committees tried to resolve disputes within their ranks, their mission concepts grew more expansive, and expensive. The Discovery program would instead be, like the older astronomy-oriented Explorer program, proposal-based. The program manager would put out an announcement of opportunity every few years, and a single principal investigator would propose an entire mission in response. While most proposals would probably be team efforts, with many scientists involved, the principal investigator would be responsible for constraining the scope of the mission to within the allowed budget throughout its life cycle. By putting a cost-cap on each announcement, Huntress would force potential principal investigators to limit their objectives to the available funds at the outset of their efforts. The Discovery program concept would restructure the way planetary science was done.

At a meeting of NASA's Solar System Exploration Advisory Committee at Woods Hole in summer 1991, Huntress had gained support for a fiscal year 1994 "new start" for the Discovery program. The first Discovery mission was to be the Near Earth Asteroid Rendezvous (NEAR). He also got committee support for one new outer planet mission, one new inner solar system mission, and one more mission that was left undefined. The inner solar system mission was to be the network. But it had been kicked off to the right of the planning schedule, with its new start not until 1997.[39]

On October 29, 1991, Huntress laid all this out for the Mars Science Working Group. In addition to its scientific troubles, the MESUR network was now so far out in the future that it was going to be difficult to maintain support for the program. So Huntress had invited two presentations on alternate, smaller, and earlier Mars missions: a Mars aeronomy mission and an early MESUR demonstration mission called Surface Lander Investigation of Mars (SLIM). Scott Hubbard presented SLIM as a 1996 launch of a single MESUR lander. This would enable demonstration of the project's technical merits early and improving its political and technical chances. Estimated at $137 million, the mission seemed to fit the emphasis on low cost. SLIM, however, didn't gain the support of the assembled scientists. By itself, it would do very little science; the point of the network had been global data, and a mission that looked like a single, far less capable Viking was not attractive. So the science working group's response to it was tepid.[40]

The next day, however, Matt Golombek rescued the SLIM concept with the Mars Science Microrover. He showed the group video of the Rocky 3 sample collection demonstration and offered the microrover as a potential addition to SLIM. The microrover could, Golombek argued, act as the sensor deployment mechanism that was missing from the concept. The microrover could easily bring an APXS to a convenient rock. It might also be able to deposit a seismometer, among other possibilities. And, Golombek emphasized, it could be relatively inexpensive. Unlike the 800 kilogram sample return rover, the microrover could weigh a few tens of kilos and cost a few tens of millions.[41]

The microrover concept drew extensive discussion at the meeting, leading to support for adding one to a SLIM-like mission. Summarizing the meeting for Wes Huntress, Mike Carr said that the microrover had drawn "considerable enthusiasm" from the group, because it would be "a demonstration of a new capability at Mars." It would also, he reflected, have strong public appeal and draw attention to the Space Exploration Initiative. He concluded that "a MESUR vehicle carrying a microrover and highly focussed [*sic*] science is an attractive possibility."[42]

Ames Research Center, however, lost the MESUR mission at almost the same time. Scott Hubbard remembers that NASA's associate administrator for space science, Lennard Fisk, came out to Ames to pin down the center's commitment to MESUR. But when Hubbard and Fisk met with the center's director, Dale Compton, he wouldn't sign up to the task.[43] Asked years later, Compton recalled,

> I don't think we could have staffed it. Even if we'd staffed it, I think if it had started to get in trouble, I don't think we'd have had the extra resources to pour into it to pull it out of troubled times. That was where we got in trouble in other projects. We'd learned the hard way that if you don't do something right and you can't pull it out of trouble, it causes you all sorts of pain. It's not just the pain of going back for more budget and arguing that it's still an important project; it's the pain of not being able to treat it properly in-house, whereas JPL has got the luxury of finding all the folks that have been working on technology and are helpful in other projects. They can just pour talented help into a project like that.[44]

Compton also had given higher priority to winning another mission concept, the SOFIA airborne infrared observatory that was to succeed the Kuiper airborne observatory that Ames had operated for many years. His decision left Fisk little choice but to transfer MESUR to JPL. At the end of October, Fisk informed JPL's director, Edward C. Stone, that MESUR was being sent to Pasadena.[45]

SLIM Becomes MESUR Pathfinder

From headquarters, Wes Huntress pressed Stone to appoint an operations specialist named Anthony J. Spear as the MESUR Pathfinder project manager. Spear had just finished a tour as project manager of the Magellan mission to Venus and was heading JPL's Discovery program concept study. Huntress knew Spear from his own career at JPL and considered Spear a renegade from JPL's dominant, excessively conservative engineering culture.[46] Spear, an enthusiastic skier in his spare time, thought a big reason for JPL's high costs was an antagonistic relationship between its flight projects and its mission assurance and safety organization, and he set out to change that. He brought in a senior mission assurance engineer at the very beginning to smooth that relationship.

When Spear took over MESUR, the mission was still structured to deliver 16 landers to Mars over three launch opportunities beginning in 1998, and the SLIM idea of launching one lander early to demonstrate the concept was still being studied but had not yet been approved. But in spring 1992, Huntress called him up and told him that he'd just lost a political battle inside NASA headquarters to keep the lunar science program in his Office of Space Science.[47] Instead, it was being transferred to Michael Griffin's Office of Exploration, and he was afraid he'd soon lose Mars as well. In order to keep Mars science from being taken away, he needed to show that Office of Space Science could get to Mars quickly and inexpensively.

So Huntress told Spear that he wanted him to do the SLIM concept in "Discovery mode," as a $150 million, cost-capped mission. Its target would be the 1996 launch opportunity. Huntress also told him that he needed to use the mission to shake up JPL. The place was too slow, stodgy, and expensive. Additionally, he liked Ames's landing concept: airbags, instead of a Viking-like landing rocket.[48]

The new version of SLIM was quickly named MESUR Pathfinder. While the Discovery program was intended to be a series of science-focused missions, Pathfinder was not. Instead, it was clearly defined as a technology mission. In other words, its function was to prove that the larger network mission's hardware would work, not that it could deliver quality science. It also would serve to demonstrate that landing on Mars could be done for far less than with Viking. But Pathfinder did have to have some scientific content, if only to satisfy the Mars scientists who had spent many years advocating for a new mission. What this component would be was the subject of some controversy over the remainder of 1992.

Huntress had decreed that Pathfinder use a Delta II rocket, the smallest active U.S. launcher, to get to Mars, so it had limited mass available for scientific instruments. The Delta II had never been used as a launch vehicle for planetary missions before. Much of Pathfinder's mass allocation for science instruments was being held open for the microrover, as it was the item that the working group scientists thought would have the greatest value. In June 1992, with Matt Golombek, Donna Shirley, and many others cheering it on, the Mars Science Microrover had performed a mock martian expedition in the Arroyo Seco. Disembarking from JPL's Surveyor lunar lander model, and weighing only 8 kg, the microrover had deposited a seismometer, made its way to a preselected rock, and taken data there using an APXS-like instrument. The demonstration got a rover added to Pathfinder; the rover, in turn, represented both a major engineering challenge and the new science capability that MESUR Pathfinder would demonstrate.

As JPL was beginning to define MESUR Pathfinder, the European and Russian space agencies were also embarking on new Mars planning. Matt Golombek served as the liaison to the European effort, which was a parallel to the MESUR network idea. Called Marsnet, it was to be a network of landers equipped similarly to those from MESUR. During 1992, discussion about the mission had focused on the complementarity between Marsnet and MESUR and on the possibility of carrying out the two missions cooperatively. Marsnet was still in its early definition phase and was scheduled for its approval review in March 1993; this placed the mission somewhat out of phase with MESUR.

Russian planning was farther ahead. The Russian government, in dire financial straights, had suggested suddenly to NASA officials in fall 1992 that their large Mars 94 orbiter, which was scheduled to carry thirty instruments and two "small station" landers to Mars, might be able to take three landers instead. But the United States would have to pay for the additional lander. This offer inaugurated a quick effort to figure out how the lander could be "Americanized" with an instrument but without significantly changing the lander itself or displacing the seismometer and meteorology package that it already had. Huntress had given this problem to Steve Squyres's MESUR science definition team to solve.[49]

Much U.S. space science had been driven by competition with the now defunct Soviet Union, and while cooperation with its successor, Russia, was politically in vogue, the fact that MESUR Pathfinder compared poorly with the ambitious Russian mission struck Pathfinder's scientists as troubling. MESUR Pathfinder, as it rapidly evolved in 1992, was not a science-driven mission. It was a technology experiment, and that drew a great deal of unhappiness from the MESUR science definition team. Squyres, summarizing the results of the November 1992 science definition team meeting for headquarters, put it this way: "Pathfinder science is especially disappointing in comparison to Viking and to what Russia will be attempting at Mars at that time."[50] Several nations were contributing instruments and even funds to help keep the Mars 94 mission going, given the Russian Federation's financial straits.[51] But it promised a comprehensive investigation of Mars, while Pathfinder was promising, at most, a weather station with a small, experimental rover.

Defining MESUR Pathfinder, 1992–1993

The initial lander concept sent down by Scott Hubbard's project at Ames was disc-shaped. It was designed to land with either side up, with instruments that could deploy in either direction. The design avoided the volume and mass impact of the Viking landers' rocket-based landing system by using airbags to absorb the final impact with the martian surface; this also promised, the designers thought, lower recurring production costs than a Viking-style system. Four of these 80 kg landers would fit on each Delta II.[52]

In mid-1991, prior to the decision to move the project from Ames to JPL, NASA headquarters had directed JPL to conduct an independent "reality check" on Ames's concept. JPL had assigned a new employee, mechanical engineer Tommaso Rivellini, to the task. Rivellini and three others had started by looking for earlier work to draw information from. The team found that JPL had done an airbag lander study back in the 1960s—Donna Shirley had worked on it briefly—and the old designs provided a useful beginning. One of the constraints Rivellini had been given in the tasking was that the lander should be able to carry and disembark a small rover, which suggested to him a cube or tetrahedron instead of the disclike craft used in both the old JPL study and the Ames plan.

Rivellini's team quickly settled on the tetrahedron as the most appropriate shape for the lander. As the closed geometric shape with the fewest number of sides, it would also require the smallest number of motors and actuators to open in order to let the rover debark. Rivellini explained later these characteristics would make it the least expensive to manufacture, which was important since the MESUR network was supposed to include between 16 and 20 nearly identical landers. Further, a tetrahedral lander would easily fit inside the volume of an aeroshell shape like that used by the Viking landers, saving the cost of revisiting entry aerodynamics to design a different shape. The microrover,

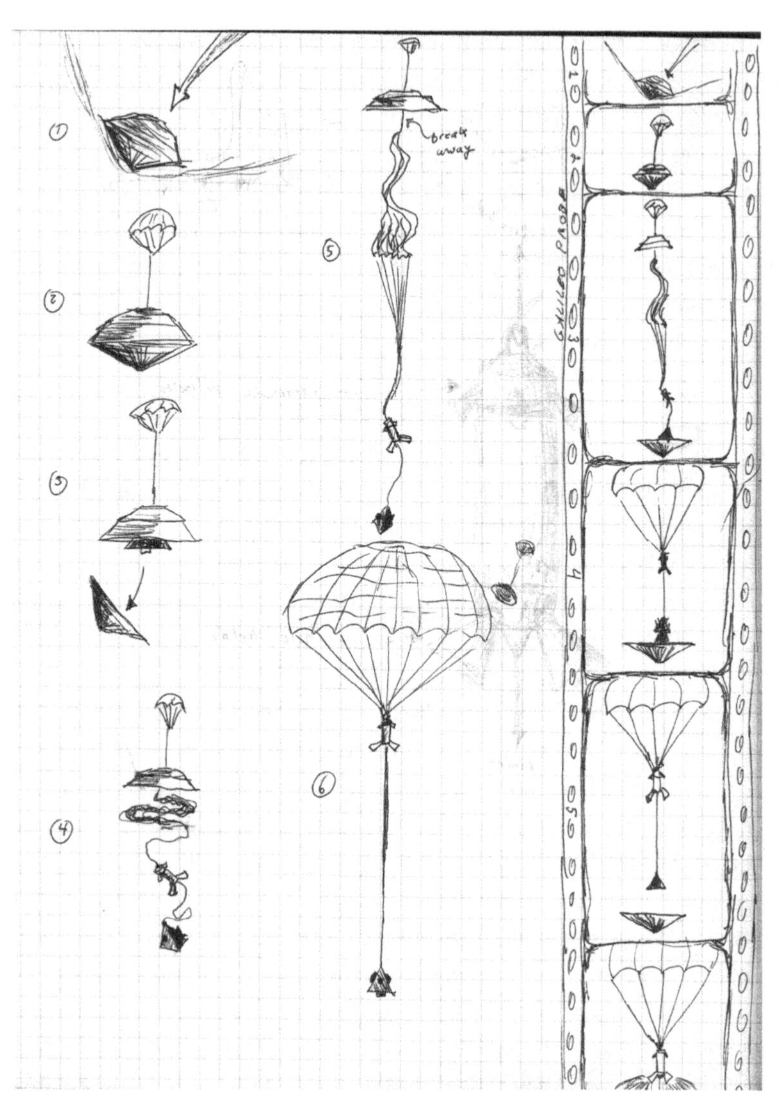

Tommaso Rivellini's sketch of MESUR Pathfinder's landing sequence. He began by thinking about the Galileo atmosphere probe's entry sequence (noted along the left edge of the "filmstrip") and expanded it. His notebooks record several iterations of the basic design concepts.
Courtesy Tommaso Rivellini.

if one should be selected for the mission, could be attached to one of the petals or the base. When the MESUR project moved to JPL, Tony Spear inherited Rivellini and his tetrahedron. It would be one of few things that didn't change much from Rivellini's initial sketches.

Rivellini's little group had then turned to the airbag concept. The airbags' chief purpose was to absorb about 60 meters per second worth of velocity on impact with the martian surface, the equivalent of hitting a wall at 134 mph. The lander would still be going this fast because its parachute would not be particularly effective in Mars's thin atmosphere; the hope was that airbags would turn out to be a more mass- and cost-efficient approach to solving this terminal descent problem than Viking's retrorockets had been, at least for smaller landers. Rivellini's 1991 effort on airbags was largely research into the relevant engineering literature and paying visits to manufacturers. Airbags are used by the U.S. military to cushion the impact of airdropped supplies, for example, so he had quickly discovered that there was an extensive airbag literature and several companies that specialized in making them. Rivellini also discovered that the Department of Energy's Sandia National Laboratory, outside Albuquerque, New Mexico, had the ability to model, build, and test small airbags. Sandia's capabilities became a key part of the project's test program. But Rivellini's early digging suggested that as long as the lander stayed relatively small, the airbag approach was a viable one.

By March 1992, when JPL took over MESUR, the project was already "short-fused." NASA intended to have the project's formal confirmation review in October 1993. That gave Spear's team only a year and a half to make all of the major technical decisions about their vehicle, demonstrate that whatever new technologies they chose had a good chance of actually working on Mars, and produce credible cost estimates. The launch period was the month of December 1996; Spear wanted to split the roughly 36-month development schedule in half, with 18 months devoted to design and procurement and the remainder devoted to the "assembly, test, and launch" phase (ATLO, in the jargon), when the spacecraft would be built and tested, in order to have plenty of time to fix problems uncovered during tests. But a 36-month development was extremely aggressive. It put enormous pressure on the team.

The relationship between the Pathfinder lander and the full Mars network was in question throughout 1992. NASA wanted Pathfinder done "in house" at JPL but also desired the larger network of landers to be executed by an industrial contractor. This presented the challenge of knowledge transfer; JPL's Pathfinder team would have knowledge that the contractor would need, but merely providing documentation would not result in an efficient transfer. NASA leaders intended to have JPL put out a request for proposals to industry that would bring contractor engineers into the Lab to work with the Pathfinder team directly, hopefully leading to a more cost-effective transition from JPL's Pathfinder to the contractor-built MESUR network. Besides the hope that an assembly line would be cheaper, later on, Administrator Goldin explained that using an industrial contractor was politically necessary to get the MESUR network approved.[53] With the end of the Cold War and the resulting reduction in military spending, aerospace contractors were complaining to Congress about funds going to NASA centers and not to them; survival of the MESUR network, Goldin implied, depended on solving this political problem.

JPL's David H. Lehman, an engineer whose experience was in attitude control and mission design, was Spear's deputy in charge of developing the network portion of MESUR. He had been given the job of reformulating the basic Ames concept of building up the network across several launch opportunities, with four landers on each Delta II launch, beginning with one launch in 1999. As the design stood, the Pathfinder lander was too heavy to permit more than two on a Delta, so Lehman intended to use microminiaturization of the electronics to reduce the mass. But the hope of adapting an assembly line approach to building the landers was quickly falling apart. Squyres's science group wanted some of the landers to go to the martian polar regions, where there was insufficient sunlight for a solar-powered lander. The polar landers would have to be powered by a different source: radioisotope thermoelectric generators.

Another challenge, Lehman realized, was finding a way of building landers durable enough to last in the inhospitable climate of Mars. Because the network had to be built up across several launch opportunities, the first set of landers would have to operate for at least eight years, waiting six years for the rest of the landers to arrive and then performing for a two-year primary mission. By comparison, Spear had planned Pathfinder for a mission of only 30 days. So there were distinct differences between Pathfinder and the network. The network landers could not be "build to print" copies of the Pathfinder lander.

These differences, and the short period of time between Pathfinder's 1997 landing and the MESUR network's first launch in 1999, caused a good deal of questioning within the project, and in the review board that Spear invited to critique his team, about the linkages between the two efforts. If the electronics, scientific instruments, and power systems were all different, and the MESUR network landers could not carry a rover, what exactly was Pathfinder a pathfinder for? The only obvious commonalities remaining were the cruise stage and the entry, descent, and landing system. So an internal review of the project by JPL a few months after the transfer ultimately recommended that the project drop essentially everything else that wasn't essential to the basic tasks of (1) getting to Mars, (2) landing on it, and (3) sending back engineering data.[54]

Underlying this review's recommendations was the realization that while NASA's Wes Huntress was demanding that JPL "take more risk," he was also adamant about Pathfinder not being allowed to fail. Failure would be more than embarrassing. It would jeopardize the entire MESUR network, which would be too far along in its design to change significantly by the time Pathfinder landed in 1997. Pathfinder's failure would almost certainly result in cancellation of the MESUR project after tens of millions of dollars had been spent on it. Steve Squyres would point out the irony of the faster-better-cheaper philosophy in a letter he wrote in late 1992 to NASA officials. "This 'New Way of Doing Business' is really more conservative in this case, as the risks of landing are not being accepted."[55]

Finally, the review panel thought Spear hadn't presented a Pathfinder mission that would actually cost $150 million. Some of the panel members did not believe that there were *any* possible $150 million landed missions to Mars, while others thought simply that Spear hadn't presented one. For example, Brian Muirhead, the deputy manager of JPL's Mechanical Systems Engineering section and at the time leading a cost-constrained shuttle-based radar antenna project, told the review chairman that such a mission was possible if Spear dropped the solar arrays; aimed for a three-day, battery-only surface mission; and sent the lander to martian high latitudes. This last requirement would make it different from the two Vikings, which landed in the martian "tropics," and demonstrate that high latitudes were feasible from an engineering standpoint.[56]

The review board's recommendation was, in essence, to get NASA to agree to stripping Pathfinder down to an entry, descent, and landing demonstration. Little to no science should be accepted by the project, and the rover should be either cut or significantly descoped in order to generate an even larger financial reserve for the project as a whole. This harsh assessment, and the potential deletion of science from Pathfinder, didn't go over well at the next day's meeting of Squyres's science definition team (SDT).

The potential deletion of Pathfinder's already small scientific content raised an obvious question with the team: why are we even sending this mission to Mars? Squyres summarized, "The landing and uprighting system planned for

Pathfinder is fundamentally new (for the US), but many on the SDT felt that more could be learned about its reliability by performing many drop tests in varied terrains on Earth than by conducting one or two landings on Mars. So we see little that *Pathfinder* would do in the way of truly critical engineering tests that could not be accomplished adequately on Earth."[57]

Squyres was right, at least as it came to practical engineering, but Pathfinder was no longer, if it had ever been, just about engineering, despite the origin of the mission. NASA's solar system exploration program manager, Carl Pilcher, countered that "Mars is 'magic' in that it holds a special place in the minds of Congress, OMB, and the public. A new start for Pathfinder in fiscal 1994 is an opportunity to start the Discovery program two years early. It will be impossible to get $150 million for MESUR so soon without Pathfinder as a Discovery mission. The lack of scientific instruments is not a problem for Pathfinder as a Discovery mission as far as Congress or NASA management are concerned."[58] Pathfinder, in other words, could be funded as an engineering demonstration as long as it went to Mars, but NASA could not get a much smaller amount of money to do engineering development here on Earth. This was, Squyres wrote back to NASA, "a strange and unfortunate state of affairs."[59]

Partly due to the scientists' opposition, Pathfinder went forward as a single vehicle mission with a science component intact. Matt Golombek got the science announcement of opportunity out in December 1992, asking for imaging instrument proposals and for science team proposals for the atmospheric structure and meteorology instruments that were part of the spacecraft's engineering subsystems. The imager proposal had a very short deadline; Spear needed to know what imager he would have to accommodate by July 1993, when his team faced a project approval review. He remained under some pressure to drop the imager throughout the next year. But the basic Pathfinder mission as defined during 1992 went forward with a single exception: the seismometer was dropped because, while the instrument was to be provided for free by a foreign government, it would cost money to integrate onto the lander. It didn't take long for Spear's team to figure out that this "free" instrument would cost too much for their budget, and Squyres's science definition team had assigned it a low priority, below both the microrover and the imager.[60]

After the November 1992 review, Spear went looking around JPL for new additions to strengthen his project team. He explicitly sought out youngsters who would be more willing to break with the traditions of JPL. But there was one important exception: the mission assurance manager. At JPL, the Safety and Mission Assurance Division was the final authority over key design, testing, and

Mars Pathfinder scientific payload

Instrument	Investigators (institution)
Imager for Mars Pathfinder	Peter Smith (University of Arizona)
Alpha-proton X-ray spectrometer (APXS)*	Rudy Reider (Max Planck Institut für Chemie) T. Economou (University of Chicago)
Atmospheric structure investigation / meteorology package (ASI/MET)**	John T. Schofield (JPL)

Source: Matthew P. Golombek, "The Mars Pathfinder Mission," *Journal of Geophysical Research*, 102, no. E2 (25 February 1997): 3953–65.

*The APXS was a derivative of an instrument flown on the Russian Fobos mission and lost. It was composed of three units: alpha and proton spectrometers provided by Reider and an x-ray spectrometer provided by Economou.

**The ASI/MET package was a facility instrument without a true principal investigator. Schofield served as investigation scientist.

safety standards, all of which could have huge impacts on a project. Spear believed that the relationship between projects and Safety and Mission Assurance was often hostile, leading to overly conservative and time-consuming design and test efforts, and he hoped bringing a senior mission assurance manager in early and including him as a key team member would smooth relations with the division. He also hired Brian Muirhead, one of his critics on the review panel, to be the flight system manager. Muirhead was tasked with delivering three of Pathfinder's four "vehicles" (cruise stage, aeroshell, and the actual lander), which gave him most of the project's money, personnel, and responsibility.

Muirhead, in turn, went looking for an electronics and software expert and turned to one of JPL's old timers, Joseph Savino, an avionics specialist hired in 1956, for a recommendation. Savino sent him to Robert M. Manning, who had once been an English major at Whitman College before switching to computer science and was working on the Cassini-Huygens mission to Saturn. Manning's first impression of the Pathfinder concept was that it was "nuts." But he also couldn't resist the challenge of doing something really new and signed only a few days after being asked.[61] Manning's expertise in software and electronics filled a major gap in his leadership team.

JPL spacecraft typically had independent, and redundant, subsystems. They are "one-fault tolerant designs." The lab's technical division delivered "black boxes" to a flight project that, ideally, would all be connected to form the spacecraft. One consequence was that JPL's institutional structure "looked like the block diagram of a spaceship," Manning observed. "For every box that's built, there was an organization for that box."[62] The Laboratory's senior management had organized the institution that way to assure that its structure was well aligned with the product it built. The idea was to make development predictable. But that rational structure was a problem for Pathfinder. JPL had never done a planetary entry project before—there was no organization to build the entry, descent, and landing "box."

In short, JPL was not structured well for Pathfinder, a problem that the team solved by "soft projectization." Spear secured the team space on the top floor of a single building, pulled most of the key engineers out of their technical divisions, and collected everyone in his office block. This was an experiment in "co-location" that was designed to improve communications and break the institution's stovepiping of subsystem tasks. It also coincided with an institutional decision to integrate the command and data handling organization and guidance and control organization in what Muirhead called a "shotgun marriage."

Pathfinder needed to make the shotgun wedding work because having separate boxes for each subsystem would make the spacecraft too big and heavy. Rivellini's little tetrahedron didn't have enough volume to fit all the electronics if they followed the traditional JPL approach. Manning convinced Spear's team that they had little choice but to integrate these different subsystems as much as possible. A single, high-performance CPU with about six times the capacity of Cassini's processors would run virtually everything on the Pathfinder lander. This enabled integrated software development and the adoption of the first operating system capable of supporting multitasking to fly on a JPL-built spacecraft. (Mars Observer had already gone that route.)

The final major step was settling on Pathfinder's airbag landing approach. For a few months, Spear kept open the possibility of using a Viking-like retrorocket. After an entry, descent, and landing peer review in November 1992, though, he decided to abandon the retrorockets and focus his efforts on the airbag lander. There was a good deal of pressure coming from NASA headquarters to do "something different." He recalled that when he presented Wes Huntress with trade studies evaluating the retrorocket descent, Huntress had been "disappointed that we even considered it."[63] Airbags became the default choice, making the new question "what kind of airbags?"

Beginning in late March 1993, Tom Rivellini, working with engineers at Sandia, carried out a series of three-eighth scale tests of airbags. One set of tests dropped a lander mock-up in ambient atmosphere conditions in the New Mexico desert, while a second set of tests took place in a high-altitude chamber to approximate Mars atmospheric pressures. A third set of tests went back to Earth's atmosphere, this time to find out whether the addition of horizontal velocity changed the airbags' performance. Rivellini's group experimented with two types of airbags. The original Ames design used airbags with blow-out vents, which allowed them to absorb much higher impacts—although only once. The vents began releasing the gas milliseconds after impact, so the vehicle's impact energy was dissipated by accelerating the gas through the vents. These bags would deflate rather than bouncing. But Rivellini's crew also experimented with unvented bags. Unvented airbags would distort on impact, but not deflate, causing the lander to bounce repeatedly like a giant kid's superball until all of the impact energy had been dissipated by friction. The drawback of the unvented approach was that these bags could not withstand the 60 m/sec vertical impact velocity that the Ames study had suggested would remain when the lander reached terminal velocity on its parachute.[64] They would burst.

Rivellini wound up recommending to Muirhead that they choose the unvented bags anyway. He explained that in his tests, the blow-out patches had not been reliable or predictable. Instead of blowing out at a specific pressure, they would blow out over a wide range of pressures, and on a few occasions not at all, which made predicting the behavior of the system impossible. Further, when the bag vents had worked acceptably well, the test body was still going too fast when the airbag had fully deflated. If it had been a spacecraft, the remaining speed it had would probably have caused it to be destroyed on impact with the martian surface. So he would need, in essence, a second set of airbags inside the first to absorb the remaining impact energy. The dual set of airbags would be much heavier than the team had expected and far beyond their mass allocation. Rivellini didn't think it could be done, at least not within the project's budget and time frame.[65]

The recommendation to use the unvented bags left the team with a substantial velocity problem. They needed to eliminate about 60 m/sec of vertical velocity (and some unpredictable horizontal velocity of up to 25 m/sec), and Rivellini was promising that his airbags could absorb just 25–30 m/sec. They had only two choices, really: make the parachute much bigger or use rockets of some sort to slow down. Muirhead asked a mechanical engineer at JPL named Dara Sabahi to study the problem. Sabahi had come to JPL after a career as a bridge builder, specializing in a special structural analysis software system NASA had developed and then commercialized.

The analysis Sabahi presented to Muirhead in May strongly favored the use of solid rockets instead of a larger parachute. The parachute would weigh much more than the rockets, and the enlarged parachute also would no longer be similar to the Viking parachutes. The team did not have the money to repeat the very high altitude tests done to quantify the Viking parachute design's performance (those had cost nearly as much as Pathfinder's entire budget), so they were depending on the validity of decades-old heritage data. Sabahi and Manning were also afraid that the giant parachute might drape over the lander, crippling it. Using rockets, however, would allow them to shrink the existing parachute design a little, resulting in less overall mass. In this approach, three small solid rockets derived from the type used on ejection seats would be attached to the backshell, the aerodynamic cover that remained suspended between the lander and the parachute.[66] The rockets would fire about 25 meters above the surface to reduce the vertical velocity to zero, followed by the lander being cut free of the backshell. Then the lander would drop to the surface and bounce to a stop.

Sabahi pointed out the rockets would also solve the parachute draping problem. If Pathfinder landed on a day with no wind, the parachute might settle over the lander, preventing the lander from opening: a hugely embarrassing end to an otherwise successful entry. Releasing the lander from the backshell while the rockets were still burning would cause the backshell to fly away from the falling lander, taking the parachute with it.[67]

The rocket approach's principal drawback was that it would require precise timing. The rockets had to fire at a specific height above the surface. Too early and the martian gravity would reaccelerate the lander to velocities too high for the airbags; too late, and the lander would already be at too high a velocity, even with rockets firing. Muirhead had hoped to keep the entry sequence control system very simple, using mechanical timers to trigger each event in the sequence. That would not be possible with the rockets. Instead, he would need a means of determining exactly where the lander was in relation to the surface in real time. The Viking project had employed a sophisticated and expensive Doppler radar altimeter to do this on its landers in 1976. Muirhead had neither the money nor the time to replicate that. Instead, embracing Pathfinder's "keep it simple" approach, and went with a plumb bob. When the plumb bob touched the ground, the release of tension on its cord would trigger a sequence that would fire the rockets, inflate the airbags, and, once a predetermined time had elapsed to allow the rockets to burn, cut the lander loose from the backshell.[68]

Approval of MESUR Pathfinder

In July 1993, Tony Spear and his expanded lander team faced their review board again, this time for the design, implementation, and cost review. This verdict would determine whether the NASA headquarters would permit their "pre-project" to move into formal project status and begin making its major procurements. To pass through this gateway, Spear's team had to convince the board that they had a sound design for their spacecraft and lander, and that they had a credible cost estimate. This board largely accepted the technical merits of the project, but with a lot of discussion over the decision to add solid rockets to the backshell. The manpower and cost estimates proved the larger hurdle to clear.

Spear had decided to scope the project's work so that he would have a large financial reserve available to use in overcoming development troubles, about $33 million of the $150 million cost cap. He and Muirhead had agreed the flight system development (the spacecraft without instruments or rover, and without the cost of operating after launch), should be $86.4 million. Mission operations they budgeted at $12.5 million, and their agreed science budget was $13.2 million.[69]

The review chairman, former Viking project manager Jim Martin, argued that the entry, descent, and landing (EDL) development seemed grossly underbudgeted and understaffed. He figured it would wind up costing close to $40 million, not the $13.6 million that Spear and Muirhead had planned on. The overrun was available within Spear's project reserve, so Pathfinder could still afford it, as long as nothing else overran significantly. However, he only saw eight people assigned to work on the EDL problem, which he and several others found ridiculously few. It didn't matter if anything else worked if the landing system failed, so he thought EDL should be the project's highest priority. The largest manpower component was actually in the integrated attitude control and information subsystem, which also had the largest share of the spacecraft budget at $30 million.[70] So the board recommended getting more people onto the EDL system immediately.[71] Its development couldn't take a back seat to anything.

The major technical issues to draw the board's attention were the addition of the solid rockets and the "heritage" of the Viking-derived heat shield. Muirhead received some rough handling over the rocket addition because it seemed to add complexity to a once-simple idea. But his team convinced the board that the rockets simplified the development problem considerably. Several board members, including Scott Hubbard, commented in their notes that the rocket decision was a good one, defusing the issue.[72] The heat shield was a larger concern. The Pathfinder team intended to use a scaled-down Viking heat shield, which the board members thought problematic because the material it was to be made of hadn't been manufactured in many years. A heat shield expert from Ames Research Center pointed out that the contractor, Martin Marietta, would have to re-create the ability to make the material. It almost certainly wouldn't be identical to Viking's.

That raised the question of whether the heritage argument was at all valid. Spear couldn't depend on the heat shield material's performance without testing it comprehensively; Ames would have to do the testing, and the project needed to budget for it.

After the review, Huntress approved the mission, which had evolved into a standalone enterprise aimed at an EDL technology demonstration with a small science element. The difficult disconnect between Pathfinder's short-lived design and that of the long-lived MESUR network landers had not yet been bridged, but that worry could be addressed by others somewhere in the future. Spear was challenged enough with the relatively simpler requirements of Pathfinder.

Conclusion

Despite the approval of Mars Pathfinder, many people at JPL thought Spear would not be able keep the project within its budget, let alone land the rather "Rube Goldbergian" vehicle safely on Mars. Spear and the team he assembled over the next few months felt themselves under constant pressure from their peers at JPL over their chances of success. Manning recalls that many of his former Cassini colleagues saw Pathfinder as a threat to them, because its public failure would bring Dan Goldin's wrath down on that mission, too. Cassini was due to launch just after Pathfinder's arrival (or lack thereof) on Mars, so this concern wasn't farfetched.[73] The undercurrent of thinking on and off the project at JPL was that Pathfinder was approved less because its reviewers believed in it than because Wes Huntress wanted it. But JPL director Edward C. Stone was on board, too. Stone, a quiet, mild-mannered physicist who had been the Voyager project scientist and a faculty member at Caltech prior to his appointment as JPL director in 1991, needed to prove to Huntress that JPL could do less expensive missions, and Pathfinder was the test case.

At the same time, the cost cap didn't prevent ambition from affecting Pathfinder. Steve Squyres's science working group couldn't settle for a simple entry, descent, and landing demonstration. They, the lander team, and NASA headquarters all believed they needed to demonstrate new scientific capabilities. It would have been embarrassing even to Pathfinder's engineers to simply deliver a "brick" to Mars. They also felt the need to demonstrate technological innovation that would impress the general public. The rover was the perfect machine to embody all of these demands—no one seems to have doubted that it would be popular, although there was concern among scientists that it would not be particularly valuable scientifically. But the rover drove the team far away from the project's original goal of a simple, mass-produced, meteorological lander.

To make things still more difficult for the team, pressure to succeed increased dramatically by the time NASA's letter of approval reached Ed Stone's office at JPL: Mars Observer had disappeared.

Attack of the Great Galactic Ghoul

In JPL lore, there is a legendary beast that lurks between the orbits of Earth and Mars. Known as the Great Galactic Ghoul, it is the product of the imagination of an engineer named John Casani who had been lead designer of JPL's early Mariner planetary spacecraft. One day in 1965, Casani had been doing an interview with a reporter. Asked why a great many Soviet spacecraft were failing to reach Mars, Casani made up the Ghoul as a joke. A few years later, the Ghoul's legend was cemented when a JPL spacecraft bound to Mars named Mariner 7 suddenly went silent, close on the heels of still another disappeared Soviet spacecraft. Casani revived the Ghoul story, and a colleague painted the Ghoul's portrait (dining happily on Mariner 7) and presented it to Casani.[1] Mariner 7 eventually reappeared and completed its mission. It had been, temporarily, victim of a battery failure. But on August 21, 1993, the Ghoul ate Mars Observer, and no one ever heard from it again. Its disappearance, after nearly a decade of development and large cost increases, threw NASA's space science directorate into confusion and put enormous pressure on JPL.

Mars Observer's disappearance came at a particularly bad time for the Laboratory. It was the first complete mission failure JPL had experienced since Mariner 8's loss in a launch vehicle accident two decades earlier. John Casani, by 1993 JPL's director for flight projects, reported to Ed Stone that the Lab was perceived as "fat, complacent, arrogant, with little regard for cost" in NASA headquarters.[2] Casani even recommended the Flight Projects Office be closed. It was seen in headquarters as a big part of the problem, enforcing unrealistically strict engineering standards that drove up cost without really reducing risk. Shutting down the Flight Projects Office would be a visible signal of JPL's willingness to reform.[3] And eventually, Stone agreed. The Flight Projects Office was closed down during 1994, ushering in a six-year period in which JPL imposed few design or testing standards on its projects.

Further, a Government Accounting Office report in late 1992 had condemned JPL business practices. NASA used the 1993 renegotiation of Caltech's management contract to impose management reforms, demanding greater accountability for labor and finances that increased bureaucracy and frustrated much of the staff.[4] Administrator Daniel Goldin's continuing attacks on the Cassini-Huygens outer planet mission, at the time the largest project at the Lab employing the most people, antagonized and demoralized staff too. ("Cassini, at a cost of $4 billion, and more than a decade in development . . . I'm terrified. This is not the way to run a program," Goldin said in October 1993.)[5]

Finally, JPL was being shrunk. When Ed Stone had assumed the directorship in January 1991 he had visited the Office of Management and Budget in Washington, where he'd been told JPL was too big and unaffordable. So he'd unilaterally decided to eliminate about a thousand positions over the next few years, starting in fiscal year 1993.[6] JPL had been run more like a family enterprise than a corporation, and many employees saw layoffs not as a necessary business practice but as a betrayal.

So Mars Observer's loss came in the midst of low morale and an unhappy, antagonistic relationship between JPL and its primary sponsor. But the loss also presented NASA's Goldin the opportunity to impose sweeping reforms on NASA's science directorate and JPL. Mars exploration was transformed in Mars Observer's wake.

Flight of the Mars Observer

Mars Observer's flight to Mars had been a relatively uneventful six-month cruise, with only one recurring problem: the spacecraft's inertial reference unit kept malfunctioning. When that happened, the spacecraft wasted control gas through maneuvers that were not really necessary. Eventually, the project team, now led by Glenn Cunningham, figured out that they had a deeply buried software fault. The hardware was fine, they determined to their great relief. They fixed the software problem with a patch in July.[7]

With great anticipation, Mars Observer's arrival sequence at Mars started on August 21, 1993, with pressurization of the fuel system. This involved firing a pair of pyrotechnically operated valves, which subjected the spacecraft to a small, but not insignificant, mechanical shock. On General Electric's Earth orbiters, these pyrotechnic shocks all occurred before the spacecraft's telecommunications system was turned on, so GE did not normally certify the telecom system for shock resistance. For Mars Observer, however, this was a problem. The transmitter's tube-based amplifier had not been qualified for the pyrotechnic shock while it was operating. So the pyro firing had a small, but not zero, chance of crippling the space craft before the mission really began.

The project team could have run tests to certify the telecommunications system's immunity to this problem, but of course they had been under continuous cost pressure. The leadership team had decided not to spend the $75,000 to do the test. Instead, their solution had been to order Mars Observer to shut down its transmitter prior to executing the pyro firing. Cunningham's flight team expected Mars Observer to be silent for 14 minutes.[8] Instead, no one ever heard from it again. The loss threw Mars Observer's engineers and scientists into disarray. Pace remembers that it took him "months and months and months to get over that."[9]

Three weeks after Mars Observer's last signal, NASA administrator Goldin asked the director of research at the Naval Research Laboratory to head an independent investigation committee. His investigation took place in parallel with investigations carried out by JPL and by Martin Marietta, which was just in the process of completing an acquisition of GE's Astro Space division. The panel's largest challenge was simply that they had no facts to analyze. Because of the telecom shutdown, there was no telemetry data available to help diagnose what had been going on aboard the spacecraft. They therefore had to work through the "fault trees," essentially charts showing what failures could produce the situation they had (no communications at all), to tell them what subsystems to study in detail. Software could be checked via the project's simulator, the verification test lab, while hardware could only be checked by studying the Mars Observer spares or by relying on hardware commonality within production batches. Other spacecraft had Mars Observer's components as well, and one could reasonably assume that a part that had failed on one spacecraft could also have failed on Mars Observer.

These activities led the board to a set of tentative technical causes. The one they considered most probable was related to the problem that had caused Cunningham's team to operate Mars Observer in the "blowdown mode" in the first place: leakage. Simulations run at the Naval Research Laboratory seemed to show that the oxidizer could have leaked into the pressurization piping. If that had happened on Mars Observer, activation of the pressurization system could have pushed the oxidizer into the fuel line, causing an explosion that might burst the piping. The unchecked high-pressure gas pouring out through the broken pipe would have sent the spacecraft into a wild spin, and probably destroyed it.[10] JPL's propulsion specialists were never able to get the titanium tubing used in Mars Observer's fuel system to rupture, though, even with much more "leaked" fuel and oxidizer than was likely to have been present, so many people at JPL did not accept the board's finding.

The investigation board identified two other potential causes. JPL's investigators had opened the spare power subsystem's electronics and found one important box was contaminated with metal particles. Further, a set of components had been mismounted so that the pyro firing shock could have brought them into contact with the spacecraft chassis, shorting out the entire power system. This would kill the spacecraft instantly, but this scenario was not likely, as the physical stress during launch was much more severe than that imposed by the propulsion system activation at Mars. The spacecraft should not have survived launch if this were a severe flaw. The panel also identified another pressurizing system failure mode that would have been immediately fatal: failure of the pressure regulator valve. In this scenario, the leaking oxidizer would have frozen the valve in its "open" position, causing overpressurization and rupture of the fuel and oxidizer tanks. This could not be ruled out because Mars Observer did not have a temperature sensor on the regulator.

Taking a larger view, the review panel contended that the primary failure of Mars Observer had been the project's reliance on "heritage," the very argument that NASA had used to create the Planetary Observer concept. In the panel's words, "many of the spacecraft systems had been so extensively modified for Mars Observer that their heritage had been lost; others, whose heritage remained intact, should have been requalified to verify that they would function properly on an interplanetary mission of three years duration (an environment for which they were not designed)."[11]

Nearly twenty years later, this conclusion still angered George Pace.

> We did not take the hardware, modify it and fly it without analysis and test. We did a preliminary design review [PDR], a delta PDR and a critical design review [CDR] on every subsystem and the system. Review boards consisted of experienced JPL and outside reviewers. We had system engineers and representatives from each subsystem on the spacecraft team interfacing with RCA/GE. NASA even asked me one time why I needed so many people. We had a system engineer and two quality engineers in residence at GE. We did component and system level functional and environmental tests like we did on all other missions. If we had done what we were being accused of, the development would have been a lot cheaper. No one said we spent too little, they were constantly complaining during development that we were spending too much.[12]

Pace believes the true cause of Mars Observer's loss was the use of titanium valves in the propulsion system instead of more traditional stainless steel. Less flexible than stainless steel, the titanium valves might have allowed hot plasma from the pyrotechnics to blow past the valve gaskets and ignite fuel in the lines. After Mars Observer's loss, a British spacecraft with the same type of valves was lost to this failure. The titanium had been used to save weight, not money, and heritage had nothing to do with it.

Regardless of the actual cause, Mars Observer's loss helped reinforce NASA headquarters' poor view of JPL. If the GAO's audit had impugned JPL's institutional management, Mars Observer's loss impugned its technological capabilities as well.

Recovering from Mars Observer

The shock of the sudden disappearance of Mars Observer was not limited to NASA, JPL, and Martin Marietta. Bill Boynton's gamma ray spectrometer team at the University of Arizona had planned a celebration for the week after Mars Observer's orbit insertion. After the spacecraft vanished, he decided to hold the party anyway, figuring that he could use the opportunity to explain what had happened, as best he knew. It could be a wake, of a kind, helping his team cope with the loss of a decade's work on the GRS. Several members of his team faced the prospect of sudden unemployment. Mars Observer's youngest principal investigator, Phil Christensen, likewise faced an uncertain future. Unlike Boynton, Christensen did not have tenure, so in addition to having to find new jobs or funding for his team, he might be out of work right along with them. Christensen had bet his career on Mars Observer's scientific productivity; he had not been publishing scientific papers due to all the work on his instrument, and now he had nothing to show his tenure committee.[13]

The personal challenges faced by the project's scientists were not helped by the public reaction. Late night television host David Letterman made the disappearance a comedy subject: "Top Ten Reasons for Losing Mars Observer"—"Space monkeys" earned number 1.[14]

Wes Huntress, who had succeeded Lennard Fisk as NASA's associate administrator for space science late in 1992, handed the question of how to recover Mars Observer's scientific objectives to Charles Elachi, JPL's assistant director for space and earth science. Elachi, a native of Lebanon who arrived in Pasadena in 1969, was a radar engineer by education who had led development of the first space shuttle science payload, the shuttle imaging radar. Elachi had heard about Mars Observer's loss from Tony Spear—while riding on a bus at Los Angeles

International Airport. The job assignment of figuring out what to do about it came while he was mowing his lawn. He was given only a couple of weeks before having to brief Goldin and Huntress on September 15, 1993, not quite a month after the disappearance.

Elachi knew that Goldin was strongly interested in finding a way to do smaller, less expensive science missions, as was already clear from Pathfinder. That meant regaining Mars Observer's science objectives without redoing Mars Observer itself—it was too big, too expensive, and, due to a lack of Titan IIIs, it would have to be reengineered for a shuttle launch anyway. A Mars Observer II would not be a good candidate for a 1994 mission, and probably not for 1996 either, due to orbit-imposed limitations on mass that could be sent to Mars during those launch opportunities. Besides, Goldin had already set in motion a policy to put all NASA science payloads into Delta II (or smaller) sized-packages; Mars Observer was far too heavy for that smaller launch vehicle.

Huntress offered Elachi a variety of suggestions on shrinking the mission: a variant of the Strategic Defense Initiative Office's "Clementine" lunar orbiter might be flown with copies of some of the Mars Observer payload in 1994; a variant of the Miniature Sensor Integration Technology satellite, also developed for the Strategic Defense Initiative Office, could fly part of the payload the same year; JPL could perform a quick industry survey to determine the level of interest in flying all or part of the payload in 1994 or 1996 using some combination of the Mars Observer spares and newly built components. None of these options would regain all of Mars Observer scientific objectives in a single mission, but they would be a beginning.

Glenn Cunningham's engineers didn't like the Clementine option, believing that the spacecraft wasn't compatible with Mars Observer's instruments. The miniature sensor integration technology satellite option wasn't much of an option either. JPL had been part of this project, but the people who had done the work had mostly been incorporated into the Mars Pathfinder project—they weren't available. Besides, the satellite had been a tiny and intentionally short-lived Earth orbiter. The spacecraft was not designed to host large instruments and would have to be reengineered for the three-year Mars mission. So Elachi and his team had quickly put their emphasis on the "industry" option.[15]

Cunningham put out a request for information to solicit the views of potential contractors on reflying all or some of the Mars Observer instruments in 1994, 1996, or 1998. About a dozen companies responded. After briefing NASA on the results, Cunningham gained Goldin and Huntress's agreement to release "Phase A" study contracts to all of them, to get higher fidelity cost estimates and preliminary designs. These had to be done quickly to have any hope of getting a launch in 1994, or even 1996, given the lateness of the year.[16]

Money, as is always the case, was also a major influence on the process. Mars Observer's operating budget had been about $34 million for fiscal year 1994, and this was the funding the reflight effort depended on. In the first week of October, however, the relevant congressional committees decided to cut this to $10 million, stating "the conferees are disappointed in the recent loss of the Mars Observer. A total of $10,200,000 has been included for a possible 1995 or 1996 reflight of the Mars mission."[17] This action ended any possibility of a 1994 reflight and left little to begin rebuilding the scientific instruments. The first post-loss meeting of the Mars Observer Project Science Group brought this point out repeatedly. Since the instrument teams knew how to rebuild the instruments, keeping the key members of those teams available was crucial to the possibility of an inexpensive reflight.[18] Losing them to other tasks would result in wasting a lot of time and money in redevelopment. Some relief came from Lockheed Martin, which had just completed purchase of GE Astro Space before Mars Observer's loss. The company refunded the entire $15 million award fee, sending George Pace two checks because, apparently, the machine that cut checks could not go higher than $10 million.[19] The refund also became available to support the reflight.

JPL's review of the industry proposals occurred the first week of November, and the Mars Observer science team met in conjunction with the associated industry briefings. There were 13 proposals for reflight: six for a complete reflight in 1996, two for two half-payload launches the same year, and five for single half-payload launches in both 1996 and 1998. The project science group's preference was for a complete reflight in 1996, which they thought would be the least expensive overall option. It would also allow the instrument synergy to remain intact. The group found the option to refly their instruments across two opportunities least desirable, because they did not believe this was possible.[20] The orbit geometry for the 1998 opportunity was less favorable than 1996, meaning that a Delta II would be able to send less mass toward Mars in 1998. The two orbiters could therefore not be identical. Either the 1998 orbiter would have to be lightened considerably or some instruments would have to be dropped. The option to refly Mars Observer's payload across two opportunities would, they thought, be very likely to turn into flying it across three opportunities. In short, recovering from Mars Observer would require another decade under this scenario.

That's what wound up happening. The request for proposals released in January 1994 asked for proposals covering only the 1996 opportunity and limited the spacecraft mass to a level that meant only some of Mars Observer's instruments could be aboard. It also fixed the spacecraft cost at $54 million plus a 15 percent fee. In February, the Mars Observer project science group met to have a "shoot out," to decide which instruments would be on the reflight. While they didn't yet have a selected spacecraft, they already had a pretty good understanding, from the December industry survey, of what mass would be available for the scientific payload. Those results hadn't been pretty, from the scientists' standpoint. Phil Christensen remembers an uncomfortable day of briefings for the team:

> By now we'd worked together well and we're friends and we were a team, and it was suddenly everybody got an hour to make their case. No one wanted to go in and say, gee, pick me and not him. Very, very awkward situation, and we were kind of in this waiting room, waiting for each one of us to get our chance. But again, the reality was it came down to mass and cost. Dan McCleese's atmospheric sounder instrument was forty kilograms plus, and Bill Boynton's gamma ray spectrometer was forty kilograms plus, and it didn't take a genius to figure out that for the mass available you could fly all the other five experiments. But if you put either the PIMRR or the GRS on, it would take the place of three other instruments.[21]

Christensen had assumed that as the junior scientist, he'd be the first dropped. But his thermal emission spectrometer was selected for the reflight, along with Mike Malin's camera. Selecting Christensen's lighter instrument also meant that other instruments could be accommodated. Those ended up being the magnetic field detector, the laser altimeter, the radio science package, and the Mars balloon relay.

In July, JPL selected a proposal from Lockheed Martin's Astro Space division to fly these instruments in 1996. The proposed cost for the spacecraft was

Mars Global Surveyor payload

Instrument	Investigators (institution)
Mars orbiter camera	Michael Malin (Malin Space Systems)
Thermal emission spectrometer	Philip Christensen (Arizona State University)
Mars orbiter laser altimeter	David E. Smith (Goddard Space Flight Center)
Radio science	G. L. Tyler (Stanford University)
Magnetometer / electron reflectometer	Mario Acuna (Goddard Space Flight Center)
Mars relay	Jacques Blamont (Centre National d'Études Spatiale)

Source: Adapted from Arden L. Albee et al., "Overview of the Mars Global Surveyor Mission," *Journal of Geophysical Research* 106, no. E10 (25 October 2001): 23291–316.

$72 million, with $51.2 million being the expected cost, a $12.8 million award fee and project reserve, and $8 million for JPL's management cost.[22] This proposal had not been the least expensive, but it had been evaluated as having the lowest risk. The proposal team, led by Claude "Bud" McAnally, had chosen to use the Mars Observer spare electronics, reducing their development risk considerably. Glenn Cunningham thought this probably had saved them $80 million or so; while the electronics would have to be retested and refurbished to overcome the criticisms from the Mars Observer investigations, it would still be less expensive than starting over.[23]

McAnally's proposal team had also saved a great deal of mass by proposing to use aerobraking at Mars to obtain the final orbit. This part of the proposal had credibility because Astro Space had just completed an aerobraking experiment using the Magellan spacecraft orbiting Venus; engineers from that project would be available for the new Mars mission. Aerobraking reduced the amount of fuel the spacecraft had to carry considerably by using the thin upper atmosphere to create drag, slowing the spacecraft down. While the use of aerobraking would increase mission risk, it had the advantage of demonstrating an innovative approach to spacecraft design and operation, something NASA leaders desired to see. The team had also gained an additional increment of mass savings by proposing the use of a composite spacecraft structure, basing it on a communications satellite.

Further, McAnally's proposal had taken advantage of a preexisting contract with JPL's Cassini project. Cassini was formally a JPL "in-house" project, but Lockheed Martin was building its propulsion system under contract. So McAnally proposed using a modification of Cassini's propulsion system for the Mars spacecraft, allowing the project simply to buy into Cassini's existing component and assembly subcontracts. Cassini had already borne the development risk, and riding on Cassini's contracts saved time that would otherwise have gone to procurement and design activities.

Glenn Cunningham was named the project manager at JPL, and he quickly moved to keep intact most of his Mars Observer team. He had to fight to keep George Pace as the spacecraft manager, though, because many people at JPL blamed Mars Observer's loss on him. "When I picked the staff for MGS, I decided there wasn't anybody here who knew the hardware better than George knew the hardware. I thought that was very important, because the Lockheed Denver people had to learn all about the hardware. They just got a boxcar full of stuff, and they had to learn all about it, but George knew all about it and knew all the history and everything."[24] Cunningham brought in M. Daniel Johnston, who had led the aerobraking experiment on Magellan, to direct aerobraking planning, and he responded to the Mars Observer failure board's criticism of insufficient project-level systems engineering by hiring as project engineer Peter C. Theisinger, who had recently been section manager for the Systems Engineering section at JPL. At Lockheed, the veteran McAnally stayed on as spacecraft manager.

Forming a Mars Program

After the formulation of this first Mars Observer reflight mission, quickly named Mars Global Surveyor (MGS), NASA still faced the issue of reflying the remaining two Mars Observer instruments, McCleese's atmospheric sounder and Boynton's gamma ray spectrometer. The members of Steve Squyres's Mars Science Working Group had made this their top priority, and NASA's Wes Huntress had agreed. At the same time, Dan Goldin wanted a true Mars program: a series of missions with one or more spacecraft launched in each 26-month opportunity. To that end the Mars Surveyor Program Office was established at JPL in February 1994 to manage the overall Mars program on behalf of NASA.

Glenn Cunningham was offered the opportunity to leave the Mars Global Surveyor project to become the new program's manager, but he declined. Instead, he recommended Donna Shirley, who was then manager of the Pathfinder rover development, for the job. She accepted. John McNamee, a no-nonsense former Texas construction manager who had gone back to school for an aerospace Ph.D., was assigned the job of working with NASA's science officials on the initial program requirements. He later became manager of the 1998 pre-project effort. From the science side of JPL's house, Dan McCleese was assigned as program scientist. Because there was strong interest at NASA headquarters in "stretching" the slim dollars available to the Surveyor program through international cooperation, JPL mission analyst Roger Bourke was appointed to sound out European and Russian interest in the new program.

For a brief period, scientists pinned their hopes for a full recovery of Mars Observer's science on the international "Mars Together" effort. Started under the Bush administration and incorporated into a set of U.S.–Russia initiatives undertaken by the Clinton administration, Mars Together was intended to be a series of cooperative missions. This effort was vital to the Mars Science Working Group because the faster-better-cheaper ground rules laid out by NASA headquarters for the U.S.-only missions were extremely demanding. In 1998 NASA wanted both an orbiter and a lander. For the 2001 launch opportunity, NASA was requiring two landers. Each Mars opportunity would have about $130 million available to spend. By itself, Mars Global Surveyor would cost about $154 million; the Pathfinder lander, $150 million.

Further, these missions were to fly on a launch vehicle that did not yet exist, known at this point simply as the "Med-Lite." Part of Goldin's program to reduce cost by reducing complexity and mass, the Med-Lite vehicle would provide about 450 kilograms mass capability to Mars in 1998, about half that of the Delta II that Mars Global Surveyor was to use.[25] Elachi had emphasized in his September 1993 sales pitch to Goldin and Huntress that a Med-Lite sized orbiter could carry only one of the two remaining Mars Observer instruments, not both. But a Mars Global Surveyor-sized spacecraft launched on a Russian Proton rocket could carry both easily. Squyres's Mars scientists really wanted the Mars Together option to come to fruition.

These new ground rules were first exposed to the Mars Science Working Group in late February 1994, where they were greeted with skepticism. The Med-Lite requirement drew great criticism. An orbiter that the Med-Lite could accommodate might not be able to carry *either* of the remaining Mars Observer instruments. That was because NASA and JPL had agreed to a new requirement that the next orbiter to Mars also would provide the communications relay functions that Mars Observer and Mars Global Surveyor had contained. This decision was made to support the relay of data from future landers back to Earth. The relay system, of course, took up precious mass. This new requirement for communications relay appeared to make meeting the science objectives impossible.

Meanwhile, under the Mars Together banner, Roger Bourke had brokered a deal with NASA and the Russian government to investigate the possibility of getting Russia to fly a copy of the Mars Global Surveyor spacecraft on a Proton rocket in the 1998 launch window.[26] "Possibility" was the key word. Not knowing whether his launch vehicle would be a large Russian Proton or a small Med-Lite, McNamee had to pursue both options in designing a mission while awaiting an expected June 1995 decision from NASA regarding the Russian ride. Either option would leave a scant three years to get the spacecraft built.

The Mars Office also had to respond to the sudden opportunity to do landed science on Mars. Here, too, the opportunity was a mixed blessing. According to NASA's ground rules, JPL had to contract out the lander project. The only current lander design effort, Mars Pathfinder, was an in-house JPL project. It could not simply be "handed over" to industry, and JPL had not been under any obligation to document it well to facilitate such a transfer. That meant the contractor would have to develop a new spacecraft. This fact, and the larger context of the Mars program, gave the scientists considerable pause. Under the plan, JPL and whoever its contractors turned out to be would be developing four different spacecraft simultaneously: Mars Pathfinder, Mars Global Surveyor, the 1998 orbiter, and the 1998 lander. "In such a cost-constrained program, four new developments in four years strikes us as unlikely to be achievable," Squyres warned.[27]

But the scientists, especially Squyres, very much wanted the landed science opportunity, whatever its engineering difficulties. So they sought a way to decide, quickly, what a 1998 lander might do scientifically. Remembering that the working group had spent years on the Mars network science strategy without being able to get it accepted, Dan McCleese suggested that they not try to plan a decade-long program and instead allow the lander program to be "driven by the science at it arises."[28] Both he and Squyres thought NASA needed to change the way it did science missions. According to Squyres, "Our approach in the past has always been to begin from top-level science objectives, and then construct a program that meets those objectives. Time and again, this objectives-driven approach has failed, for the simple reason that it produces inflexible and ultimately unaffordable programs."[29] So they recommended instead that, since NASA would be developing a sequence of small landers, it set "broad characteristics" for a given flight opportunity and then request proposals from individual scientists for complete missions, individual instruments, or both. In other words, McCleese and Squyres wanted the Surveyor program to operate the way the Discovery program was supposed to, but focused on Mars exploration. Because the time horizon for choosing the 1998 lander mission was short, McCleese suggested, and the Mars Science Working Group agreed, that they hold a workshop that summer to define the most desirable theme for that mission so that the project could move forward.[30]

In early May 1994 the workshop convened at the Pasadena Hilton, a frequent venue for such things due to JPL's lack of conference facilities at its own little spot in the San Gabriel foothills. Because the 1998 orbiter was going to be devoted to recovering the rest of Mars Observer's payload, the workshop focused solely on the 1998 lander. More than 160 people attended, impressing the organizers with the range of interest in Mars and the diversity of potential instruments. To Squyres and the rest of his working group, however, there was already a clear demand for a specific scientific theme for the 1998 lander, and already mature instruments available to carry it out. This theme was the "evolution of Martian volatiles and climate."[31] Just as Earth's surface contains many volatile materials (e.g., either solids and liquids that become gases at relatively low temperatures), the martian surface was expected to do the same. Some of these volatile substances, like water and carbon dioxide, directly affect planetary climate. As had been clear from the Viking orbiter imagery, Mars once had plenty of water, and part of a "volatiles" mission's goals would be to contribute towards an understanding of what happened to it. Depending on the specific instrument collection, such a mission could also contribute to a climate history. Mars had clearly undergone a dramatic climate shift at least once in its past, and such a mission might also be able to answer questions about when this had happened, and whether it had happened more than once.

There were already instruments available to look into some of these questions. One of McCleese's former postdoctoral students, University of California, Los Angeles, scientist David Paige, had been working with members of Tony Spear's Pathfinder team and with Bud McAnally at Lockheed Martin to propose a polar Pathfinder for the Discovery program. Its instruments were all derivatives of common geophysical instruments used on Earth. Paige had made clear at the May meeting that 1998 and 2000 were the only two near-term availabilities for the more scientifically desirable martian south pole. And orbital dynamics strongly favored a 1998 launch.[32] So Squyres's working group recommended to NASA that the RFP set the mission theme as Evolution of Martian Volatiles and Climate.

Finalizing the 1998 Missions

While Mars Science Working Group was figuring out what science the lander mission might do, Donna Shirley's office at JPL and the science directorate at NASA headquarters were trying to figure out how to write a RFP for the 1998 orbiter and lander that could encompass both possible scenarios of the orbiter ("big" Mars Together or "small" Med-Lite) and a lander that could reach any spot on Mars with a worthwhile payload. This effort was made more challenging by the reality that Shirley and McNamee had no single point of contact at headquarters. Instead, at various times during 1994 and 1995, five different officials provided sometimes-conflicting instructions. This led to misunderstandings, bad feelings, and even an attempt by Shirley to resign in August 1995.

Much of the difficulty revolved around the JPL Mars Office's inability to get NASA officials to understand the technical and financial difficulties the new program faced, even with the relatively independent voices of Steve Squyres's working group making the same points in their quarterly reports back to headquarters. The orbiter's inability to host *either* McCleese's atmospheric sounder or Boynton's gamma ray spectrometer—if it also had to carry the data relay system—was interpreted at NASA headquarters as not being able to carry *both* of them, for example. The funding expectations of JPL and NASA were also not aligned with each other. An internal JPL review of McNamee's mission plans in September 1994 produced unanimity that the "baseline missions will be extremely difficult to achieve"; the panel thought he would be at least $30 million over his cost cap.[33] NASA, of course, expected JPL to deliver within the funds NASA had available.

Another point of conflict was McNamee's insistence that he be allowed to award the spacecraft development contract for both vehicles to a single contractor. He was convinced that having to manage two different contractors for the orbiter and lander would put him well outside his cost cap, simply because it would require two management teams at JPL. But he wasn't able to get NASA headquarters to allow the RFP to specify this approach. Instead, he was required to alter the guidelines so that contractors could bid on the orbiter, lander, or both. But if contractors bid on both, they had to provide separate proposals for each, as well as additional material specifying the financial savings from doing both.[34]

Lack of funds for the Mars program office at JPL simply exacerbated all these difficulties. Shirley had $5 million allocated for fiscal year 1995 to define the 1998 lander and orbiter; pay for the Mars Together effort; fund an ongoing study effort for the international version of the Mars network mission, InterMarsNet; and provide financial support for Boynton's and McCleese's science teams. This wasn't even enough to pay for solid Phase A and B studies of the 1998 missions, let alone the other activities. She commented later that this left JPL "totally dependent on the contractors,"[35] as she was unable to afford independent studies to validate contractor claims.

But by fall of 1994 the picture was clearing somewhat. Mars Together did not seem practical for the 1998 opportunity. Roger Bourke's efforts had produced a lot of information about the Russian program, but little of it was positive. He became convinced that Russia could not afford the 1998 mission plan in addition to the ambitious Mars 96 mission that was already well into its construction phase. In early November, NASA administrator Dan Goldin visited his Russian counterparts and came away convinced of this as well. The "big" Mars Together orbiter option disappeared.[36]

The Med-Lite version of the Mars Surveyor 1998 RFP was released on January 3, 1995. It allowed for costs up to $106.3 million, including the contractor's fee. Recalling that Lockheed Martin had bid $54 million for the larger Mars

Global Surveyor orbiter just a few months earlier (and which was being partly built from Mars Observer's spare parts), this was an optimistic price tag for two spacecraft. The scientific instruments for the lander were to be selected via a NASA announcement of opportunity to be released later in the spring, after the spacecraft orbiter contract had been awarded. Initially, the orbiter instruments were to be simply the Mars Observer instruments left behind by Mars Global Surveyor. But beginning in January, NASA began insisting that a small camera be incorporated on the orbiter. By February, the camera had become a "must fly" instrument, even if it meant kicking one of the two Mars Observer instruments off the spacecraft. This insistence only worsened McNamee's mass and money problems, of course. It also inspired ill-will between Shirley's office and headquarters. The program executive believed that there had been an agreement to have a camera on the 1998 orbiter from early in the program, but there was no documentation of such an agreement. It wasn't found in the program requirements documents either.[37] The camera requirement was eventually added to the announcement of opportunity, with a $3 million cost cap and an 8 kg mass limit.[38]

In March 1995 JPL awarded a single contract for the two vehicles to Lockheed Martin Astronautics in Denver.[39] John McNamee recalls that their proposal had two main strengths. The company had made the best case for synergy between the two spacecraft, both in technical terms—such as using similar flight avionics—and in management terms. It also had pledged substantial internal investment in the project, promising to devote about $11 million in internal research and development funds (IRAD) in the project.[40] There were other advantages to working with Lockheed Martin again. The 1998 projects would be able to draw on the company's aerobraking experience with Magellan and Mars Global Surveyor, which reduced overall risk. The proposal also relied on miniaturized avionics that it had developed under a different NASA program, the Small Spacecraft Technology Initiative. These were essential for staying within the 450 kg mass limits of the Med-Lite vehicle.

The Lockheed Martin proposal effort had been led by Parker Stafford, a guidance and control engineer with experience going back to the Viking landers. The orbiter Stafford's team had designed for the 1998 opportunity was significantly different from Mars Global Surveyor. Besides different avionics, the 1998 orbiter, named Mars Climate Orbiter for its primary payload, had only a single solar array composed of three panels. The radiator for Dan McCleese's atmospheric sounder was the driver of this design choice. The radiator always had to "see" deep space in order to keep the instrument at the proper temperature. Solar arrays generate heat, and Mars radiates heat, so the radiator could not be on a side that would see either Mars, a solar panel, or the Sun. Simply because of its size, it also could not be on the same side of the spacecraft as the propulsion system. So the designers had adopted an asymmetric configuration that satisfied all these conflicting requirements. To protect the solar array's drive mechanism from large torques during engine firings and aerobraking, the array would be moved into a restraint that would support the mechanical loads. The vehicle's orientation would also be changed so that when the array was stowed for these events, it would be "behind" the spacecraft. This would permit the airstream flowing past the spacecraft during aerobraking to see a relatively symmetric, and aerodynamically stable, configuration.

The lander concept had three legs, and used a propulsive final descent, like the Viking landers. Studies had indicated that for a lander of the mass being sought, the rocket-based approach of getting down to the martian surface was likely to turn out lighter than an airbag lander like Pathfinder. This determination reflected the Pathfinder team's own experience to this point. Tony Spear's project had experienced very high mass growth during 1994. They could afford it because the MESUR network lander had originally been sized so that two could be carried on a Delta II, not just the one they were actually building. Lockheed's designers and McNamee's management team knew they had no such luxury on their Med-Lite. So they chose a path that seemed likely to result in less mass and less mass growth.

There was one key difference between Viking's propulsive descent design and the new lander that drew quite a bit of attention early in the project. Viking had used throttleable rockets for its final descent to the surface. A Doppler radar had fed its flight computer with the altitude and horizontal and vertical velocity data it needed to regulate its dozen directional thrusters. But the Mars Polar Lander's thrusters were not throttleable. They would be turned on and off by the flight computer to regulate the descent instead. Sometimes this is called a "bang-bang" mode of operation, and it has a problem known to any plumber or home improvement enthusiast: water hammer. Since fluids are incompressible, quickly opening or closing a valve in a fluid system initiates a shockwave that travels through the fluid into the tanks, from there to all the other pipes in the system. The resulting pressure transients could cause some of the thrusters to misfire, although Lockheed's analysis indicated that the transients would probably not be significant enough to cause this.

After the Phase A contract was negotiated, NASA's Med-Lite selection team gave the Mars 1998 team a gift: they chose a rocket with much more capacity than the 450 kg originally specified. The 1998 missions were assigned to a pair of McDonnell-Douglas Delta 7325 rockets, which could send 565 kg to Mars in that year. Each additional kilogram of mass added to the spacecraft required an additional 0.8 kg of propellant for the Mars orbit insertion burn, so the increase gave them an additional 55 kg of useful spacecraft mass with which to work. In June, McNamee informed NASA headquarters that he could now accommodate the camera, communications relay, and one of the two remaining Mars Observer instruments on the orbiter—but not both. He also explained that even the larger launch vehicle left him with a mass margin that was "uncomfortably low for designs at this level of maturity. Historical data suggests that developmental spacecraft like the 1998 lander and orbiter require mass margins of 25–30% at the beginning of a project's construction phase ['Phase C' in the jargon] to insure achieving an acceptable mass for launch."[41] If McCleese's sounder was selected, the mass margin would be 16 percent; if Boynton's lighter spectrometer was chosen, a 20.5 percent margin would result. From McNamee's engineering perspective, choosing the spectrometer made more sense.

In July, Wes Huntress notified McNamee that McCleese's sounder had been chosen for the 1998 orbiter. McCleese's second pressure modulated infrared radiometer (PMIRR) was to have its optics made in Russia, a small extension of Mars Together. While the gamma ray spectrometer had been rated lower risk, headquarters' selection had been predicated on risk to the follow-on 2001 mission. With the inability to get both instruments aboard McNamee's orbiter, NASA had told Shirley to plan for a 2001 orbiter instead of the second 2001 lander originally specified in Huntress's ground rules. But because of relative positions of Earth and Mars in that launch window, the Delta 7325 would have a much lower mass capability. Accommodation of the sounder on the 2001 orbiter would be even more difficult than it already was on the 1998 orbiter. From this slightly longer-range perspective, putting Boynton's spectrometer into the lower-capability year made more sense than saving McNamee some trouble in 1998.

The science announcement of opportunity released in the spring of 1995 had set August 15 as the deadline for proposals to supply the orbiter's camera and for the lander's payload. In November, NASA selected two cameras proposed by Mike Malin: the Mars Surveyor orbiter color imager (MARCI) and, for the lander, the Mars Surveyor descent imager (MARDI). Officials also chose the integrated lander package suggested earlier by David Paige at UCLA, the Mars volatile and climate surveyor (MVACS). Paige's proposal included four major elements: a mast-mounted stereo camera like Pathfinder's (and by the same principal investigator, Peter Smith); a two-meter-long robotic arm with its own camera and a scoop to acquire soil samples; a meteorology package; and a thermal and evolved gas analyzer, which could examine the concentrations of various volatile compounds in the soil samples.[42]

Paige's proposal for the lander mission was to send it to 71° south latitude, arriving in the late spring. After the end of the southern summer, it would probably not receive enough sunlight to keep itself functional. So like Pathfinder, it would have a very short primary mission: 86 days. Most of the instrument assemblies were to be built by the University of Arizona and delivered to JPL for integration. Paige's instruments were expected to cost $20 million to build and to have a mass of 17 kg. The mission would be called Mars Polar Lander.

These instrument selections didn't quite complete the two missions' basic definitions. NASA's Goldin and Huntress still wanted to gain Russian participation in the 1998 lander and kept open the possibility of adding a Russian instrument to the lander. A few options were under discussion: an electromagnetic sounder for probing soil depth, a Mossbauer spectrometer for examining mineral composition, and a laser instrument for measuring atmospheric dust. Shirley and McNamee opposed their addition, however, because they had developed the project budget around a single principal investigator who would be paying for payload development and integration to the spacecraft out of his budget. They didn't have more money to pay to integrate other instruments.[43] Nonetheless, NASA added the lightest Russian instrument in December, a lidar for measuring atmospheric dust, after McNamee confirmed that the lander's mass margin was sufficient for it, but not the others.

NASA also wanted the lander to carry a pair of "microprobes" to Mars. These were being developed in a separate NASA program, called New Millennium. The program's purpose was to develop new, high-risk space technologies. The microprobe project's manager was Sarah Gavit, who had been a systems engineer on Magellan while working for Lockheed Martin and had then moved to JPL as a fault protection engineer on the Cassini mission.[44] Her project was to demonstrate an entirely passive entry and landing capability on Mars, for $25 million. The probes were impactors: they would simply drop to the surface of Mars, hitting the ground without deceleration beyond that provided by atmospheric friction. This meant withstanding up to 60,000 g's acceleration. The probes were supposed to provide basic meteorological data, relaying it to Earth through the balloon relay on Mars Global Surveyor. They would also be able to sample the soil for water content.[45] Because the probes would be simple and inexpensive,

Mars Surveyor 1998 payloads

Spacecraft	Instrument	Investigators (institution)
Climate Orbiter	Pressure modulated infrared radiometer	Daniel J. McCleese (JPL)
	Mars color imager	Michael Malin (Malin Space Systems)
Polar Lander	Mars volatiles and climate survey (MVACS)*	David Paige (UCLA)
	Mars descent imager	Michael Malin (Malin Space Systems)
	LIDAR	V. S. Linkin (Space Research Institute)
Deep Space 2**		Susan Smrekar (JPL)

Source: Adapted from NASA 1998 Mars Missions press kit, December 1998, www.jpl.nasa.gov/news/presskits.cfm (accessed 7 February 2012).

*The MVACS was an instrument package consisting of a surface stereo imager (SSI), the robotic arm and a robotic arm camera, a meteorology package, and a thermal and evolved gas analyzer. The SSI's co-investigators were Peter Smith (University of Arizona) and H. Uwe Keller (Max Planck Institut für Aeronomie). The robotic arm and camera and meteorology package were provided by JPL. The thermal and evolved gas analyzer's principal investigator was William Boynton of the University of Arizona.

**The Deep Space 2 microprobes contained four facility instruments: a sample collection and water detection experiment, a soil thermal conductivity experiment, an atmospheric descent accelerometer, and an impact accelerometer.

the hope was that they could be the basis of an affordable Mars meteorological network. They were also to be tiny, small enough to be held in two hands.

Gavit remembers that McNamee was a tough negotiator over the probe addition. The deal they finally struck was that the microprobes couldn't have any electrical connection to the lander at all. This meant that once attached to the spacecraft, they could not be turned on or tested. So after launch, the first JPL would hear from them would be when they landed on Mars. The probes also could not require the lander to be in any particular orientation to Mars on release, so they had to be self-righting. And the New Millennium program had to pay for the cost of integrating them to the lander. Further, the deal they struck did not guarantee that the Polar Lander would actually carry them. At this point, the lander's mass margin was about 16 percent, too low for comfort on a new spacecraft design, so McNamee would only promise to decide whether to fly them after the lander passed its thermal vacuum test series. By that time, the lander's true mass would be known, and the availability of excess launch capability could be determined accurately. It was not much of a promise.[46]

Mars Climate Orbiter and Mars Polar Lander were to be launched in December 1998 and January 1999, respectively. To manage the development of the two vehicles, McNamee had five full-time people at JPL, and seven more staff members relocated to Denver to work directly with Lockheed Martin's team. He also planned to use "bits and pieces" of people at JPL, as Shirley put it, as part-time consultants. He had about $3 million to pay for these to bolster his small team. He intended to "hire" them using four-hour work orders to JPL's technical divisions, a novel arrangement for the Laboratory. It was all he could afford.[47] It also made the idea of "soft projectization," which Tony Spear was pursuing on Pathfinder, largely irrelevant. Soft projectization was supposed to achieve better results by gaining personal commitment to the project's success from the team members through "cradle to grave" involvement in the mission. McNamee's tiny full-time team might achieve this, but his consultants would have no such commitment.

Strategy beyond the 1998 Missions

In parallel with the development effort on Mars Global Surveyor and the effort to define John McNamee's 1998 missions, Donna Shirley's office was also trying to formulate a longer-term strategy. A science strategy came together relatively quickly during 1994 and achieved NASA's blessing in February 1995. But by the middle of 1995, the science strategy initially adopted had been downgraded in importance, replaced with a new mandate to prepare the way for human missions to Mars. This led to the removal of strategic planning from Shirley's Mars Exploration Program Office and its relocation to NASA headquarters.

The Mars program's chief scientist at JPL, Dan McCleese, had advocated establishment of a scientific strategy for the program that was aimed at understanding the martian hydrologic cycle, something of interest to both scientists and astronauts. Water on Mars and Earth circulates between the surface and the atmosphere, changing phase and location depending on local temperatures; the water cycle on Earth affects weather and climate. It probably did on Mars, too. The surface or subsurface reservoirs of Mars's water were still missing, and finding them was a scientific priority.

Water is also fundamental to life. In fact, the Mars Underground that had evolved during the 1980s had been organized by three scientists very interested in the intersection of life and water (see chapter 1). The research submarine *Alvin*'s 1979 discovery of "chemosynthetic" organisms, ones that live off the chemistry of deep sea vents instead of sunlight, had been the scientific impetus for organizing the Mars Underground. These "extremophiles," as they're sometimes called, formed an entire ecosystem powered by geothermal energy. The Viking landers had left virtually no doubt that the martian surface was sterile, and until these sea vent communities were discovered, that seemed to have settled the issue of present-day life on Mars. The Viking scientists had had no reason to believe that life could exist independent of the Sun. And for most of the 1980s, in fact, nobody drew the connection between the sea vent communities on Earth and the potential for subsurface martian life. In 1992, some of the Mars Underground members published the first paper detailing their ideas about subsurface martian habitats, proposing that the Tharsis region seemed a likely place to look for "hotspots" that could still be hosting life.[48]

At the same time McCleese was promoting the water strategy, NASA's Office of Exobiology was finalizing a report that called for an explicitly exobiological strategy for the exploration of Mars. Published in April 1995, it proposed a strategy that would evolve over five phases, the first of which focused on "the role of water, past or present, and on the identification of potentially fruitful sites for landed missions."[49] A global reconnaissance phase from orbit would make the first effort at finding Mars's water, then in Phase 2 landed missions would characterize those sites chemically. In Phase 3, more landers would search for organic compounds. Phase 4 was the Holy Grail of Mars science: sample return. Scientists wanted nothing more than getting martian rocks back to their labs on Earth. Finally, astronauts would be sent "to establish a detailed geological context for any exobiologically significant observations made previously."[50]

If one wanted to send humans to Mars for whatever reason, one needed to find water. The expedition members would need it. Basing the near future of Mars science around the martian water cycle unified what could be divergent scientific and exploration interests. In addition to being a sound scientific strategy, it also seemed sound politically.

In January 1995, McCleese pitched the water strategy to Huntress. In February, Huntress took it to Dan Goldin, who approved of it as the scientific basis for the ten-year Mars Surveyor program plan. Their agreement became known as the "Goldin-Huntress contract." The resulting mission sequence included the orbiter-lander pair in 1998; two water-oriented small "neolanders" in 2001; an international Mars network, much like the earlier MESUR network mission, in 2003; and a 2005 Mars sample return.

At the same time McCleese was getting NASA to sign on to the water strategy, Carl Pilcher at NASA headquarters chartered another scientists' committee that came to be called by its acronym, MELTSWG (pronounced *MELT-swig*), the Mars Exploration Long Term Strategy Working Group. This committee, chaired by Geoff Briggs at Ames Research Center, had some overlap with Steve Squyres's science working group (including the presence of Squyres) but took as its charter "the establishment of an orderly framework for the scientific exploration of Mars by humans—focusing on the big question of whether life originated there."

SEARCHING FOR WATER ON MARS

1996		1998		2001	2003		2005	← TIMELINE
PATHFINDER	MARS GLOBAL SURVEYOR	MARS WATER SURVEYOR	MARS POLAR LANDER	WATER NEOLANDERS	INTERMARSNET	COMM/AERON	MARS SAMPLE RETURN	TABLE II: MISSION GOALS
	●							ESTABLISH GLOBAL TOPOGRAPHY & STRATIGRAPHY
	○	●						CHARACTERIZE SEASONAL CYCLES OF WATER
		●	●					IDENTIFY SOURCES & SINKS OF NEAR-SURFACE WATER
	○	●			○			UNDERSTAND ATMOSPHERIC TRANSPORT
	●	●	○					DETERMINE POLAR ENERGY BALANCE
	●							MAP DEPOSITS OF CARBONACEOUS ROCK & EVAPORITES
[illegible]	●			●	●		●	WEATHERING: MINERALOGY & SOIL CHEMISTRY*
[illegible]				●	●		●	INVENTORY LAYERED TERRAINS, SURFACE DEPOSITS*
		○	●					INVENTORY PERMANENT POLAR CAPS
		○			○	●		QUANTIFY VOLATILE ESCAPE & SURFACE-AIR EXCHANGE*
				○	○		●	ISOTOPIC ANALYSES*
							●	DATE AND ANALYZE SURFACE MATERIALS IN DETAIL

KEY: ○ INITIATES OR CONTRIBUTES TO GOAL; ● MAKES MAJOR CONTRIBUTION TO GOAL; ▭ (shaded) ORBITER; ▭ LANDER; *AT MANY SITES

Mission sequence for "following the water" of ancient Mars, January 1995.
From "Searching for Water on Mars," H2OonMars_Jan95.pdf, Dan McCleese materials, Historian's Mars Exploration Collection, JPL.

Briggs hoped that by providing a scientific strategy leading to human missions the committee could "help speed the process of putting human space exploration back on the national agenda."[51]

This group offered two options for the post-2005 future. These choices represented different phasings of the same basic set of missions, starting with mobile robotic laboratories and leading to robotic sample returns. The committee assumed that the mobile laboratories would gather samples and cache them on the surface. In later opportunities, landers would collect samples and shoot them back to Earth. The pace of the two plans was relatively slow, with Briggs's group assuming that there would be six or so sites on Mars worthy of detailed exobiological examination before deciding where to return samples from—or send astronauts to. One option had sample return in 2016; the other, 2020.

These scientists' ambitious ideas, however, did not go over well at headquarters. Responding in July, Carl Pilcher commented that "the view point expressed is from a fairly narrow scientific perspective. The Mars exploration program is not the property of scientists. It is really owned by the people who pay for it, and their desires (e.g., to experience the excitement and danger of Mars exploration vicariously through human explorers, not just through telepresence) have to be a fundamental part of our planning." The phasing of both options was "way too plodding." The program would "bore people to tears with this approach."[52] In other words, Pilcher didn't think a sustained, cautious robotic exploration program was supportable politically. Humans had to get to Mars sooner rather than later. In fact, Briggs had noted in his summary that the human program might very well leapfrog these plans. "Human exploration," he thought, "carried out for reasons only loosely connected to science, is likely to overtake a robotic exploration program—especially its later phases of sample gathering and return."[53]

Indeed, that seemed to be the case. In May 1995, NASA formally removed the JPL Mars program office's strategic planning authority. It wasn't clear immediately why this was being done. But by the end of August, it became obvious: Goldin had decided to press for a White House decision in 2004 to send humans to Mars, and he intended the Surveyor program to provide some essential human-specific data about Mars. The intensity of certain kinds of radiation in Mars orbit and at the surface was the most important of these data needs.

At headquarters, Steve Saunders of the Mission from Planet Earth Office was appointed to the strategic planning task, leading Donna Shirley to try to resign on August 25. Giving the inability to resolve the poor communications with headquarters that had marked the 1998 mission decisions and the removal of JPL's planning role as her reasons, she told Wes Huntress and JPL director Ed Stone that she didn't want the job any more. Shirley's letter also caused Saunders to call Dan McCleese, and explain why headquarters had removed Shirley's planning authority. "Goldin's focus is now on humans on Mars," he told McCleese. "If we only had to worry about the robotics piece, things would be very straight-forward. The integration of technology for human exploration must be factored into NASA's planning now. Therefore, JSC clearly has a major role in all aspects of the program and in program planning."[54] It wasn't clear what JPL's role would be in the new roadmap being drawn up.

Ed Stone rejected Shirley's resignation, but he nonetheless appointed engineer Norm Haynes, then manager of the Deep Space Network, to take charge of the Mars Exploration Program Office. Shirley remained as Haynes's deputy, retiring late in 1998. However, a change of leadership didn't resolve the problem of communications with NASA, where several people continued to have various pieces of the "Mars pie" and to issue conflicting instructions and information.

Finally, as it became clear at NASA headquarters that McNamee's Mars Climate Orbiter could only carry one of the two Mars Observer instruments, Huntress redirected planning of the 2001 mission toward an orbiter-lander pair. That way Boynton's spectrometer could finally be reflown. In addition to Huntress's promise to get all of the Mars Observer instruments reflown, the gamma ray spectrometer was essential to the water strategy. Its map of surface ice distribution, if it identified any, would be an important guide for future missions.

Conclusion

As Mars Observer's science team had feared, complete scientific recovery from the loss of the spacecraft would take until at least 2001. Observer's disappearance had triggered a broad set of initiatives aimed at not only recovering its scientific content but also creating a new way of doing business that would reduce mission size and costs while enabling the formation of a true Mars program that would result in a sustained presence at the Red Planet. Questions of strategy were not entirely resolved during 1995, but that was the consequence of the late-breaking competition between advocates of robotic exploration and human exploration of Mars.

There were casualties of this programmatic realignment. One was the large MESUR network idea. Goldin had decreed that there would be no more billion-dollar missions, and MESUR was well over a billion. While Tony Spear's MESUR Pathfinder was still alive, its original raison d'être had vanished. No longer a Pathfinder for the full network, the mission was transformed into JPL's entry into the world of "faster, better, cheaper." Also gone was the grandiose Mars Rover sample return idea. Instead, over the next few years, sample return would get replanned around much smaller budgets. The International Mars Network working group would continue planning MarsNet for a few more years, but it too would not come to fruition.

Finally, the turmoil at JPL over the direction "faster, better, cheaper" was taking was not confined to the Mars Exploration Program Office. Rob Manning, chief engineer on Mars Pathfinder, was a member of the Mars Surveyor 1998 project review board and was offended by that project's implicit criticism: that Pathfinder was too expensive by about half. He considered the pulsed-mode lander's entry, descent, and landing to be more complex and difficult to test than Pathfinder's; it should have been more expensive to develop, not less. "The only free variable was risk. They would simply save money by not testing and by minimizing analysis. Just build it and go."[55] The review board understood that was McNamee's only choice. But it was what NASA wanted. During the faster-better-cheaper era, the customer, NASA headquarters, got what it wished for.

CHAPTER FOUR

Engineering for Uncertainty

During their development phases, between 1994 and 1996, both Mars Pathfinder (MPF) and Mars Global Surveyor (MGS) engineering teams confronted a feature of Mars that makes it an especially challenging destination. Mars has much greater variability in its atmospheric density than does Earth. Because both projects intended to use the atmosphere as a decelerator, their vehicles had to be engineered to cope with variation. And because the telecommunications lag between Earth and Mars is several minutes, their machines had to be able to deal with those rapidly changing conditions on their own.

The need to engineer their spacecraft for uncertain conditions led both projects to make changes to the design as their understanding of Mars conditions, and of their vehicles' likely performance, changed. It was an iterative process of design, analysis, test, and redesign. Pathfinder's project reviews, negative during 1994 and part of 1995, also led the team to reconceive their entry, descent, and landing (EDL) verification and test program around extensive numerical simulations. In parallel, engineering considerations affected the decision regarding where on Mars Pathfinder should be sent.

Pathfinder

Mars Observer's loss and its aftermath presented Tony Spear and his little band of renegades with a couple of changes. First was the disappearance of the "network" portion of MESUR. Without the network, their MESUR Pathfinder project was no longer merely a pathfinder for the larger MESUR—it was the whole thing. The disappearance of the network allowed Spear's group to focus on Pathfinder itself, uninhibited by the needs of the network. More problematic in the short run was the loss of Mars Observer's spare electronics to Glenn Cunningham's Mars Global Surveyor project. Spear and his spacecraft manager, engineer Brian Muirhead, had counted on about $6 million worth of spares from Mars Observer being available. Making up for the loss of these spares now had to come out of the project contingency fund. Spear regained some of the money by pursuing additional descoping of the project up front, but not all of it.[1]

Pathfinder was buying many of its electronic parts from the Cassini-Huygens project to save time and money, but chief engineer Rob Manning also wanted some of Cassini's specialized designs for Pathfinder. While the two missions were very different in their needs and goals, like all spacecraft they had many things in common: they needed to produce and distribute power, and they had to communicate through the Deep Space Network, to name two. The late 1980s had seen the development of integrated circuits that could be customized, called "application specific integrated circuits." The Cassini spacecraft manager, Tom Gavin, had had his engineers design several of these. For Pathfinder, they were probably overkill. But many of the engineers that David Lehman, who took over the attitude control and information management subsystem development after MESUR network was dropped, brought into Pathfinder came off Cassini jobs, and they knew the designs already. More importantly, they knew what they would need to do to adapt Cassini's designs to Pathfinder's particular needs. So the project's leaders saw modifying Cassini's designs as a time and cost saving measure.

One thing they did not want from Cassini was its data processing system. Cassini had adopted a military specification processor (the MIL-STD 1750) and the Defense Department programming language that went with it, called ADA. "We weren't big ADA fans," Manning remembers, "even though a lot of people got very good at that very difficult programming language on Cassini."[2] Instead, he wanted to use C programming, running under a commercial multitasking operating system on a commercial microprocessor. These were much less expensive to buy and could make up an inexpensive testbed that he could use for software development and simulation studies long before the space-qualified hardware would be available.

These choices also led directly to a much more centralized data processing architecture for Pathfinder. The limited processing capabilities of pre-1990 microprocessors had led JPL spacecraft engineers to employ distributed computing architectures on their spacecraft, with each major subsystem having its own processors. This made integration of all the subsystems expensive but resulted in extremely reliable spacecraft. Manning didn't think this was necessary any longer. Commercial microprocessors available in 1995 had far more processing power than spacecraft actually needed—one could do everything. And microprocessors were extraordinarily reliable, so redundancy was not really necessary on a mission as short as Pathfinder's. One consequence that Manning didn't fully foresee, though, was that the centralized architecture made spacecraft design much more reliant on software than it had been, and the small Pathfinder software team, under Glenn Reeves, would struggle to keep the software development on track.

Spear, Muirhead, and Manning faced their critical design review in July 1994. By then, they had to make all of the major design decisions, finalize the detailed requirements for subsystem performance, produce most of the detailed design drawings and interface documents, and demonstrate that they had done enough testing of their new, special-to-Pathfinder technologies (e.g., the airbags) to have some confidence of success. They also had to convince the reviewers that they had a reasonable plan for testing their spacecraft before shooting it off to Mars. This all turned out to be more difficult than they expected, with the result that they postponed their critical design review until September and even then barely passed it. The events of 1994 left some of the project's leaders wondering if they really could pull it off. But they also resulted in one of Pathfinder's major innovations.

In March 1994, about the same time that mechanical engineer Tom Rivellini was out at the Sandia National Laboratory doing the first set of prototype airbag tests, Manning reached an unhappy conclusion. He realized his systems engineers could not write a set of EDL requirements that could be met within the project's cost cap. Traditionally, JPL's design requirements were based on "worst-case scenarios." Each subsystem's performance was specified in terms of its own worst case, with little consideration of other spacecraft elements. Worst case for the heat shield would be entering the Mars atmosphere on a cold, high-density atmosphere day, while worst case for the parachute was the opposite: deploying on a hot, low-density day. These two conditions could not actually occur at the same time on the real Mars—it would either be cold or warm at the time and place of entry, not both. But this process of "worst-case rollup," as it's often called, is how systems engineers traditionally built robustness into their designs.

Worst-case analysis applied to the spacecraft's internal communications, too. Pathfinder's complex EDL sequence required that certain subsystems provide specific data to other subsystems at particular points in the sequence. Data from accelerometers and the radar determined when the flight computer ordered ejection of the heat shield, deployment of the parachute, and lowering of the spacecraft on its bridle. Worst cases applied at every step of this chain. What if the accelerometer data arrived at the moment the computer suddenly experienced a fault and went into its 40-second-long reboot mode? Answer: the spacecraft would crash, which Manning referred to as "lithobraking."

When Manning applied all of the worst-case evaluations to all of the different subsystems, he had an impossible spacecraft.[3] Or if not impossible, then as expensive to develop as Viking had been. Either way, he was sunk.

The solution was suggested by one of the navigation team members, Sam Thurman, an Oklahoman who had moved to JPL from the Charles Stark Draper Laboratory in 1989. Navigation at JPL is done probabilistically. The last time navigators know where a spacecraft is in absolute terms is the instant before launch vehicle ignition on the pad. After that, they have probable locations. So navigators were used to doing probabilistic assessments. Thurman's suggestion was simply to do what navigators did: use Monte Carlo analysis to develop the requirements. In a Monte Carlo analysis, one could subject a computer model of a system to a large number of randomly chosen cases, generating a statistically relevant database on its behavior. For example, a simulated parachute could be subjected to hundreds of different atmospheric and deployment conditions quickly and inexpensively. Using the old Viking lander data, Manning (or, rather, scientists he could recruit) could develop a probability distribution for the atmospheric density on entry day. This probabilistic approach eliminated the need to design to impossible conditions (such as the simultaneously hot and cold atmosphere), while still permitting the requirements to encompass the vast majority of likely cases—98 or 99 percent of them.

The decision to develop the subsystem requirements around probabilities caused Manning to have his various subsystem cognizant engineers find or make computer simulations of their subsystems. Bobby Braun at Langley Research Center, for example, who had been recruited by Pathfinder's mission manager, Richard Cook, into the navigation team because he had modified a space shuttle atmospheric entry simulation for Mars back in 1993, could already simulate Pathfinder's entry conditions down to around the parachute deployment altitude.[4] And the parachute contractor, Pioneer Aerospace, already had a simulation for the parachute. So a complete simulation covering entry to touchdown could be cobbled together fairly quickly.

In preparation for the project's critical design review, Spear scheduled a peer review of the EDL system for September 1–2, 1994.[5] But once again, the review was difficult for Rob Manning and Sam Thurman, who were still struggling with the problem of how to design and test Pathfinder's overall EDL system. To validate the performance of Pathfinder's unique EDL approach, Manning had planned to do a series of drop tests. But during the late summer, a series of demonstration drops at China Lake of the combined parachute, backshell with rockets, and a dummy lander had not produced good results. The impacts damaged the test equipment, although not fatally. This test series was supposed to have led to a new set of drop tests beginning in mid-1995, this time with the lander containing the engineering model electronics. Any damage to these would have been more than expensive, it would have been catastrophic, Manning recalled thinking.[6] The engineering model electronics were also the project's software development testbed, and risking them was out of the question. So the team went into the early September peer review thinking that their plan was not very good. And the review panel agreed with them.

In addition to the risk imposed on the hardware, the reviewers did not think they would learn much from the test drops. The Viking project had carried out a series of drop tests to verify the strength and opening characteristics of the parachute, but not to verify that the lander worked—there was no full-up final descent test. The earlier Surveyor lunar lander program, carried out between 1960 and 1967, had done such tests, but the Surveyor engineers had had to make a series of compromises (e.g., a one-sixth-scale lander to simulate the Moon's one-sixth Earth gravity) that meant the test wasn't entirely realistic. And, they'd had a series of failed tests that destroyed test landers, too. The Viking veterans on the review board knew all this. So Pathfinder's reviewers opposed doing a similar sort of full-up drop test. But without this full-blown system test, the Pathfinder team had no way to determine whether the individual elements of their complicated EDL would actually work together as a system. They had nothing demonstrating the linkages between the pieces.

Thurman and Manning realized they had not taken their simulation ideas far enough. What they needed was to wrap all of the subsystem simulations into a single, larger simulation that could start from the moment of aeroshell separation from the cruise stage and run all the way down to the surface. As long as the physics of the individual component models were valid, then the physics of the master simulation would be as well. This master simulation—they wound up calling it the "Mother of All Simulations"—would serve as *the* system test.[7] In other words, there would be no real-world full-scale system test—there would only be the simulation.

The team recognized that there were advantages of simulation beyond removing the risk to the testbed. The physics of the models could be made accurate for Mars. One of the challenges of getting realistic tests of the full range of potential Mars conditions in Earth tests is that Earth has far less variation in key parameters than Mars. The large martian atmospheric variability produced quite a lot of variation in the timeline of the descent to the surface; in turn, that range of variation had already driven the team away from a mechanically timed descent toward a radar and software-controlled one. Using Monte Carlo techniques on a valid EDL simulation would permit them to understand the impact of this variability on their vehicle's performance much more fully than a handful of system drop tests ever could. They could run many thousands of entry simulations and address a wide range of variability for the cost of one drop test.

One major challenge in using numerical simulation to understand a design's performance lies in the question of how well the simulation reflects the actual vehicle. Or as Thurman would point out years later, convincing yourself that the model is physically valid was the key to the process.[8] Simulations run on invalid models are worse than useless, as they waste resources and provide false confidence in the result. So after the key insight that they could use Monte Carlo simulation to assist in the system design and testing, Manning and the team's mechanical guru, Dara Sabahi, rebuilt the test program around validation of the physics of the various subsystem models. This new program was iterative in nature. Thurman gathered or had made models of all the different subsystems, then used the simulations to design tests that could be run on the testbed or on hardware, like the parachute or heat shield, which would provide insight into the physics of the models. Test data would be compared to the model output, and the models would be fixed if they did not agree. A second round of comparison between model output and test data could then be held, if necessary, to complete model validation.

During the several months that it took Manning, Thurman, and a handful of others to reconstruct the EDL program, Brian Muirhead was wrestling with a couple of other problems. The mechanical design of his spacecraft was not going smoothly. Pathfinder was the most mechanically complex spacecraft JPL had ever attempted to build. The parachute, bridle, airbags, airbag retraction mechanisms and the spacecraft structure, with its multiple petals and the actuating mechanisms that went with them all belonged to Division 35, Mechanical Systems, and the division was behind schedule in getting its designs done. Part of the problem was that Pathfinder was being designed in a period of transition at JPL, when the Lab was switching its design and manufacturing over to a new computer-aided solid modeling system. Pathfinder was the first vehicle to be fully designed with the new system, and the switchover was disruptive. It was also the first flight project to be carried out entirely under the metric system at JPL. Finally, the Laboratory's mechanical section was having troubles lining up staff to work on Pathfinder.[9] The little project was not seen as attractive work in comparison to the much larger, longer-lived Cassini.

Muirhead also faced the challenge of relentless mass growth. Pathfinder gained about 50 kilograms between July 1993 and March 1994, and almost another 75 kg by the critical design review in September.[10] Unlike any previous JPL project, the key mass constraint for Pathfinder was its entry mass, not its launch mass. Entry mass was critical because it affected the heating rate the heat shield would experience and the timing of the parachute deployment. So did the entry velocity and entry angle. While the Pathfinder team was reusing the old Viking heat shield material, and was much lighter than Viking had been, it had a much higher entry velocity because it was entering Mars's atmosphere directly from the Earth-Mars transfer orbit, not from Mars orbit. Since the heat shield's capabilities were relatively fixed, and the entry velocity was also essentially fixed, in order to accommodate the mass growth, Richard Cook had to find a way to narrow the range of probable entry angles.

There were several causes of the mass growth. The early jump in mass was a product of integration of the rover. The spacecraft tetrahedron had to be enlarged to accommodate it, making the design heavier. Tom Rivellini's three-eighths-scale airbag tests in March 1994 indicated that he needed to add a second abrasion layer to the bags to better protect the gas-tight bladders inside; this produced another substantial mass increase. He gained some of that back through redesign, however, switching from a three-lobed to a six-lobed configuration. This actually shrank the bag area significantly, although not enough to cancel out the extra layer's mass. A more detailed thermal analysis of the lander during entry conditions caused them to add mass to the lander structure so that it would act as a larger heat sink, slowing the internal heat-up rate. There were still other, lesser, causes. Muirhead had to start issuing mass challenges to other subsystem engineers in hopes of recovering some of his eroding mass margins.

Reconstruction of the EDL test and verification program was just beginning when the critical design review (CDR) was held in September.[11] The review focused on the EDL system maturity, or, rather, the lack of it. Muirhead commented later, "We had known going into the CDR that entry/descent/landing was in trouble, and didn't need to be told by the board to focus in it. They had told us anyway."[12] Muirhead and Spear also had managed to convince each other that they had plenty of time left to get the mechanical interface control documents completed—but they didn't. These documents specified how each mechanical component interacted with every other component and were required for a successful review. Earlier in the year, due to the expanding scope and complexity of the job, Muirhead had added a senior engineer who was more cognizant of mechanical systems, but the new leader of the effort had not made up quite enough time.[13] All in all, September was a bad month for the little project.

The poor critical design review results caused the Discovery program manager to call for another review in January, this one called an "independent readiness review." The review board was essentially the same group Spear had inflicted on his team back in 1993, headed by Jim Martin. The appointment of an additional review produced a great deal of concern in the project team. It was not hard to imagine the mission was headed down the road to cancellation. To survive, the Pathfinder team had to pull together by mid-January a credible EDL test program.

Sam Thurman became the head of the simulation effort. By the end of October, Thurman, Ken Smith, and programmer Chia-Yen Peng had major chunks of their simulations running. On January 5, Spear had Thurman, Manning, and Sabahi undergo a "pre-review" of the EDL effort with John Casani, Duncan MacPherson, and Israel Taback. Those three delivered a "trouncing," in Manning's recollection. The team still did not have a clear idea about how they were going to mesh the simulation with the test program. Since they did not have a clear idea, their presentation was, unsurprisingly, muddled. They still had not made the simulations central to the test program, clearly linking each real-world test to a piece of the simulation that needed validation. Manning remembers that the pre-review "caused Sam and me to re-think how to tell the story. Up until then we had not been that assertive about the notion that EDL at the system level is only validated by flying a validated flight system computer model in a virtual Mars environment."[14] Their new goal became making the Mother of All Simulations the "truth model." Once the test program was completed, they would require that the flight-quality hardware be brought into conformance with its simulation, not the other way around.

On the EDL-focused second morning of Jim Martin's January review, Martin had declared at the beginning "we'll know in four hours whether this project will be cancelled or not."[15] For this review, Manning's troika had prepared a two-page chart centered on the complete EDL sequence of events. Above a drawing of the entry sequence, at each stage the chart presented the relevant system simulations; below the sequence diagram, it named the relevant subsystem and system level tests. This diagram became the major focus of discussion, as it graphically represented the interrelationships among all the simulations, tests, and the entry sequence itself. It told the "story" that had been missing before and allowed the reviewers to identify potential gaps in the program. As a result, Martin's panel declared that the board's previous concerns had been "largely mitigated."[16] This was the first positive review Pathfinder had received in nearly two years.

Several of Pathfinder's leaders credit Martin's review in January for keeping the project alive after its poor technical showing in late 1994. Martin also helped them resolve another increasing problem: the number of reviews being piled on them. While the reviews they had experienced had forced them to improve their test program, they were expensive in terms of time spent preparing. So Martin argued that the NASA-level formal reviews needed to be curtailed during the next year, with the next one to be held at the end of December 1995 or early January 1996.[17] At that point, the EDL test program was scheduled to be completed, the spacecraft would largely be assembled, the flight rover delivered, and the first complete set of flight software finished. It was going to be a busy year for the team, and another review would not help them overcome their challenges.

The Trouble with Simulation

The one component of the EDL system Tony Spear's Pathfinder gang knew they could not simulate effectively in the virtual world was the airbags. In addition to requiring lots of expensive supercomputer time to run, these numerical airbag simulations contained too many simplifying assumptions to be credible. Sandia's airbag model did not include seams in the bags, for example, which would obviously matter. So while nearly everything else could be modeled effectively, the airbag designs could only be tested in the real world.

Sandia did not have large facilities that could be pumped down to Mars atmospheric pressure for full-scale testing, so after a national search the Pathfinder team settled on using the NASA Lewis Research Center's Plum Brook station to simulate the landing. Plum Brook's vacuum chamber had been built for the Apollo program in the 1960s. It wasn't quite tall enough to reach the proper impact velocity from a static drop, so Rivellini came up with a bungee-cord accelerator that would do the trick. Because some of the tests needed to replicate impacts with a horizontal velocity component, he also had the Plum Brook operators rig up a platform that could be raised to various angles. Geologist Matt Golombek provided a Mars relevant rock distribution for the tests; the team bolted sharp, angular volcanic Earth rocks to the platform to simulate what they expected to find on Mars.

Initially, since his three-eighths-scale tests had gone well, Rivellini expected to do two test series. The first tests, done in March 1995, included a vertical velocity component only, and the test bags seemed to perform well. Rivellini remembers that he had an inkling of problems to come, though. After one of the tests, some of the lights in the vacuum chamber exploded from overheating, and the glass shards went through the abrasion layers "like a hot knife through butter."[18] What would happen when the bags got dragged along sharp rocks?

There was a six-week delay before he found out. The Plum Brook chamber operators needed the time to raise the platform and bolt down all the rocks. On the fourth test, done with about half the maximum impact velocity they expected, the test bags tore. "But it wasn't catastrophic," Rivellini remembers. Test number 5, on May 10, 1995, was. Done at full speed, it "unzipped the airbag from pole to pole. It just ripped wide open. It was just this catastrophic scene. It was great fun to watch on the video, but it was horrifying to experience professionally. That was really the watershed test that told us we were basically screwed."[19] They didn't have an airbag design, and they were not even two months from the beginning of assembly, test, and launch operations.

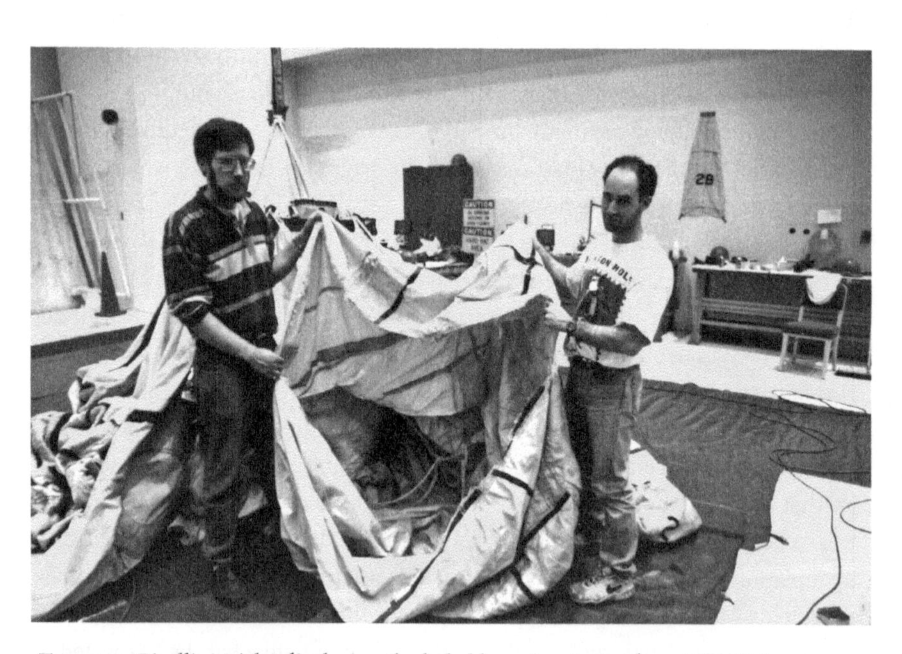

Tommaso Rivellini (*right*) displaying the hole blown in a test airbag at NASA Lewis Research Center's Plum Brook station.
Courtesy Tommaso Rivellini.

Rivellini and Skip Wilson of ILC Dover, maker of the airbags, devised a new set of tests that were done at Dover's facility, in which they strapped various potential new materials and layer designs to the top of a 55 gallon drum and inflated them. Then they dropped rocks on the fabric from the top of a water tower. They also used a mechanical stress testing machine to force rocks through various fabric constructions. At the same time, the team used the Mother simulation to reexamine the probable range of impact conditions. Shaving a few meters per second off the maximum impact velocity would make a big difference.

They went back to Plum Brook in July for a second set of full-scale tests. This time, instead of a set of identical bags, ILC Dover had made bags of several different constructions based on the experiments they had done earlier. Instead of trying to prove that they had a single good design, they were trying to figure out which of the different constructions looked to be the most robust. This test series produced a lot of good data but not much hope. "By any stretch of the imagination, it was catastrophic performance of the second airbag set we put out there. Pretty much all of the concepts showed catastrophic levels of damage during that test." The project was completely dependent on the airbags, of course, so Brian Muirhead asked Rivellini to call him after each test. The tests were done late at night to save electricity costs, so Rivellini was calling with bad news, every night or two, after midnight in Pasadena. "It was a pretty bleak and dreary time for us."[20]

But some good information did come out of the second test series. They learned they could afford to reduce the pressure inside the bags and still provide sufficient cushioning for the lander. They had put too much margin in the bag "stroke," the amount the bags could compress toward the lander at the moment of impact. This was a function of pressure, and they could sacrifice some of this margin to reduce the stress on the fabric. They also learned that in every case, multiple abrasion layer bags had performed better than single-layer, even if the single-layer bags had seemed nearly bulletproof in thickness. For the third test series, all the bag sets were fabricated with two, three, or four abrasion layers of various materials, some thick, some thin.

The third series of airbag drop tests was carried out in conjunction with the first full-scale inflation test for the bags, which was done in a different vacuum chamber at Plum Brook. Rivellini remembers this as a welcome break from the drop testing. The drop test series had not been encouraging, but the inflation test, in October, went well, with the speed at which the big bags inflated surprising the viewers. "Everything worked perfectly, the airbags went perfectly, the gas generator worked really well, the performance over time worked very well, so it was very successful, and that was really important for morale because we were really bumming at that point."[21]

They dropped that set of airbags, known as the full system development set, too, on October 17. This first test looked like another catastrophic failure—a jet of sooty air sprayed out of the lander. But that turned out to be from damage to the "boot," a cone of fabric that served to connect the main airbag structure to the lines that held it to the lander, and sealed the gases in as well. It was too short and had torn loose. It just needed lengthening to accommodate the bag movement during impact. Surprisingly, the bags themselves were intact. They had only small tears in the outermost abrasion layer. So Rivellini had them dropped again without repairing the abrasion layer damage, at the highest velocity they expected Pathfinder to impact at, 25 meters/second. He recorded in his lab notebook, "very good and hard impact, NO bladder damage at all."[22]

Rivellini was running out of time. He wanted to do more tests to build a better database, but to make the launch date, he had to adopt a final design and have ILC Dover fabricate the flight bags. He adopted the inflation test bags' four-layer design as the final, modified so that the only the bag faces had the four layers, reducing to three layers close to where the bags met, and tapering off to two close to the lander itself. This was an accommodation he'd negotiated with Muirhead to stay within mass limitations; Pathfinder was getting close to its entry mass limit. ILC Dover built two sets of bags, one to go to Mars and one for a set of qualification tests to be done out at Plum Brook in April 1996.

The airbag development was probably Pathfinder's most visible challenge, and it probably produced the most angst within the project. But it was not the only challenge. Muirhead had originally intended to trigger the airbag deployment and retrorockets with a plumb-bob hanging below the lander on a long tether. This scheme had been discarded relatively early in 1994s in favor of a small radar altimeter designed for military aircraft and cruise missiles. The altimeters were already on hand; JPL's mechanical section had bought them as instrumentation for the test program. The problem that frustrated Sam Thurman and a few others through late 1994 and well into 1995 was that the swinging motion of the lander seemed to make the altimeter act strangely.

The altimeters chosen for the test program were small and light. They had two operating modes: a "search" mode, in which they sent out radar pulses rapidly and looked for a stable ground return, and a "lock" mode, in which they pinged more slowly, analyzed the return to provide altitude data to the guidance software, and expected to see small, but not large, changes to the radar return in each cycle.

Pathfinder would be descending, quite rapidly, so the flight software needed rapid updates to its altitude in order to calculate when the airbags should be inflated and when the braking rockets in the backshell should be fired. If it happened to be landing on a windy day, the lander would also be swinging—and the simulation studies done late in 1994 and early 1995 indicated there was a good chance the swinging would cause the altimeter to lose its lock on the ground one or more times. When that happened, the altimeters' default mode of resending the last known altitude to the flight software caused the software to misestimate its descent velocity, resulting in cases where the simulation fired the braking rockets after the spacecraft had already descended below the Mars surface—that is, crashed.

It took the Pathfinder team months to understand what was happening. Sam Thurman reflected years later that the manufacturer's representative did not fully realize how the altimeter actually worked, causing early confusion. Some of the early drop tests did not help find the cause of the strange behavior because one of the four altimeters bought for Pathfinder was itself faulty and operated differently than did the other three, producing another element of confusion.

During July 1995, Thurman held a series of parachute-borne altimeter drop tests to characterize the altimeter's performance. The test range used, at China Lake Naval Air Warfare Center, had optical trackers and laser range-finders to produce precise trajectories for the test bodies. The tests were done right after dawn, when the colder air provided the helicopters that did the drops with slightly more lift. These independent data sets were then compared to the altimeter's data to examine its performance and to help understand the conditions in which it lost its lock on the surface. A digital terrain model of the area was used in this analysis process too, since the test range was not flat. Thurman recalled,

> We wanted to be able to predict the behavior we'd been seeing in the tests, rather than just assuming the altimeter worked perfectly when it became clear that assuming it worked perfectly was the wrong thing to do. . . . Ultimately after a dozen drop tests, we got to where you could take the altitude and attitude of the unit as a function of time during a parachute drop out at the [China Lake] Naval Air Warfare Center over desert-like terrain. You could feed those things in a relatively simple model of the surface it was looking on into Scott's [Schaffer, the radar

> systems engineer] program, and it would produce an indicated altitude history that was pretty close to the one we got from the actual test unit.[23]

Muirhead could not afford to pay for redesign of the altimeter, and the project did not have the time to wait for a redesign in any case, so Pathfinder's team "solved" the altimeter problem by programming around it. After using the drop-test data to improve the Mother simulation, Thurman's group used the simulation to estimate how far the lander had to swing off vertical to lose radar lock on the martian surface and how often that was likely to occur. Then they used the simulation-derived estimates to write an algorithm that could keep track of the swinging using the spacecraft's accelerometer data to project when the altimeter might be in danger of losing lock. That let them generate an independent estimate of the descent rate to replace the altimeter's until the altimeter relocked on the surface. This analysis and programming effort extended well into 1996, after the flight altimeter had already been delivered and integrated to the lander.

At the end of January 1996, Rob Manning and Thurman faced Jim Martin's review board again, this time with their test program nearly complete. Manning told the panel that the "test and validation program has shown that EDL is going to work."[24] The project EDL team had tested to performance margins exceeding its specified requirements. Rivellini's test of the full scale development bag set in October had been at a velocity that covered about 98 percent of the probable impact velocities (according to the Mother simulation), and the airbag development had been the project's laggard. So the project team had developed confidence in their design; apparently, that confidence convinced the reviewers as well. Martin's reviewers blessed Spear's engineers, and the Pathfinder team moved out from the dark cloud they'd been under since their difficult critical design review.

Mars Pathfinder was first fully assembled in JPL's Spacecraft Assembly Facility at the beginning of March 1996; at nearly the same time, Rivellini was out in Plum Brook doing the final qualification drop tests for the airbag design.[25] Pathfinder's electronics had already been undergoing testing for months. The small programming team had also been stomping out software bugs with the hardware's assistance since late in 1995. This was not its final assembly, to be sure; several flight components were to be delivered and integrated at Kennedy Space Center. The flight parachute, airbags, flight rover, and batteries all had to be installed in Florida. But several environmental tests had to be run on the complete spacecraft at JPL, using "flight-like" components standing in for the flight-hardware that could not be completed in time.

The rover team delivered the flight rover to the spacecraft assemblers on January 23, 1996. The previous year, the flight rover had been given the name Sojourner Truth; Sojourner's nonflight twin became Marie Curie. Sojourner had been proposed by Valerie Ambrose, of Bridgeport, Connecticut, who had won a competition Donna Shirley had arranged with the Planetary Society.[26] The initial integration of Sojourner with the lander did not go so well. The lander and Sojourner did not communicate initially. This had happened before, with the testbed "lander." An investigation showed that while the faulty circuit board in the testbed had been repaired, its twin in the flight lander had not been. Later in the test program, problems with the commercial modems that Sojourner and the lander used to communicate emerged, too. The modems experienced substantial frequency changes as their temperature changed, which meant that if the lander modem and rover modem were at different temperatures, they quite literally could not hear each other. Traditional spacecraft radios—in fact, most radios for any use—resolve this problem either by putting the crystal that produces the radio frequency into a small constant-temperature "oven" or by adding additional circuitry to compensate for the frequency drift. The rover's small team hadn't realized the problem existed until after they had already delivered Sojourner, and while they knew how to fix it in principle, it was too late to implement a fix to the rover's hardware.[27] They had no additional space in the rover to add the necessary components. So the project launched knowing that they had some risk of not being able to communicate.

The extensive test program revealed other design flaws as well. The cruise stage's solar array ran far too hot during the spacecraft system thermal vacuum chamber testing. This turned out to be the result of an inadequate representation of the solar arrays in the cruise stage's thermal model. It was also fatal to the project if it couldn't be fixed. The high temperature caused the array to generate less voltage than the spacecraft needed to operate; due to the locations of some of the heater controls, some of the fuel lines were too cold. Just as the rover communication problem could not really be solved, it was too late to fix this properly, too. Instead, the trouble was mitigated by JPL's recently retired thermal blanket expert, Hugh von Delden, who had sworn never to come back to work after retiring but did anyway, and his old assistant, Andy Rose. The two men rearranged the hand-made Mylar and Kapton blankets covering the cruise stage once, while it was sitting just outside the test chamber, and had it back in testing within a few days. This improved the situation, but not enough. So von Delden and Rose made more changes, this time knowing that their work could not be tested again before launch. The week the effort consumed was all Brian Muirhead had to give them before having to move to other tests. They would not know if they had succeeded until Pathfinder reached space.[28]

Then in May, Sojourner failed its "stand-up" test. In order to fit the rover into the tight confines of the lander, the mechanical engineers had made the rover suspension collapsible. When the rover was ordered to stand up, the rear wheel motors were supposed to force the front wheels against stops on the lander petal; the resulting compression would slowly force the rover body upward, and the suspension members would lock into place. During the test, the rover almost rose far enough for its suspension to lock—almost. A couple of weeks of investigation led to the conclusion that a subtle software fault was the culprit. Once discovered, the problem was easy to fix. Changing a parameter that the software used to evaluate the position of the rover's instrument arm was all that had to be done.[29]

In July, right before they shipped Pathfinder off to Florida, Spear's crew ran the spacecraft through a five-day "end-to-end" test. With the components of the spacecraft hooked together by cabling, they ran the vehicle through its entire life cycle: launch, cruise, EDL, and surface operations. Their goal was to ensure the flight hardware and flight software did what they expected it to in each mode and to remove any doubt about potential incompatibilities within the complex vehicle's subsystems. This trial went acceptably well, but because the flight software still needed work, Muirhead decided to run the test again at Kennedy Space Center.[30]

Landing Site Selection

The process of selecting a landing site took almost as long as it took to build Pathfinder, with the final decision being reached in March 1996. Matt Golombek had kicked off the process back in April 1994 with a workshop at the Lunar and Planetary Institute in Houston, Texas. About 80 interested scientists had attended, to help analyze the potential of (or to make explicit pitches for) about 20 sites.

The lander's tentative design placed strong constraints on the choice. Pathfinder had to go to a tropical latitude, between 0° and 30° north, for its solar arrays to provide enough power; preferably it would be sent to somewhere between 0° and 15° latitude, to maximize power and science return. The chosen site also had to be below the "zero datum," the martian equivalent of mean sea level on

Earth, because the parachute's minimum deployment altitude was only 5.5 kilometers above that zero line. The line had been established using Viking orbiter data, but the Vikings had no altimeter; as Golombek put it, Mars's altitude map had been "kluged together" and wasn't very reliable. So to ensure that the parachute opened well above the real surface, they had to choose a site that was well below zero altitude.

Pathfinder's airbags, of course, also defined a rock size limitation. Any site chosen had to have a fairly low probability of rocks larger than one meter in diameter (they also assumed that most rocks would be partially buried, so only the top half-meter of such rocks had to be worried about). Richard Cook's navigators also could only promise a landing accuracy defined by an ellipse 200 km long and 70 km wide, which meant that the entire region within the ellipse had to meet the latitude, elevation, and rock size constraints.[31]

At the April 1994 landing site selection meeting, scientific opinion clustered around three types of landing sites. Many, but not all, of the geologists favored selection of what they called "grab bag" sites. Outwash plains on Earth hold rocks of many different types, washed there from a wide variety of locations. Geologists believed that would be true on Mars as well, so targeting Pathfinder and its rover to sites that looked like outwash plains would provide them with different rock types to sample with the rover's alpha proton X-ray spectrometer. The scientific drawback of this idea was that the scientists would not be able to tie the individual rocks back to wherever they originated, but that was not seen as a major drawback, as they knew so little of Mars's composition. A minority of geologists, however, wanted to send Pathfinder to sites in the martian highlands. These made up a majority of the planet's terrain, overall, but only a small number of such sites fell within Pathfinder's landing capabilities, mostly because the highlands were too high in altitude. Finally, meteorologists at the workshop sought to land in areas that were mostly composed of dark-colored materials. Surface color affects surface heating and local atmospheric circulation, and the Viking landers had both gone to light-colored areas. A single Pathfinder landing could not tell meteorologists anything about global circulation, but it could inform understanding of the local meteorological effects of surface coloration.

Golombek took the workshop's list of twenty sites back to JPL to begin analyzing them. As project scientist, he had to generate a shorter, prioritized list to present to the project science group and to NASA headquarters officials for final selection and approval. First, he dropped ten proposed sites that did not fall within the lander's engineering constraints. Then he gathered all the available photographic, radar, and thermal data from previous Mars missions and Earth-based planetary radar research to help with further paring of the list. He eliminated six more sites based on these data, as they appeared either too rocky or too dusty (there remained some fear that the lander might sink deep into a dust-covered surface). This left him four sites that fit all the engineering and safety restrictions: two lowland "grab bag" sites and two highland sites, one of which was also a favored "dark" site.

He presented the resulting four sites to the Mars Pathfinder Project Science Group in June. After a great deal of discussion of each location's relative merits among the project scientists, he sought a vote to prioritize the sites. Not everyone in the group wanted to choose the site so early in the project's life, as some new data would be gathered using Earth-based radar and the Hubble Space Telescope during the following year that might alter the relative value of the sites. But Cook interceded in the discussion to explain that the engineering team needed to know the landing site elevation and latitude in order to finalize design of the lander. Parachute size and deployment timing would be affected by the chosen elevation, and latitude could affect the size of the solar arrays, battery, and amount of power required to keep the lander electronics warm overnight. So the group voted on the four sites, making an outwash site on Ares Vallis their top priority.[32]

Jim Martin's review board challenged their choice at the September 1994 critical design review. The preferred site at Ares Vallis, Golombek recalls, was the original Viking Lander 1 site that had been abandoned when the Viking Orbiter imagery had appeared to show an unacceptably rocky site. So Martin thought the site had been chosen primarily for scientific reasons, not safety ones, and Pathfinder's requirements specified safety as the primary concern. Pathfinder was not supposed to be a scientific mission, after all. As an engineering demonstration, a safe landing and transmission of the resulting engineering data back to Earth were the primary goals.

Martin sought an independent review of the project science group's choice from Michael Carr, one of his Viking veterans and a leading expert on martian geology. Carr met with Golombek and other project personnel in late February 1995 to review the selection process and the data that Golombek had used in making the selection, and came to the same conclusion. Carr concluded that the site was about as safe as the first Viking lander's final landing site had been—mostly flat with some very low, rolling hills and lots of small rocks that were well within the airbags' tolerance. While Carr didn't favor the site geologically, he told Martin that the project "has done all it can, within its limited resources, to ensure that the site is within acceptable bounds of safety."[33]

The immediate challenge to the site selection resolved, Golombek's next step was site certification. The Viking project team had required this second step to demonstrate the safety of their chosen landing site by gathering more data about it. But Jim Martin's Vikings had done this second step while the two orbiters were already at Mars providing new data to inform the certification process. The failure of Mars Observer and the fact that Pathfinder would not orbit Mars prior to landing left Golombek with no new orbital data, so the best he initially thought he could do was gather more radar data on the site. But two of Mars Observer investigator Phil Christensen's graduate students, Ken Edgett and Jim Rice, suggested that he could also look for an "Earth analog," a site on Earth that resembled the Ares Vallis site.[34] Visiting such a site could serve several purposes. It would help scientists gain a three-dimensional perspective on the two-dimensional Mars photographs that they had; it would show the Pathfinder project's engineers what the landing site would really look like (Golombek had a sense they thought of landing sites as paved lots); if they invited science writers along, the trip could gain the project publicity while also helping to get across to the public the challenges it faced; and inviting science teachers along would help fulfill the project's educational and outreach goals.

The three found an analog in the Channeled Scablands region of Washington state. This is an outwash plain created by catastrophic floods at the end of the Pleistocene epoch, whose sources were a set of giant glacial lakes. The region contained braided channeling and rocks and boulders strewn about randomly by the flood waters. It was also relatively close to Spokane, facilitating visits. So Edgett and Rice organized a pair of field expeditions that were held in late September 1995 in company with related educators' and public events. The field expeditions included guided overflights of the area, designed to help the expedition members understand the relationship between how things looked from the ground (the perspective the Pathfinder lander would have) and how they looked from the air (which Pathfinder could not "see").

Edgett, who headed Arizona State University's public outreach effort, also organized a competition to select kindergarten through twelfth grade teachers to accompany the expedition. Edgett asked interested teachers to propose how they would use the experience in classroom and public activities afterward and to explain how they would obtain additional financial support for the trip (Pathfinder's tiny education budget could not fund everything) and, especially, for related activities afterward. Out of 63 proposals, he and Rice selected 11 teachers. Later they added a twelfth, rewarding a Spokane educator who helped them organize a teacher workshop held September 27, 1995. The teacher workshop and public open house featured several of the Pathfinder engineers and scientists. JPL mechanical engineer Howard Eisen stole the show at the open house with a rover that demonstrated its terrain-climbing features by driving over children lying on the floor.[35]

Several Pathfinder engineers accompanied the field expedition, including Cook, Manning, and Rivellini. The first day in the field was, as the final report put it, "quite disturbing."[36] The expedition field guide took the group to the rockiest site in the region, dubbed the Monsters of Rock. It was littered with huge boulders carried and deposited by the flood waters, including one the size of a house. Howard Eisen was able to demonstrate that the microrover would be able to operate among these rocks, but Rivellini and others didn't think the lander could survive such a site. The site was not representative of the entire region, however—Golombek had chosen it because it was the worst-case scenario. They hurriedly replanned the second day's field excursion to present a set of other sites nearby that were much less rocky in order to assuage the engineers' concerns.

That second night, the group discussed the two days' field excursion over dinner. Most impressive about the field excursion to the Pathfinder engineers was the range of potential landing conditions that the Scablands provided, from the Monsters of Rock site to areas that were entirely free of rock. The "rock abundance" averages that the team used for planning purposes had tended to disguise the true variability of rock distribution and offered them a somewhat misleading sense of security about the Ares Vallis site. But the trip also ultimately gave the Pathfinder engineers more confidence in their design. The recognition on the second day that extremely rocky sites like the Monsters of Rock were actually rare even in a catastrophic flood plain caused Cook to comment that there was "a fairly high likelihood that we can handle the Vallis site on Mars if the Channeled Scabland is a good analog."[37] Tom Rivellini was actually quite relieved by the trip. He discovered that rocks deposited by great floods tended to be rounded and partly buried, which made them much less hazardous to his airbags than the ones he'd had bolted to the test platform. So his airbag tests seemed pretty conservative.

Golombek held a second workshop after the field trip, which resulted in affirmation of the Ares Vallis site, and it was again approved by the project science group in November 1995. Jim Martin asked for more independent reviews of the landing site safety, this time from both Mike Carr and Henry "Hank" Moore, another Viking veteran. Both again agreed, with Carr arguing that while one of the other sites appeared slightly safer in the light of radar data gathered during 1995, it was scientifically uninteresting. Pathfinder's science requirements were only 10 percent of the total mission, he pointed out, but to his mind, even this small value placed on science tipped the balance between the two sites toward Ares Vallis. He still did not think it was any less safe than the final Viking 1 site. Further, after seeing Tom Rivellini's videos of the successful drop tests in December 1995, he considered the Pathfinder lander more robust to rocks than the Vikings had been.[38] Whatever small additional risk the Vallis site might impose, it was worth it. Matt Golombek and Richard Cook briefed NASA officials March 1, 1996; they, finally, agreed.[39]

Mars Global Surveyor

Glenn Cunningham's Mars Global Surveyor had to launch within a month of Mars Pathfinder, despite having started its development phase much later. Fortunately, its development was almost entirely problem-free. Reuse and reconditioning of the Mars Observer spares eliminated much of the risk that projects typically faced; the team did not have to worry about avionics deliveries not meeting the schedule or showing up but failing their basic acceptance testing. Further, Cunningham and engineer George Pace took out of the Mars Observer experience that the "hands-off" contracting mode that JPL had imposed on them had not worked well, so they pursued almost the opposite policy. They sought a "badgeless" management, with several JPL staffers living out in Denver to work full time with Bud McAnally's Lockheed Martin engineers. Pace had an office in Denver that he was in about half time.[40] This effort to forge a single project team even extended to the instrument teams. Where magnetometer designer Mario Acuna had not been allowed to interact with the Mars Observer contractors directly to clean up the spacecraft's magnetic signature, on MGS he could work directly with the solar array vendor to ensure they would not interfere with his experiment.

The one extraordinary issue that caused the team some angst had to do with the project's unique aerobraking needs. MGS was the first spacecraft designed to use aerobraking to reach its final orbit. Global Surveyor would fire its engine once, to reach an elliptical orbit, and then use atmospheric drag to slowly convert the orbit to a much less elliptical one. Once atmospheric drag had removed enough velocity, another short firing of the main thruster would be used to "circularize" the orbit. The aerobraking process was supposed to take three months, with drag passes occurring every other day early in the mission but increasing to several times per day as the orbit shrank.

Atmospheric density is not a constant, particularly at Mars. Instead, it changes with the planet's diurnal cycle, becoming less dense during the day and more dense during the night. It varies with temperature in the same way, warmer days producing a less dense atmosphere than colder ones. Seasonally, it varies about 25 percent. The density of the upper atmosphere is also affected by solar flares, which can produce an instantaneous heating and expansion, resulting in a density drop. Dust storms have a similar blooming effect on the upper atmosphere. All of this was known qualitatively, and for Earth it's known quantitatively as well. Meteorologists continually measure the near-cousins of density, pressure, and temperature, for weather forecasting.

Nobody measures these quantities for Mars—Dan McCleese's atmospheric sounder was supposed to, but of course that instrument had disappeared along with Mars Observer. And the total range of density variation on Earth is quite small compared to Mars. In an important sense, Global Surveyor would be flown blind. The operators on Earth would not know with any certainty what the density would be prior to committing to a drag pass. So the spacecraft had to be designed to tolerate a wide range of variation. To ensure the design was sufficiently robust, Cunningham asked JPL scientist Richard Zurek to organize a panel of Mars atmosphere specialists. Their conclusions about the martian range of density variation would be the basis for the design.

Atmospheric density during the drag passes affected several aspects of spacecraft design. The highest densities the spacecraft would encounter established the requirements for structural strength and thermal tolerance. The lowest densities, however, primarily affected schedule. The lower the density, the less drag each pass would generate, leading to more passes being needed to reach the desired orbit. Alternatively, McAnally's engineers in Denver could design some mechanism that would allow increasing the surface area of the spacecraft on particularly low-density passes to add more drag. But that would make the spacecraft more complex, heavy, and expensive. So it was not really an option. Instead, they had to pick a fixed surface area that would generate all the drag they were going to get. The flight team also could use the spacecraft thrusters to make small maneuvers to "chase the atmosphere," as system engineer Pete Theisinger put it, and this ability also allowed them to adjust for changing densities.[41]

When Lockheed Martin's team had proposed the initial design for MGS, they had assumed the martian upper atmosphere's density was likely to vary by about 30 percent to three standard deviations—in other words, it was highly unlikely that it would vary by more than 30 percent. Zurek's group of atmosphere specialists argued that this was actually much too low—a realistic range of variation for Mars's upper atmosphere was more like 70 percent. Cunningham's engineers decided to take 90 percent as their new variability number after reviewing the scientists' arguments, adding 20 percent to compensate for navigational inaccuracies.[42] So the spacecraft's design margins were suspect in early 1995.

The largest problem this gave the team was in thermal design. Having to make substantial changes to the spacecraft hardware would be expensive. But there was another way out of the problem: they could make the drag passes shallower to stay within the existing hardware's capabilities. Doing this would greatly lengthen the time they spent in aerobraking, though. They figured it could add another year to the process, costing an additional $15 million to sustain the manpower necessary to handle the orbit analysis that had to be done after each pass in order to safely command the next pass. So initially, raising the drag passes did not appear to be a terribly satisfying option. One of McAnally's engineers suggested a third option. They could add extensions to the solar arrays that would increase the surface area, producing more drag to compensate for the altitude increase. The project could implement the higher-altitude drag passes and still stay on its original aerobraking schedule. His idea was just to add flaps to the ends of the arrays, not more solar cell area, so the change would be cheap: a half million dollars, the company promised Cunningham.

The flap addition was a great idea, but it put Cunningham in a bit of a bind. He had been adamant in establishing a project ground rule that in order to stay within his budget there would be no changes to the spacecraft or its instruments, and he would have to break his own rule to add the flaps. After a meeting with his systems engineer Peter Theisinger, George Pace, Bud McAnally, and mission designer Sam Dallas, he decided that the flap addition was worth it. In a memo he sent out to his engineering team, Cunningham wrote: "While I have continue[d] to advocate the reduction of mission performance or mission return in order to keep to our cost paradigm, I find the aerobraking margin issue to be extraordinary. The significant financial impact of the inability to achieve the 2 PM orbit the first time, the necessity to protect the solar arrays from excessive aerodynamic heating . . . [and] the large uncertainty in the orbit-to-orbit atmosphere combine to make sufficient margin in aerobraking a necessity."[43] He told Pace to have the flaps added.

The project's spacecraft critical design review was only a month after the flap addition decision, the first week of May 1995. It went very well, with only the subject of mission operations causing some concern in NASA headquarters.

Mission operations drew attention because to save money, Donna Shirley had proposed having all of the Surveyor program spacecraft operated by a single team after launch. This operations team would be set up as a quasi-independent Mars Surveyor Operations Project. Because MGS would be the first Mars Surveyor program spacecraft, Cunningham had argued that his operations team should simply become the Surveyor Operations Project after the fall 1996 launch. Pete Theisinger would be tasked with figuring out how to incorporate the ground systems for John McNamee's two Mars Surveyor 1998 spacecraft into the operations project between 1996 and 1998. When the Mars Surveyor 2001 spacecraft launched, they too would have to be added. The program manager at headquarters did not believe that Cunningham's mission manager had a good handle on this, or that Shirley's office really understood what it would cost to implement and operate.[44]

This change was actually quite a radical proposal. Typically at JPL, each spacecraft was operated by its own dedicated team, some of whom came directly from the vehicle's development team. The lab's operations specialists had always considered that important because every spacecraft is different. They have different thruster arrangements, different solar array configurations (or none), different command dictionaries, different responses to commands, different instrumentation. The operators needed to know the characteristics of each individual spacecraft well in order to successfully operate it, and only the builders knew these details. The new operations project would have to do a very good job of retraining the MGS operators for McNamee's follow-on spacecraft to operate them successfully. What would that cost? It wasn't clear in 1996.

Some of the concern at NASA headquarters about cost also came from the reality that operating Global Surveyor during aerobraking was going to be complex. M. Daniel Johnston, the aerobraking designer, intended to draw on analyses from two computer models, a Mars atmospheric circulation model at Marshall Space Flight Center called MarsGRAM and an aerothermodynamic model at Langley Research Center, to help predict the behavior of the atmosphere and its impact on the spacecraft. He would also draw on observations from Phil Christensen's spectrometer, which could provide a temperature profile of the portion of the atmosphere visible to the spacecraft. These data provided an indirect measure of density but would only be available early in the aerobraking process. As the spacecraft's orbit shrank, the instrument would not be able to see useful parts of the atmosphere. An interproject agreement with Mars Pathfinder was to bring in atmospheric opacity information as a measure of dust in the atmosphere while Pathfinder was active. Finally, Global Surveyor had accelerometers aboard for use by its attitude control system. These would tell the operators what the spacecraft's deceleration actually was during each pass, another indirect measure of density. The operations team would have to collect and analyze these bits of data to plan out the next pass. The command sequences for the passes, finally, would actually be developed out at Lockheed Martin's Denver facility, where the majority of the spacecraft operations team would be located.

So operations during aerobraking would take quite a bit of analysis and coordination, over and above the normal "caretaking" functions that every spacecraft team had to do (power and thermal management, monitoring attitude, etc.). The follow-on 1998 orbiter would also have to go through the same process. Fortunately, Global Surveyor was scheduled to finish aerobraking before the 1998 orbiter started. Unfortunately, the 1998 orbiter was a much different spacecraft, so the flight team would have to be extensively retrained to handle it properly.

Cunningham's MGS project started its assembly, test, and launch operations phase in August 1995, a month after Pathfinder did. Before the spacecraft left Denver for Kennedy Space Center, the Lockheed Martin–JPL team put the spacecraft through the same kind of end-to-end test that Spear and Muirhead led for Pathfinder. The mission operations facility that Lockheed had established commanded the spacecraft through all of the tasks it would perform on its flight to Mars. These tested their command sequences and the vehicle's response and served to train the spacecraft's operators as well.

Conclusion

Mars Global Surveyor rocketed skyward on November 7, 1996, becoming the first planetary launch for the Delta II 7925 rocket. Mars Pathfinder followed it in a dramatic night launch, at 1:58 a.m. Eastern Standard Time on December 4. Both launch vehicles appeared to provide perfect rides into space. But both spacecraft had troubles immediately after separating from their upper stages. One of MGS's solar arrays deployed improperly, imposing vibrations on the spacecraft. Analysis done by the Lockheed team in Denver suggested that the array had deployed too quickly, damaging a locking mechanism on its hinge. The unlocked array was swaying a little as the motors that kept the two arrays aligned to the Sun slowly moved them.

But this problem didn't seem severe. The team's analysis showed the hinge was strong enough to withstand the aerobraking forces without the latch if the arrays were rotated to face the direction opposite of how the engineers had originally intended them to during aerobraking—with the solar cell side facing the airstream instead of away from it. The arrays had been designed to withstand this anyway, as there were certain scenarios in aerobraking where this might occur. So other than causing some of the aerobraking procedures and command sequences to be replanned, the flaw did not appear to have an impact. During the first part of the cruise to Mars, the flight team tried several maneuvers intended to get the array to lock, but without success. They also hoped that once they started aerobraking, the aerodynamic forces would lock the array for them, and the trouble would go away.

Pathfinder's troubles seemed more alarming. Neither of the redundant Sun sensors on the rear of the cruise stage could find the Sun. Pathfinder was a "spinner," in space jargon, spinning around a Sun-aligned axis as part of its stabilization scheme. Inability to find the Sun was potentially fatal to the mission. The midcourse trajectory corrections and the final orientation of the spacecraft prior to cruise stage separation at Mars had to be managed by the on-board attitude control system, and it needed to know where the Sun was in relation to itself in order to perform these acts correctly. These operations could not be controlled

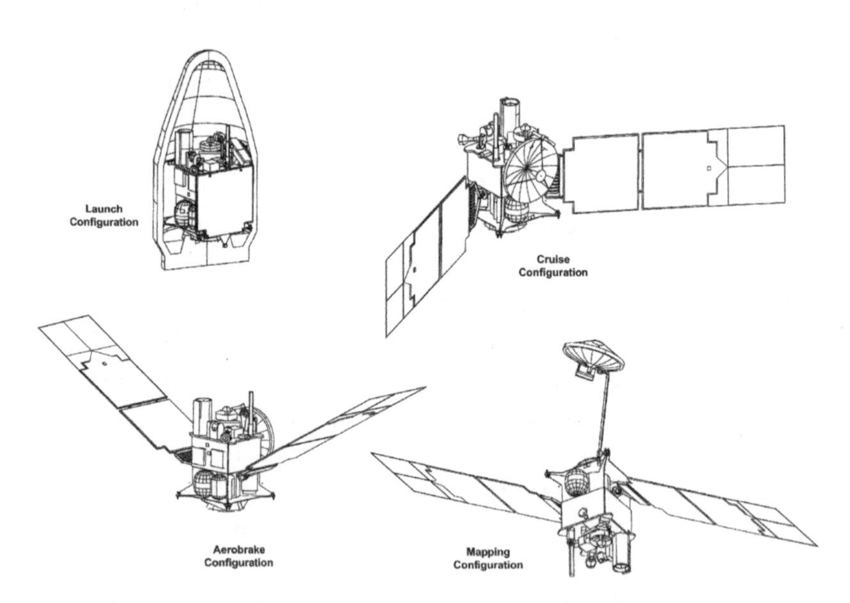

Mars Global Surveyor configurations. Aerobraking required the solar arrays and high-gain antenna to have different positions than they would in mapping orbit. The "nadir deck" with the instruments points up in the launch, cruise, and aerobrake configurations and down in the mapping configuration.

manually by the team back on Earth. Analysis of Pathfinder's engineering data showed that one sensor in fact saw nothing at all, but one had found the Sun perfectly well. Yet its output signal was so low that the flight software was discarding the signal as noise. Something indeed was wrong. The attitude control engineer figured out how to program around the problem, allowing the software to accept the data. But then the team couldn't upload the software patch. What should have taken a few minutes to do took many hours; the telecommunications engineers eventually concluded that the spacecraft receiving antenna's view of Earth was being blocked by another part of the spacecraft on each revolution. Dropping the data rate to a thirtieth of that normally used finally got the new code to the spacecraft. JPL and the launch vehicle manufacturer eventually concluded that the Sun sensors had been damaged by debris from the rocket's pyrotechnics.[45]

The troubles for Global Surveyor and Pathfinder, though, were nothing compared to Russia's Mars 96. When it launched on November 16, the spacecraft's upper stage misfired, stranding it in Earth orbit. It reentered over the South Pacific and disintegrated, effectively ending Russian planetary exploration for at least a decade. MGS's balloon relay would never receive the balloon data it was designed for.

Glenn Cunningham held a little ceremony to return about $6 million to NASA, his underrun from the development phase of the project. Spear was right on budget after several years of hearing that his task wasn't possible; Brian Muirhead saw no value in giving money back if it could be used to reduce risk. Feeling somewhat vindicated, Spear held a two-day "Lessons Learned from Faster Better Cheaper" conference in April—months before his team's ultimate test on Mars would occur.[46] JPL's delivery of an in-house-built, operational spacecraft for $170 million was sufficient cause for celebration. John Casani paid up on his bet with Dave Lehman, delivering the fine red wine to a party for the Pathfinder team at Lehman's house.

Mars Global Surveyor and Pathfinder costs (millions of U.S.$)

Project	Development	Total through end of primary mission
Mars Global Surveyor	148	270
Pathfinder	171	265

Source: Mars Global Surveyor Arrival press kit, September 1997, and Mars Pathfinder Landing press kit, July 1997.

Note: Mars Pathfinder includes the $25 million cost of the Sojourner rover. Costs are in real year dollars.

But despite the post-launch celebration, the team understood that Pathfinder still had to work on the planet's surface. Rob Manning's EDL team spent their flight to Mars running simulation sets and squashing software bugs to ensure that it would.

CHAPTER FIVE

Mars Mania

On August 7, 1996, three months before Mars Global Surveyor's launch, NASA's Space Science Enterprise had held a press conference to announce the discovery of organic material in a rock from Mars. The suspect rock, meteorite ALH84001, had been found in Antarctica in 1984. Studied by a research team headed by David McKay of NASA's meteorite laboratory at Johnson Space Center, this piece of debris was traceable to Mars by four lines of evidence, including the presence of trapped gases in glass bubbles that could have come from nowhere else. The organic compounds were of a type often associated with the decay of biological matter, and the rock also contained small amounts of carbonates, which can form through both biologic and nonbiologic processes. Most interestingly, the rock contained tiny magnetic grains like those manufactured by some Earth bacteria.[1] In short, McKay's group thought they had found strong evidence for a living Mars—at least in the very remote past.

NASA headquarters had known about the discovery since April, when McKay's group had approached Wes Huntress, associate administrator of space science, about the paper they had submitted to *Science*. Accepted in July, the paper caused Huntress to approach NASA administrator Dan Goldin about it, and Goldin had contacted the White House. This evidence of life on Mars got Goldin audiences with President Bill Clinton and Vice President Al Gore prior to the paper's formal August 16 release. Publication was greeted by a media storm; President Clinton asked for a bipartisan "space summit" to discuss the finding's implications for the future of the space program.[2] Held in October, the summit helped Goldin to stem the decline of the NASA budget. He had been told earlier in the year to expect around $12 billion from 1999 forward; instead, the White House would halt the slide around $13 billion. In fact, it didn't drop even that far. The approved fiscal year 1998 budget was $13.6 billion.

McKay's group, whether or not they truly had found remnants of ancient microscopic Martians, thus dramatically altered the environment in which JPL's Mars Exploration Program Office operated. The program suddenly got more money, and a slight name change—to Mars Exploration Directorate—but it also received new mandates: it *would* produce a Mars sample return mission in 2005. And it *would* provide data supporting a near-term human mission to Mars. NASA's demands were not negotiable, and they threw the program office's own plans into disarray. At the same time, the rock from Mars dramatically increased public attention to the Mars effort. When Mars Pathfinder bounced to a stop on the Ares Vallis outwash plain the following year, more than a billion people were waiting for its first images.

Mars Surveyor Program Strategy Redux

In February 1996, JPL's Dan McCleese was given the task of creating the implementation strategy for the 2005 Mars sample return mission that had been the subject of much discussion the preceding year. He chartered a Mars 2005 Science and Implementation Committee composed of both engineers and scientists. Many of the scientists were drawn from universities, with the engineers coming from Mars Pathfinder and from JPL's "Team X," a recently formed organization designed to quickly generate and cost advanced mission proposals. Between February and May, this group tried to construct a sample return strategy that could be carried out within the Surveyor program's resources and schedule. Their conclusion was that it could be done, but not without substantial changes to the existing program.

Team X had been asked to study the simplest option, a direct-entry, direct-return architecture. The spacecraft would, like Mars Pathfinder, enter the Mars atmosphere directly from its transfer orbit from Earth. It would then take a single, 100-gram sample of soil and within four hours launch that sample back to Earth. The mission would do no other science beyond taking a photograph of the landing site. The two-stage ascent vehicle would put the sample container on a 2.5-yearlong trajectory toward Earth's Moon—in case it failed en route, the ascent vehicle and its sample container would not hit Earth. Twenty days before arriving, if the vehicle was still healthy, JPL would direct the ascent vehicle to turn toward Earth, release the sample container, and then turn back to its lunar impact trajectory. This could be done, they thought, for $300 million, but it would have to fly on a Russian Proton. The mass that had to be sent to Mars greatly exceeded the capabilities of the Delta II/III series and even of the larger Ariane

5. While the team had used Lockheed Martin's polar lander design as part of its cost basis, the landed mass on Mars had to be far larger, on the order of 1,400 kg.[3]

From a technological standpoint, this scheme appeared to be the simplest possible mission architecture. It minimized the number of flight vehicles that had to be built and integrated as well as the number of interfaces that had to managed. The mission design used only existing, flight-experienced equipment, excepting the sample collection chain and the Earth-return vehicle (if you accepted, as Team X did, that the Mars Polar Lander would land successfully in 1999, validating the design). Even the ascent rocket was built out of existing rocket engines (derived from the space shuttle's maneuvering thrusters). So it minimized development needs.

But these facts did not necessarily make it a desirable mission, or one that was low risk. From the standpoint of desirability, McCleese and the other scientists on the committee faced the problem that the "grab sample" would be whatever the lander happened to have landed on. It might well have very little scientific value, and some members of his group thought such a mission would not be worth the cost and effort. The plan also carried political risk. What would happen to the Mars program's political support if the returned sample were as sterile as the Viking data had suggested the surface was? A sterile sample could conceivably end both the Mars robotics program and kill Goldin's drive for human missions in the twenty-first century. The counterargument was that even a boring sample would contain chemical information and permit the development of an absolute date for the sample. Human knowledge of Mars's surface age was based on analogy with the lunar surface, which had been accurately dated via rocks returned by Apollo astronauts. The lunar analogy could be wrong, and the returned sample might at least provide an indication of its correctness. Ultimately, the committee concluded that while a grab sample should be a required contingency measure for the mission, a rover should be the mission's baseline sample collector.[4]

The engineers on the committee were not thrilled with the mission, either. What Team X had done was scale up Lockheed's Mars Polar Lander enormously, assuming that whatever scaling problems might arise could be fixed easily. But as the Mars Pathfinder team had discovered with their airbag nightmares, that isn't necessarily a valid assumption. Things don't always scale linearly or easily. So the engineers contended that if NASA was serious about getting a sample return mission to Mars in 2005, then the Mars program had to prove the landing capability in the 2003 opportunity. They wanted to send a "test mass" at least equal to that of the ascent vehicle, about 700 kg, to Mars. It did not have to carry a scientific payload—in fact, the engineers preferred being able to send a "dead on arrival" lander just to demonstrate the entry, descent, and landing concept—but risk reduction required sending a vehicle much larger than even the 600 kg Viking landers had been. The scientists, of course, could not back a no-science mission, so the hammered-out compromise was to try to send a big long-range rover in 2003 that might collect samples for the 2005 mission to pick up. This would serve both the engineering need for demonstration and the scientists' need for science, but at considerable additional cost. The group hoped that NASA's Human Exploration and Development of Space Enterprise (HEDS, the part of NASA that carries out astronaut missions) would pay for the 2003 rover because there probably was not enough money in the Mars program's own budget for it.[5] But at the very least, they argued, NASA should offer to carry instruments from other countries on the putative 2003 lander.

Cost also drove the group to recommend that the 2001 mission plans be substantially curtailed. At this point, NASA leaders were still planning to have the 2001 project send an orbiter and a lander to Mars, although what those vehicles would do beyond flying Bill Boynton's long-delayed gamma ray spectrometer was unspecified. But the Mars program clearly could not afford to fly a lander in 2001 at all if it were to pay for the scaling up of Polar Lander to more than twice its mass for 2003 and then scale it up again and develop the technologies that the sample return mission itself required for 2005. The technology development costs alone were estimated at $200 million. Without a substantial budget increase, the only way NASA could afford the 2003–2005 sample return sequence was by raiding the 2001 mission's planned budget.[6]

Dan McCleese presented the mission sequence proposal to the Committee on Lunar and Planetary Exploration (COMPLEX) in June 1996, intending to present the final report in August. Instead, the rock from Mars intervened. During the first week of August, Huntress sent McCleese new instructions to do a quick study of what kind of robotic exploration program was needed to answer whether life ever existed on Mars. Team X made a crash study of options for gathering evidence and the Mars Exploration Directorate team presented some of them to Huntress.[7] Neither set of plans found immediate favor, so the following month, Norm Haynes, who had succeeded Donna Shirley as the chief of the Mars Exploration Directorate, offered Huntress a somewhat different series of options. Huntress had not been willing to sacrifice the 2001 lander to the sample return mission, so under the new options the 2001 lander would carry a "demo" rover costing the same as Pathfinder's Sojourner. The big, long-range rover was out because NASA's HEDS directorate was not willing to pay for it, so instead the sample return rovers Haynes presented were approximate "clones" of little Sojourner, too.

Goldin, taking advantage of the rock from Mars excitement, demanded from Johnson Space Center a study of a human mission to Mars to be launched in 2011. The Mars Exploration Directorate at JPL had to support the study, another task for the already busy little group. This planning effort meant collecting data relevant to Mars's ability to support human explorations; for a 2011 or 2013 mission, the Johnson leadership needed these data no later than 2004. That was the year they hoped to get a presidential directive supporting an astronauts-to-Mars strategy. That meant, in essence, JPL's 2001 mission would have to provide all the necessary data. A 2003 launch wouldn't arrive at Mars in time.

In mid-October 1996, McCleese presented again to the Committee on Planetary and Lunar Exploration. His carefully drawn water strategy was no longer central; instead, the enthusiasm over the rock from Mars within NASA headquarters had resulted in "following the water" being replaced by a Viking-like "search for life." He presented four options, but the focus on life was the same in each. As he put it later, "This was a NASA-driven directive that no scientist liked. Sometimes you take one for the Gipper."[8] It didn't go over well with COMPLEX. The committee members were not about to sign on to such a narrowly focused program merely because of the brouhaha surrounding ALH84001. Instead, they stood by their previous recommendation of a broader goal of characterizing Mars's environments, past and present. "NASA should focus its Mars program, and sample return missions in particular, on the more comprehensive goal of *understanding Mars as a possible abode of life*, a goal that is fully compatible with previous recommendations."[9] Viking had fallen victim to a narrow focus on life and often was viewed as a failure because it had not found the microbes it had been sent to discover. And former JPL director Bruce Murray's argument from the Viking era—that if you don't understand the geochemical environment you're sampling, you can't make a definitive interpretation of your biochemistry data—still resonated within the committee. So COMPLEX wanted to see a broader examination of Mars's geochemical and geological evolution than McCleese had proposed. In a similar vein, it made no sense to these senior scientists to send the 2001 lander with an "inert" payload as the sample return sequence proposed. Why send a spacecraft to Mars at all, they argued, echoing Steve Squyres's complaint about Pathfinder, if it was not going to produce any scientific return?

Thus, a year before JPL formally established the Surveyor 2001 project office, there was already a conflict brewing over what the project should, or even could, entail. From a budgetary standpoint, with no new funds, JPL's engineers thought they could afford an orbiter but not a lander. Technologically, however, if cost was less important, a 2001 mission could be, and should be, used to demonstrate some key technologies that a sample return mission would need, such as aerocapture, precision landing, and a high-mass-capability EDL system. Scientists wanted scientific return from whatever the 2001 mission turned out to be, which meant funding instruments that the Mars Exploration Office did not think it could afford. And NASA's human spaceflight organization needed the 2001 lander to provide still other kinds of data.

These conflicting goals could be resolved one of two ways: either strict prioritization and removal of some of the requirements or an infusion of quite a bit more money. The first didn't happen. Over the next 18 months, both NASA and JPL senior management refused to limit the requirements being imposed on the 2001 and 2003/2005 mission sequence. However, there was a little more money to be found. The Surveyor program got a budget increase from $111 million in fiscal year 1997 to $145 million the following year.[10] The cash infusion seemed to be enough to cover part, thought not all, of the increased cost imposed by the drive toward a 2005 sample return mission. Norm Haynes and Donna Shirley knew they still did not have the money to develop the sample collection, transfer, and handling chain with its (still undefined) planetary protection requirements, to develop an Earth return vehicle, or to miniaturize certain technologies that were fundamental to all spacecraft so that more mass would be available on the sample return missions for all the mission-specific gear. The Mars Exploration Office's inability to bring its resources in line with its requirements would continue to worry its officials, even while they experienced great success with the Mars Pathfinder and Mars Global Surveyor missions. Those two enterprises simply increased the pressure on them.

Mars Pathfinder Shifts from Development to Operations

During its cruise to Mars, the rover and spacecraft teams for Pathfinder practiced operating their vehicles. Continually. These operational readiness tests (ORTs) employed the flight system testbed, the engineering model rover (named Marie Curie), and a lander mockup placed in a "sandbox" at JPL to train the teams in their jobs once the vehicle landed. These tests were a crucial part of the project. In addition to their training function, they were also the first real opportunity the engineers had to identify and correct flaws in the plans they had made for operating on Mars.

Key to the early readiness tests were the two testbeds. After the Sun sensor problem had been satisfactorily resolved, Rob Manning had turned to using the flight system testbed for debugging the entry, descent, and landing software. That software turned out to have some serious problems. Manning remembered years later that as launched, Pathfinder would have failed entirely. It had a subtle timing error that caused the EDL software to stop responding to inputs from the spacecraft's own sensors after one day's run time.[11] Prior to launch there had not been time to let the EDL software run on the testbed for a full day; there were simply too many demands for other tests. But the plan had been to order the EDL software to start running one day prior to landing. That way, if the Pathfinder team missed the communications window on landing day itself for whatever reason, the spacecraft would still make the landing attempt on its own. But this bug would have prevented that; Pathfinder would have plummeted to the surface without ever noticing that it had reached Mars.

When Donna Shirley had been promoted to head JPL's Mars Exploration Directorate, a robotics engineer named Jacob Matijevic had replaced her as rover team leader. Matijevic's rover team used the Marie Curie rover testbed to begin doing their own set of rover operations readiness tests in January. His group found fairly quickly that they had gone too far in adhering to the faster-better-cheaper mentality. They did not have enough people to help the science team plan the rover's movements and experiments, monitor the rover's engineering status (its "health"), manage the interface with the lander and lander operations team, and build, test, and transmit all the necessary command sequences. Matijevic had to hire more people.[12] Initially, his team operated on its own, practicing driving the little rover around JPL's "Mars Yard," a big outdoor sandbox with rearrangeable rocks. In April, they joined the lander team for the first ORT aimed at landed operations.

This first landed ORT was mildly disastrous. Brian Muirhead had appointed one of the junior systems engineers, David Gruel, to the job of imposing problems for the operations team to solve. For this ORT, he had arranged for the lander to be positioned on top of a steep sand dune, with the lander petals arranged so that the rover egress ramps would each have to extend down at a steep angle to reach the surface. The rear ramp he intended to be too steep for the rover to roll down, while the other was passable but would let the rover down into a small trough. He expected that the rover team, using the lander's stereo imager, would be able to determine relatively quickly that the forward ramp was the safer, to get the lander team to deploy it, to generate the rover command sequences to move off, and to have it all done on the first "day" on Mars. But instead, it took the engineers three long, painful Mars "sols" just to get Marie Curie to the sandbox surface. And they were so inefficient that it took the team five Earth days to perform these three sols of work.

Matt Golombek laughed about this a decade later. "It was horrible! We would've killed the lander, because the rover blocked so much of the solar array area." The rover blocked about a third of the total array area, leaving not enough electricity to operate the lander without imposing a drain on the battery. So as the team struggled with the rover egress problem, the lander's rechargeable battery would deplete and be close to dead after three sols of trying; it put a rather hard limit on how long they could spend trying to get the rover off. It took them this long in part because trying to figure out the ramp angles using nothing but the stereo imager on the lander's mast and a couple of telemetry data streams proved far more difficult than the engineers had expected. Rover control system engineer and chief driver Brian Cooper initially chose the wrong ramp. He thought from the images that the forward ramp would put the rover into a spot it could not maneuver out of, and he asked the lander team to use the spacecraft petal actuator motors to try to get the lander better positioned for deployment of the rear ramp. Once the rover was deployed (by Jake Matijevic going out into the sandbox with the ramp and attaching it manually while the cameras were off), it became clear that Cooper had made the wrong choice. The ramp was obviously too steep. So then the rover and lander teams had to build and uplink the command sequences to deploy the forward ramp and move down that one instead.[13]

"It showed us the weakness of our software tools" for operating the vehicles, Golombek remembers. Rover systems engineer Andy Mishkin added that the ORT exposed the immaturity of the operations team and its procedures. From a strictly operations perspective, almost nothing had gone right. The teams had been so consumed with getting their hardware to the launch pad they had put very little time into development of the procedures necessary to guide operations decision making. So in addition to making poor decisions, their decision-making process was inefficient. If time on Earth is money, on Mars time is power. The team couldn't afford to be so inefficient on "real Mars" because their vehicles couldn't support the energy drains that the egress operations imposed.[14]

Mars Pathfinder and a New Passion for Mars

While the Pathfinder team was training, public enthusiasm for the Mars missions started to grow. No one had expected this; in fact, most of the team had felt largely ignored by JPL, NASA, and even the aerospace trade press until just prior to landing. Pathfinder was a low-budget technology demonstration mission, not a scientific showboat, so no one at JPL had seen it as having the media potential that the very telegenic Viking and Voyager missions had demonstrated more than a decade before. Further, many at NASA headquarters, and even at

JPL, had taken the political hostility in Washington toward the old Space Exploration Initiative's plan to send humans to Mars as an indication of lack of public interest in Mars. That turned out to be incorrect. Pathfinder landed in the midst of a media circus and lived its short life on the surface in a fishbowl.

One reason for the interest in the mission was proximity. Pasadena was home not only to JPL but also to one of the nation's largest space advocacy groups, the Planetary Society. Founded by Carl Sagan, Bruce Murray, and Louis D. Friedman to promote planetary exploration, the society boasted 100,000 members by the late 1990s. It had scheduled a "PlanetFest" celebration for the weekend of Pathfinder's landing, renting out the Pasadena convention center a few blocks from the converted house it used as its headquarters and working with JPL to set up live feeds from the mission support area (MSA). This setup would enable far more people to "be there" as Pathfinder landed—or crashed—and experience the thrill of another planet by telepresence. It wasn't unusual for JPL to hook up an off-lab venue to accommodate more spectators than could be hosted onsite; its own Von Karman auditorium could hold only about 200 people and was reserved for news media during landings. The Viking project had wired the auditorium of La Cañada High School, just outside JPL's main gate, for its two landing days, enabling some of the project's families and other VIPs to witness the events. The Planetary Society hoped to attract many more people to witness Pathfinder's descent.

The society had scheduled lectures and panel discussions to occupy the three days of PlanetFest. The festival opened on Thursday evening, July 3, 1997, with a symposium on "Mars into the Next Millennium," kicked off by NASA administrator Dan Goldin. Donna Shirley participated in a panel discussion on women in science and engineering. Amateur astronomer Thomas Bopp lectured on his co-discovery of the comet Hale-Bopp, which had made its closet approach to Earth that March. David McKay of the ALH84001 meteorite analysis team lectured on the Mars rock discovery and exhibited a fragment. Recognizing the rapidly growing Hispanic population of the region, the society held a full day of Spanish language presentations on July 4. And, with the first generation of planetary explorers rapidly graying, the society had also set up "A Child's Universe," a planetary science educational playground for a new generation.[15] The society's efforts brought thousands of people into Pasadena for the weekend, and its advertising for the event reached many more.

NASA's Goldin had also helped prepare for a very public return to Mars by requiring a new emphasis on education and public outreach within the science directorate. This order had been met with mixed feelings. Many scientists considered public outreach to be outside the proper domain of scientists—science "popularizers" were held in low regard in scientific circles. Carl Sagan, while famous, was never elected to the National Academy of Sciences partly because he was viewed as a "mere popularizer."[16] At the same time, other scientists thought this view very short-sighted. Since most science in America is publicly funded (and essentially all planetary science is), getting the general public interested seemed vital to the long-term health of the science. Phil Christensen at Arizona State University was a supporter of outreach, winning a proposal competition to have graduate student Ken Edgett establish a Mars K–12 education program. Edgett, in turn, organized Mars sessions at the National Science Teachers Convention in April, a Mars Day at the Arizona Science Center for July 2 and a teacher workshop for September.

While these educational initiatives were aimed at children, they also had the effect of exposing adults to planetary science. Parents had to transport their children to these activities, so they also served as indirect advertising. Educational activities helped expand the community of interest in Mars and Mars Pathfinder.

JPL had also established a Web site for the Lab in 1995 and a separate site for the Pathfinder project in 1996. The availability of the Web allowed the project to engage in an experiment in public outreach that was a little controversial. The three-person Web site staff would be allowed to put the imagery returned by the lander and the Sojourner rover up on the Web immediately. There would be no delay for scientific analysis, which also meant that Peter Smith of the University of Arizona, the stereo camera's principal investigator, would not have the period of exclusive use that had been traditional for NASA experimenters. Smith was concerned that he could be scooped by someone else with his own data. But Wes Huntress had backed the idea. Assuming it landed safely, Pathfinder would send data back to Earth once per day, every day. A daily Web site update would allow visitors to follow the mission's events essentially in real time.

Spear had also supported the development of a World Wide Web application that enabled the general public to plan simulated missions for the Sojourner rover. Developed by an engineer named Paul Backes at JPL, it was originally intended for use by the science planning team. Backes had thought that science team members might want to stay at their home institutions to plan the rover mission and developed a Web-based tool to enable it. He discovered that, in fact, they didn't want to stay home. While it was expensive to come to JPL and stay throughout the mission, the reality was that scientists found value in being together at the scene of the action. So instead, his software tool, called the Web Interface for Telescience, was used at JPL to help plan the actual Sojourner rover's mission. And, a parallel installation allowed Web-connected members of the public to plan missions for a virtual Sojourner using the real mission imagery.[17]

Rob Manning remembers that all of these efforts became real for the project engineers when the national press suddenly appeared at JPL the week before landing day. "They absolutely descended upon us, and this little project became this big deal."[18] Tony Spear got a sudden directive for a VIP briefing for Wes Huntress and Dan Goldin on how Pathfinder was supposed to work. Spear charged Manning with that task, so Manning had to spend the afternoon of July 3 explaining the entry sequence with the aid of his big EDL test chart and a video of some of the testing. He fielded questions about why his EDL system was single-string (money and volume), what vulnerabilities it had (winds and big, sharp rocks), and why it had never been subjected to a full-scale demonstration on Earth (pointless). He recalls leaving the briefing convinced more than ever that the system would actually work; his audience wasn't so sure. But by this time, reviewers' unhappiness didn't matter. The vehicle would reach Mars at 10:00 a.m. the next day, and 10:00 p.m. that night in Pasadena was the last moment at which the vehicle could have its software or course "tweaked." JPL's visitor parking lot and the sides of Oak Grove Drive, outside its main gate, were lined with television trucks waiting for the big event. Wes Huntress's experiment in faster-better-cheaper Mars landings was going to be very public.

Spear hosted a big party at his house in La Canada, almost within sight of the Rose Bowl, the night of the July 3. He left to meet with Richard Cook and Matt Golombek to decide whether to make any last minute changes during the one window of telecommunications left before Pathfinder entered the martian atmosphere. They decided not to do anything.

The original Pathfinder design had provided no means of knowing what was going on during the entry sequence until the vehicle had landed, deflated and retracted its airbags, and unfolded itself. Then it would power up its telecommunications system, find Earth, and transmit the engineering data the spacecraft had generated during the entry and landing sequence. But this design hadn't flown with Wes Huntress. Pathfinder was supposed to be an engineering demonstration mission, primarily, but JPL would not get any engineering data at all if the spacecraft failed. Huntress made Spear's crew figure out some way of sending back real-time data. He had also given them another $6 million to do it with.

This task had been difficult for the little project. Unlike the nuclear-powered Viking landers, Pathfinder ran on batteries from the moment it separated from its cruise stage until it was opened up on the surface and bathed in the weak, rising Sun. The battery charge was a limited resource, and running a transmitter consumed precious watt-hours. The team's solution was to leave the transmitter on and connected to a low-gain, omnidirectional antenna. After each major event in the entry sequence, the transmitter would vary its carrier frequency according to a predetermined scheme—a kind of semaphore. The Deep Space Network would detect that, and the network operator in the control center could look up the prearranged code on a chart. Pathfinder could send 16 discrete codes this way. So it could be followed in almost-real-time, except during its hypersonic entry when a plasma cloud formed around the spacecraft, blocking transmissions. (The distance between Earth and Mars on July 4 imposed a ten minute communications delay.)

Pathfinder's entry sequence started with venting coolant. Then the computer ejected the cruise stage, which was pushed away by springs so that it would not follow the lander into the atmosphere too closely. Inside the MSA, JPL's version of "mission control," Rob Manning took over the director's chair. There was nothing to direct, of course; the sequence was fully automated. And the ten-minute signal lag was longer than the total entry sequence, so if anything went wrong between hitting the atmosphere and hitting the surface, there was nothing anyone could do about it. Manning's job was to interpret the signals sent back by the lander for the assembled team—and for the television cameras staring in at them. (On JPL's video master, an unidentified voice comments that he'd never seen so many television cameras at the Lab before.)[19]

The mood in Pathfinder's small MSA was pensive throughout the descent. Manning occasionally broke in on the microphone to explain various bits of detail to the television audience. They weren't sure, for example, that they'd actually be able to follow Pathfinder's transmission all the way to the surface on their monitors. They'd installed a much more sensitive recorder, at the Deep Space Network station outside Madrid, Spain, that would actually be receiving the transmission, to capture the signal to tape for later analysis. But, in fact, they were able to follow it in Pasadena, and as the entry sequence ran and the signal kept coming, more and more team members drifted over into the corner of the room that contained the telecommunications terminals. They clearly got the signal announcing parachute deployment, sparking a moment of clapping and cheering, but the team didn't quite realize when they'd landed. A sotto voce "yes!" by Manning off camera was their clue; Muirhead looked over at him in surprise, and then Manning said, "That's a very good sign, everybody."[20] The mere fact that the lander was still transmitting was proof enough that it had landed; by that time, Manning knew, it would have been smashed to bits against the surface otherwise.

The room erupted around him in cheers and hugs. Spear stood in the middle of the room shedding tears of joy as Wes Huntress, Jim Martin, Dan Goldin, and Ed Stone all came in and congratulated him for having done what a lot of people had thought impossible: landing on Mars for a tenth the cost of Viking.

Pathfinder's airbag retraction, petal opening, and self-righting processes were slow, so it was not expected to get its high-gain antenna into position to transmit to Earth until after 4:00 p.m.in Pasadena. That was when it would send its first images back to Earth. So after learning that it had landed successfully Spear and Manning went down to the Pasadena Convention Center to explain the landing events to the overflow crowd in person. Goldin had decided to rename the landing site Carl Sagan Memorial Station in memory of the recently deceased famous astronomer. Sagan's widow, Ann Druyan, was a Planetary Society

Video frame from the Pathfinder mission support area. Standing (*from left to right*): Matt Golombek, unknown, Tony Spear (*arms raised*), Sam Thurman. Partly obscured by Spear's arms are Dave Spencer (*left*) and Dara Sabahi (*right*). Rob Manning is seated. The Pathfinder project could not afford to hire JPL's own photo lab to photograph the landing event, so this video is the only imagery JPL owns of the event.
Frame from Mars Pathfinder Landing Day Commentary Coverage, 4 July 1997, AVC-1997-183, part 2. Courtesy NASA/JPL-Caltech.

board member and was present to receive the honor. Tony Spear was supposed to make the announcement but forgot in all the excitement; Goldin pulled him back on stage, and they made it together. Spear also announced that Pathfinder had carried to Mars a microchip with the names of 100,000 Planetary Society members on it. The chip had been prepared for the Russian Mars '96 mission; launched November 16, 1996, it had reentered Earth's atmosphere and crashed somewhere in South America or the adjacent Pacific Ocean. Spear had let the Planetary Society install a duplicate chip on Pathfinder's lander.

The first image back, a panorama of the lander itself from Peter Smith's camera, revealed that one of the airbags was blocking the ramps the Sojourner rover needed to get off the lander. While this was a significant problem, it was also one that Dave Gruel's gremlin activities had prepared the rover team for. One of the ORTs had included exactly this problem, so the command sequence necessary to raise the affected pedal and reactivate the airbag retraction mechanism to try to pull the airbag further under the lander already existed. After some minor modifications the operations team sent the commands up just after 6:00 p.m. A second set of images returned around 8:00 p.m. confirmed that the bag had retracted as they'd hoped and the ramps were clear.

At the same time, some members of the project had set up the real lander's situation in the testbed's sandbox to examine the ramp deployment. Their simulation convinced them that the ramps would provide Sojourner with a path down to safe terrain. The flight director sent the ramp deployment command shortly after 10:00 p.m. By then, it was after dark at the landing site, so the next day's imagery would provide confirmation.

The blocked exit ramps were not the only problem the team had to cope with their first day on Mars. Sojourner had suddenly stopped communicating with the lander in the early afternoon, leaving the rover team scrambling to figure out what had happened. This was not the first time the rover and lander refused to communicate; it had happened during the systems testing, but both rover and lander teams thought they had fixed the problem. The available telemetry did not leave them many clues as to the cause, and Jake Matijevic made the rover team quit shortly after midnight to get some sleep before the next day's tasks. Pathfinder's first day on Mars left the lander team ecstatic and the rover team anxious—their mission might be over already.

All of these activities were very public. JPL's Media Relations division had set up three remotely operated cameras in Pathfinder's MSA. They provided continuous coverage to the Lab's auditorium, where reporters were congregated, to

NASA's own "NASA TV" channel, and out to PlanetFest in downtown Pasadena. One of the cameras was pointed right at the rover control station, so the possibility that the rover might have failed got transmitted immediately.

But the piece of tape that all the evening news shows that night chose to run was the team's emotional response to getting the signal that indicated Pathfinder was sitting upright on Mars right after 10:00 a.m. Replayed throughout the next day too, it helped drive traffic on the mission Web site to unprecedented levels: 32.8 million hits on July 4, 40 million on July 5, both new records for the still-young World Wide Web. The team members' informality and joy drew comment from newspapers, too. The *Los Angeles Times* praised their "infectious enthusiasm" for restoring a sense of wonder about space exploration, delineating a theme common among newspaper reports of the mission's arrival.[21] One analyst compared them to the military-like image of space exploration conveyed by Jim Martin's Viking team twenty years before. The cool militarism of the old Cold War space program had been replaced by a new model: "nerds in love."[22]

The imagery Pathfinder sent back the next morning showed the ramps had deployed successfully. And to the surprise of some, anyway, Sojourner sent back a complete memory load of telemetry too. Whatever the previous day's problem had been had fixed itself, at least temporarily, and the telemetry showed that it had executed all the commands it had been given as well. Sojourner appeared to be fine. So while the rover engineers suspected that they had an intermittent problem that could resurface later, they dropped the issue to get the rover moved off the lander. First, the team transmitted the command for it to "stand up" and latch its suspension, along with a set of commands to photograph it afterward to provide visual confirmation that it had succeeded. Then they ordered it to move down the rear ramp, sending along with that command others designed somewhat surreptitiously to allow production of a short "rover movie" of the vehicle moving down the ramp to the surface.[23] These succeeded, too, and the rover movie clip went out on the Web site right after being assembled.

The mission success criteria that guided the operations team's first week on Mars included a 360° panorama of the area around the lander. Smith's camera took the exposures over the first two sols, and the panorama was reassembled on one wall of the MSA. Matt Golombek took a very nonscientific approach to naming the rocks in the image: he let the team members choose names by sticking Post-It notes to the picture. One of the first rocks visible, right behind Sojourner after it had moved down the ramp, got the name Barnacle Bill for its knobby appearance. Others became Scooby Doo, Bamm-Bamm, and Pooh-Bear, reflecting perhaps too many Hanna-Barbera cartoons watched in childhood (or college dorms). The names stuck. Appearing quickly on the Web site and in newscasts, they became the identification of those rocks in scientific publications as well.[24]

The rover science group couldn't resist the urge to have Sojourner put the rover's alpha proton x-ray spectrometer (APXS) instrument onto Barnacle Bill immediately—it only required turning 70° and moving backward a third of a meter. The rock was nearby and had an obviously interesting composition (if not-yet known). Brian Cooper had the command sequence built for transmission at the end of sol 2's operations for the rover to execute on sol 3. Sojourner hit the rock on its first try and returned its first spectra from the rock that evening.[25] They had been incredibly lucky to have such an excellent target nearby.

Sojourner completed all of its mission requirements by the end of its fifth sol on Mars, two days early. But it continued to operate normally, although the 14-hour days its human operators were putting in took their own toll. A miscommunication between the Deep Space Network, the lander team, and the rover team caused the project to miss a communications pass on sol 7—a major sin in JPL's little world—and losing that day's data. A more formalized procedure was

Sojourner with Mars rocks "Barnacle Bill" (*left*) and "Yogi" (*upper right*).
Image PIA00660. Courtesy NASA/JPL.

worked out to prevent a similar error in the future, but it had been a product of having too few people to do all the necessary work. The lander's primary mission ended August 8, but it also continued to function, and NASA supplied more money to keep the project operating.

The lander's weakness was its rechargeable battery, which the team knew could not be charged effectively more than about 80 times. Its principal function had been powering EDL and morning wake-ups; the solar panels provided enough power during daylight to power the lander and recharge the battery. The rover's first potentially fatal weakness was also a battery, in its case, a nonrechargeable battery intended to keep the little vehicle alive overnight. When it failed, the rover would either "wake up" with the Sun—or not. If not, that was the end of its mission. The rover's battery failed on sol 56, but the rover survived the loss, waking up with the Sun each morning and carrying out the orders stored the afternoon before.

The remaining Pathfinder team hoped to do the same with the lander once its rechargeable battery failed. On sol 83, it appeared to the operations team that the lander's battery might provide one more night of data from the meteorology experiments, and they sent up a command sequence that would have it activate those instruments after local midnight and send the data back in the morning. The lander missed its communications pass the next day, and the team never heard from it again. Their post-hoc analysis was that the battery failed during the night, resetting the lander's internal clock. The clock was essential to things like pointing the high-gain antenna at Earth, but the lander should still have powered up with the sunrise and tried to use the low-gain, omnidirectional antenna to regain contact. The fact that it did not suggested something else had failed when the battery died. One of the battery's functions was maintaining the lander's internal temperatures via automatic heaters, so it was possible that a particularly cold night caused an electronics failure. The team would never know. Without the lander, the still-operating Sojourner could not communicate with Earth, so its mission was also over. It was programmed to drive around the lander until it regained communication, so it was stranded on Mars, sentenced to rolling around in a circle until it, too, failed.

Pathfinder proved to the NASA establishment that "faster, better, cheaper" worked. Tony Spear's little band had spent a small fraction of what Jim Martin's Vikings had to achieve a Mars landing. Led by Shirley and then Matijevic, the rover engineers had introduced a technology not even possible in the 1970s for an equally tiny sum. They had also proved that Mars, or at least Mars Pathfinder, was popular. The Pathfinder Web site's peak traffic day, July 8, had garnered 80 million visits. A Mattel Sojourner rover "action toy" had sold out across the United States. Goldin's bet on Mars, and on faster, better, cheaper, had been vindicated.

Tony Spear left the project on July 11, becoming project manager of an effort to develop radiation-strengthened microelectronics for a future mission to Pluto. Brian Muirhead succeeded him as project manager; simultaneously, Muirhead was also managing another technology demonstration project, "Deep Space 3." Rob Manning went on vacation for a while, found he couldn't get another project job, and became chief engineer of the Mars program. Sam Thurman had already left, moving over to John McNamee's Mars Surveyor 1998 project as the project engineer, providing systems engineering oversight of both vehicles. Jake Matijevic's rover engineers went en masse to the Mars Surveyor 2001 rover project that was just being established.

Matt Golombek and the rest of the science team worked on the Pathfinder data for another two years. Their "big picture" assessment of the landing site was that they had been correct in evaluating it as a huge outwash plain from some ancient martian catastrophic flood. The surface features implied a flood of hundreds of meters in depth. The rock composition revealed by Sojourner's APXS was andesitic, which was a bit of a surprise. Andesites are common on Earth in the Pacific rim, a product of volcanism along the subduction zones at the edges of crustal plates. The known martian meteorites, such as ALH84001, were mafic rocks, which result from rapid cooling of magma. Andesites are a product of a much slower cooling that enables chemical differentiation. The existence of andesites meant that Mars had a complex tectonic history.[26]

Mars Pathfinder, finally, had triggered a cultural response in the United States not seen since the Apollo Moon landings. Some of the response probably was due to presenting Mars to a new generation for the first time. Twenty years had elapsed since Viking; anyone born after 1977 had probably never seen a picture of Mars. Pathfinder also had the elements of a great television-age story: a small band of renegades from the dominant culture; a period of trial-and-tribulation—what Rob Manning called the Six Minutes of Terror—during EDL; new, and more media-friendly, technologies; and overwhelming success. It also had a great cast of characters. Spear, Shirley, Muirhead, Golombek, and Manning were all very expressive people, a far cry from the American cultural image of engineers and scientists as cold, rational, human calculators. They'd made their mission a very human one, and this resonated with the public.

They would also be a very tough act to follow.

Aerobraking with a Broken Wing

On September 9, as the Mars Pathfinder mission was winding down, Glenn Cunningham's Surveyor Operations Project was preparing to put Mars Global Surveyor into orbit. This was the day they had to pressurize the spacecraft's fuel system for its orbit insert burn on September 11. Pressurization was the point at which Mars Observer had failed, so the team was nervous. But nothing untoward happened. Cunningham told reporters afterward "to see this event pass us successfully today is a really great relief."[27] The burn itself went off perfectly as well. By evening on the 11th, Global Surveyor was safely in a 48-hour Mars orbit.

The aerobraking plan called for another short burn of the main thruster on September 16. This would place the orbit's "periapsis," or its closest point to Mars, in the upper atmosphere. Each orbit through the upper atmosphere would impose drag on the spacecraft, reducing its velocity and slowly lowering the highest point of the orbit (known as the "apoapsis"). The flight team had to manage this process carefully, with firings of the attitude control thrusters needed every other day to control the spacecraft's position within a planned "corridor" in Mars' atmosphere during each drag pass. A team led by M. Dan Johnston and Dan Lyons at JPL had developed a series of software tools to help predict the spacecraft's trajectory, as well as the impact of Mars's atmosphere on the spacecraft.

The high variability of the martian atmosphere meant that the flight team needed a daily assessment of its behavior, and the flight team relied on a network of scientific consultants organized by Richard Zurek at JPL for this. This Atmospheric Advisory Group used information from the accelerometers, from Christensen's thermal emission spectrometer instrument, and the wide-field mode of Michael Malin's Mars orbiter camera, to make assessments of the atmosphere's behavior for the flight team. This group was chiefly concerned about the potential for a large dust storm forming in the spacecraft's area. Dust storms on Mars change the density profile of the Mars atmosphere substantially and could endanger Mars Global Surveyor; Zurek's group would examine each day's data for signs that a storm was forming.

So aerobraking was an operationally complex process. Out in Denver, Lockheed Martin engineers monitored the spacecraft's "health" and built the command sequences necessary to control the craft. At JPL, a navigation team planned and monitored the aerobraking process, feeding Lockheed Martin's engineers the navigation information necessary for the daily command uplink. Zurek's advisory group operated by telephone and e-mail, as it was geographically dispersed, with many of the members at various universities. While Global Surveyor was in its early 48-hour orbits, gathering and analyzing all the data in time was only a little challenging, but as the orbits shrank it got increasingly so. Meanwhile, as the flight team gained experience, their procedures began to smooth out; the team leaders only hoped that their adaptation would be quick enough.

Dan Johnston recalls that as aerobraking began, the team did not really know where Mars's atmosphere was.[28] With no other spacecraft orbiting Mars, they had little data with which to work. They had planned for this confusion by making the early phase of aerobraking a "walk-in." They would make small adjustments to the periapsis each orbit until the accelerometer data indicated that they had "found" the proper atmospheric density. Then they would begin the more aggressive primary aerobraking phase. How aggressive they could make it was governed partly by the vehicle's thermal margins and partly by a three-day "walk-out" requirement. In the event of problems, say a loss of communications, the spacecraft had to be able to survive for three more days of drag passes before exceeding its design margins. This buffer was intended to give the flight team time to evaluate the situation and try to regain control before the spacecraft was destroyed by excessive drag. The spacecraft also could be moved out of the atmosphere at any time by another short burn of the main thruster.

The first few aerobraking passes during the walk-in phase revealed an atmosphere somewhat thicker than expected, but well within design margins.[29] The spacecraft also scored its first scientific coup, demonstrating conclusively that Mars had a magnetic field. At the surface, its strength was about 1/800th of Earth's.[30]

Two weeks after Mario Acuna's magnetic field announcement, something odd happened to the spacecraft. The –Y solar array, which had not fully opened due to the damper mechanism problem shortly after launch, suddenly almost fully opened during a drag pass. At least that's what the flight team thought initially. Telemetry back from the spacecraft showed that the array had moved about 14° toward its latched position, only about 2° from the point at which its latching mechanism should lock it into place.[31] Reflecting on this later, Bud McAnally, Lockheed Martin's program manager, said, "We thought that when we got into aerobraking that the force of the aerobraking would actually push the array right back into position and we'd be forever locked in good shape."[32]

Two orbits later, that rosy scenario disappeared. A sudden doubling of atmospheric drag forced the array into, and then past, its latching point. This could only mean the array structure had been damaged and was failing. George Pace had called Bobby Braun at Langley Research Center, who had worked on Pathfinder's atmospheric modeling, and asked him to put together a small team to determine whether aerobraking would still be possible after discovery of the solar array's deployment anomaly right after launch. His team discovered that something was interfering with Global Surveyor's Mars horizon sensor; their simulations showed it could only be the solar panel extending into the sensor's view, where it certainly should not be.[33] The –Y array was clearly broken.

Charles Whetsel, who had succeeded Pete Theisinger as project engineer when Theisinger took on the task of preparing the Surveyor Operations Project to operate the McNamee's two Surveyor 1998 spacecraft, called Dan Johnston and Cunningham.[34] Cunningham faced what Johnston called "one of the toughest decisions I've seen anyone make."[35] Should he order Global Surveyor moved out of the atmosphere to safety or continue aerobraking? There wasn't enough information yet to diagnose the problem, let alone understand its impact on the mission. Cunningham decided to let the spacecraft continue aerobraking for three more orbits to collect additional data, but he had the flight team raise the periapsis from 110 to 121 kilometers. Cunningham expected this maneuver to reduce the drag back below the point at which the –Y array had started to wobble.

There was some obvious risk in this. If the array's problem escalated rapidly, it could fail during one of the next few passes. That would be catastrophic to the spacecraft and end the mission. It was also possible that reducing the drag force on the –Y array would cause the motion to stop, in which case the aerobraking might continue at a slower pace. But at the end of the three orbits, the telemetry back was clear: the array was still "flapping in the wind," as it were. On the worst orbit the array moved 15° to 17° past its latched position.[36] The titanium hinges would not have allowed this much motion. There had to be a structural fault.[37]

After examining the data from the last three orbits with his operations leaders on Saturday, October 11, Cunningham ordered the flight team to move the vehicle out of the atmosphere entirely.[38] With the array unstable even at this much-reduced drag, his engineers needed time to understand what had happened and what might be done about it. His intent was a two-week hiatus before resuming aerobraking.

Out in Denver, Bud McAnally's engineering team set up an identical array that had been used for mechanical qualification tests to perform diagnostic tests he hoped would determine what had happened to the flight array. The evaluation team's attention focused quickly on the "yoke," a triangular structure composed of a graphite composite and aluminum honeycomb material that connected the inboard segment of the two-panel array to the gimbals that allowed the panels to rotate while tracking the Sun.

At the same time, the navigation and mission operations teams had to develop an understanding of the effect of aborting aerobraking on the mission's ability to achieve its scientific goals. This assessment wasn't long in coming. They could no longer reach the 2:00 p.m. orbit the instruments had been designed for. The schedule margin they had had at launch had been eroded when they had replannned aerobraking after the initial solar array anomaly appeared the previous December. A two-week delay made that orbit impossible. On October 17, Cunningham told Wes Huntress and Carl Pilcher at headquarters the bad news, while also pointing out that other Sun-synchronous orbits were achievable. What now had to happen was a reexamination of the mission's scientific goals vis-à-vis achievable orbits. Malin, for example, needed an orbit that would provide high enough Sun angles for good lighting but also low enough to provide shadows, which were essential for interpretation. Finding a new final orbit that would satisfy that constraint—as well as others set by other instruments, by spacecraft systems, and by Deep Space Network communications capabilities—would be a fairly complex process.[39]

The mechanical tests out in Denver quickly revealed a probable secondary failure that would explain the telemetry. When the array had deployed and its damper mechanism, which was supposed to slow the deployment, had failed, the too-rapid unfolding had cracked the yoke and weakened its structure. The yoke was composed of a honeycomb matrix with composite face sheets on both sides. The tests had resulted in one of the two face sheets cracking, and one cracked face sheet did not eliminate all the structure's strength. The crack reduced it by some amount, and McAnally's mechanical team set out to determine how much strength the yoke structure was likely to still have.[40]

During October and through November, the mechanical team subjected the damaged array to mechanical loads intended to simulate 4,500 aerobraking passes, about five times the number actually needed to achieve a circular orbit. The array finally broke, ending the test series in December, but the answers the project staff needed had been evident since the end of October: the damaged array was likely to survive aerobraking. In a status report to Wes Huntress on October 28, Cunningham reported that he was preparing to slowly walk Mars Global Surveyor back into the atmosphere beginning on November 7. The navigators would design aerobraking orbits to impose about one-third of the maximum drag force of the original aerobraking design, with the consequence that aerobraking would now take much longer than originally anticipated. There were additional consequences. Cunningham needed to reestablish the disbanded mission design group to carry out a complete reanalysis of the available final orbits, and he would have to augment the operations staff to handle the additional aerobraking work. The aerobraking staff also would have to be kept on the project much longer, which meant he needed more money to pay them. He eventually asked for $1.7 million.[41]

On October 31, project scientist Arden Albee convened a meeting of the project science group at JPL to discuss the science options available given the new orbit realities. Cunningham told the assembled members that if he simply continued aerobraking along the current plan, the spacecraft would reach an 8:30 a.m. orbit. He already assumed they would declare this unacceptable because it would leave Phil Christensen's spectrometer essentially useless. His instrument needed the surface it was viewing to be radiating strongly in the thermal infrared for its mineral detection abilities to work, which meant the surface had to be warm. It wouldn't be at 8:30 a.m. local Mars time. Cunningham was right, and the group immediately dismissed the idea.

Almost as quickly, the science group converged on the opposite of their original orbit, a 2:00 a.m. orbit.[42] This meant that while Mars Global Surveyor had been intended to take data while moving from the north pole south to the equator, crossing the equator at 2:00 p.m. local Mars time, in the 2:00 a.m. orbit it would take data while moving from the south pole northward to the equator. Mars would be receiving the same Sun angles as in the original orbit, so the result for the scientists was no net loss of capabilities. The principal drawback to this orbit choice was that Global Surveyor could not reach it prior to solar conjunction in May 1998, at which point Earth and Mars would be on opposite sides of the Sun. Conjunction made communications with the spacecraft impossible, so aerobraking would have to stop with a move out of the atmosphere again. Then, the flight team would have to wait about six months before restarting aerobraking so that the relative positions of Earth, Mars, and the Sun needed for the 2:00 a.m. orbit would be established. Aerobraking would resume in November 1998 and not be finished until March 1999. Project leaders ultimately designated this period of waiting as its "science phasing orbit."

The costs, meanwhile, continued to rise, as the aerobraking team still had to be kept on the payroll. Because the project was only funded to January 2000, the science data collected in the mission's final mapping orbit would be severely reduced unless NASA headquarters authorized an extension. At their meeting, the project science group members decided that since they were already going to be asking NASA for quite a bit to support the extended aerobraking, they shouldn't also ask for the mission extension—at least not yet. Arden Albee also pointed out that the Mars Surveyor Program's budget was a zero-sum game and that a mission extension would have to come from one of the other Mars projects' budgets. The same, of course, was true of the additional aerobraking period's costs.[43]

The orbit choice also meant that the operations project staff would still be aerobraking Mars Global Surveyor and mounting a science campaign at the same time it had to prepare for and operate John McNamee's two spacecraft. These commitments "oversubscribed project (human) resources."[44]

There also were benefits to the long delay in aerobraking imposed by the 2:00 a.m. orbit choice. As Mars Global Surveyor's orbit currently stood, its low altitude drag passes were all over the northern hemisphere of Mars. This was giving Mario Acuna's magnetometer experiment high spatial resolution detail on the martian remanent magnetic field over that hemisphere. The new profile needed to reach the 2:00 a.m. orbit would bring the low altitude passes into the southern hemisphere. Acuna remembers this as "winning the lottery."[45] He would get far more interesting data, it would turn out, than the original orbit would have allowed.

Cunningham's team started walking Global Surveyor back into the atmosphere on November 7, 1997. The Mars atmosphere was a little more dense than expected, so they proceeded more slowly than they had intended. Then they got a big surprise over Thanksgiving: down near the south pole, a dust storm grew very rapidly across the hemisphere, producing a suddenly swelling in the upper atmosphere that resulted in a 200 percent increase in density at the vehicle's altitude. Well beyond its design margin, this event forced the team out of the atmosphere again for the duration of the storm. Scientifically, it was exciting, and from an operations perspective it had put a premium on responsiveness in the flight team.

The December science group meeting revealed disappointment among the assembled scientists. They could already claim important advances in knowledge about Mars, but not many people were paying attention. While Mars Pathfinder had been hugely popular with the nation's major media outlets, Mars Global Surveyor was being ignored. Mike Malin had released the first images from his camera in mid-October, but as far as large media outlets went, only the aerospace trade journal *Aviation Week and Space Technology* had chosen to publish one.[46] Phil Christensen was surprised that JPL hadn't used the storm to try to sell an "exciting new story about drama and adaptability at Mars."[47] The group resolved to try to gain control over the press release process, hoping to get better exposure for the mission. But they would never get the attention that Pathfinder had received.

Conclusion

Taken together, Mars Global Surveyor and Mars Pathfinder had shown the world (and, more importantly, NASA), that "faster, better, cheaper" worked. For a total cost of around a half billion dollars, JPL and Lockheed Martin had delivered two working spacecraft to Mars. That was less expensive than Mars Observer had been and represented a great success by any reasonable standard—with the important caveat that Mars Global Surveyor would not have been possible on its budget or schedule without the large prior investment that Mars Observer had represented. But most people, even inside NASA, quickly forgot about this important qualification. Pete Theisinger would comment much later that Mars Observer had probably given a $100 million gift to Mars Global Surveyor just in systems engineering value and nobody, including himself, had really understood that in 1997.[48] Instead, JPL and Lockheed Martin had made Mars look easy and cheap, so pressure on them to deliver the 1998 Surveyor missions on time and on budget, and to carry out the 2001 missions for even less money than the 1998 missions, grew.

CHAPTER SIX

The Faster-Better-Cheaper Future

During the year following the exciting Mars Pathfinder landing, JPL's Mars Exploration Directorate struggled to keep its commitments to NASA. John McNamee's Mars Surveyor 1998 missions, the Mars Climate Orbiter and Mars Polar Lander, would only make it to the launch pad on time and on budget by dropping some key testing. In turn, the experience caused Lockheed to submit a much higher bid for the next pair of missions, the orbiter and lander pair for the 2001 launch opportunity, which threatened the funds available for the 2003/2005 Mars sample return project. George Pace became project manager of the two 2001 missions, which were initially known only as Mars Surveyor 2001. Pace immediately found himself in a struggle to contain the seemingly ever-expanding demands being placed on the small pre-project team by NASA. By the middle of 1998, Pace's project would be dramatically descoped to keep it from eating the sample return project's budget.

Throughout this period, NASA officials could not understand why JPL could not deliver the lower costs promised by faster, better, cheaper. Their position was that Lockheed Martin, as a private corporation, should have been less expensive than not-for-profit JPL. So the two 1998 missions had been budgeted for less money than the in-house Pathfinder had been. And NASA officials had expected the two Mars Surveyor 2001 missions, which they perceived would be derivatives of McNamee's 1998 missions, to be even less expensive. This disconnect between what NASA officials believed was necessary and what actually had to be done to deliver the four spacecraft to Mars led to a frustrating period for the Mars Exploration Office at JPL, which seemed unable to overcome its communications problems with NASA headquarters.

The Surveyor 1998 Missions

Unlike Mars Global Surveyor, the Mars Climate Orbiter design included a single, large solar array extending from one "side" of a cubical spacecraft. The radiator for Dan McCleese's atmospheric sounder was the principal driver of this design choice, according to Lockheed's chief engineer for the vehicle, Steve Jolly. Jolly, originally a guidance engineer with a wide variety of experience on missile, space shuttle, and classified defense programs, had joined engineer Parker Stafford's design team right after Lockheed had won the contract. The radiator always had to "see" deep space in order to keep the instrument at the proper temperature. Solar arrays generate heat, and Mars radiates heat, so the cooler could not be on a side that would see either Mars, a solar panel, or the Sun. Simply because of its size, it also could not be on the same side of the spacecraft as the propulsion system. So the designers had adopted an asymmetric configuration that satisfied all these conflicting requirements. To protect the array's drive mechanism from large torques during engine firings and aerobraking, the array would be moved into a mechanical restraint. The vehicle's orientation would also be changed so that when the array was stowed for these events, it would be aerodynamically "behind" the spacecraft.

The Mars Polar Lander design was a hybrid of ideas from both Mars Pathfinder and Viking. It used direct entry, like Pathfinder. It also used an expendable cruise stage like Pathfinder's, although not identical. The Delta 7324 launch vehicle provided less launch mass and a smaller diameter aerodynamic shroud to protect the lander during its ride through Earth's atmosphere, so the lander had to be smaller and the Pathfinder-derived heat shield and aeroshell had to be scaled down. The design also used the Pathfinder version of Viking's parachute. Lockheed's trade studies for the proposal, redone in more detail prior to the March 1996 preliminary design review, had indicated that Pathfinder's airbags were not the most mass-efficient solution for Polar Lander's mission. Instead, Lockheed's team had proposed using a Viking-like propulsive descent.

There was one key difference between Viking's propulsive descent design and Mars Polar Lander that drew quite a bit of concern, though. Viking had used a throttleable rocket for its final descent to the surface. A Doppler radar had fed its flight computer with the altitude and horizontal and vertical velocity data it needed to regulate the thrust of its dozen directional thrusters. But Polar Lander would use thrusters that were not throttleable. They would be turned on and off by the flight computer to regulate the descent instead. McNamee did not

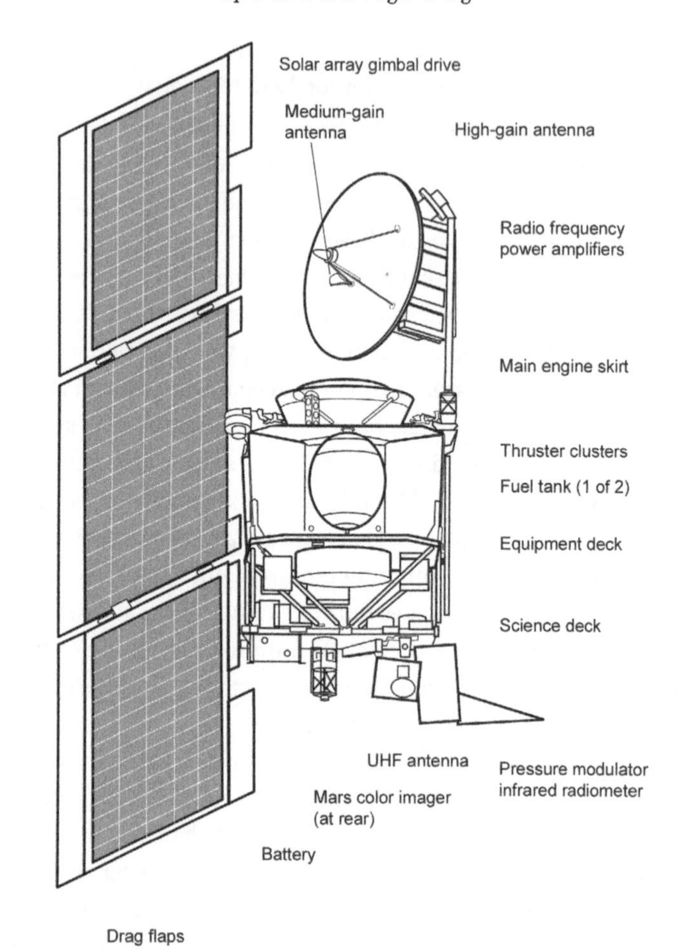

Mars Climate Orbiter (MCO). On this image, the nadir deck, labeled "science deck," points downward.

have the money to do a "hot firing" test of the descent system, so it was an unknown risk throughout the project.

Because Mars Polar Lander was using a rocket-based descent, it needed more information than Pathfinder had required in order to manage the landing. Pathfinder's airbags would absorb whatever horizontal velocity the vehicle had at the moment of impact, up to around 20 meters/second. Polar Lander's legs, though, would only be able to absorb a fraction of that, 2 or 3 m/sec, which could easily be imposed by winds. So the vehicle needed to know what its horizontal velocity was in order to use its thrusters to control it. For this, Lockheed's engineers intended to use a Doppler radar. Doppler radars derive velocity from the

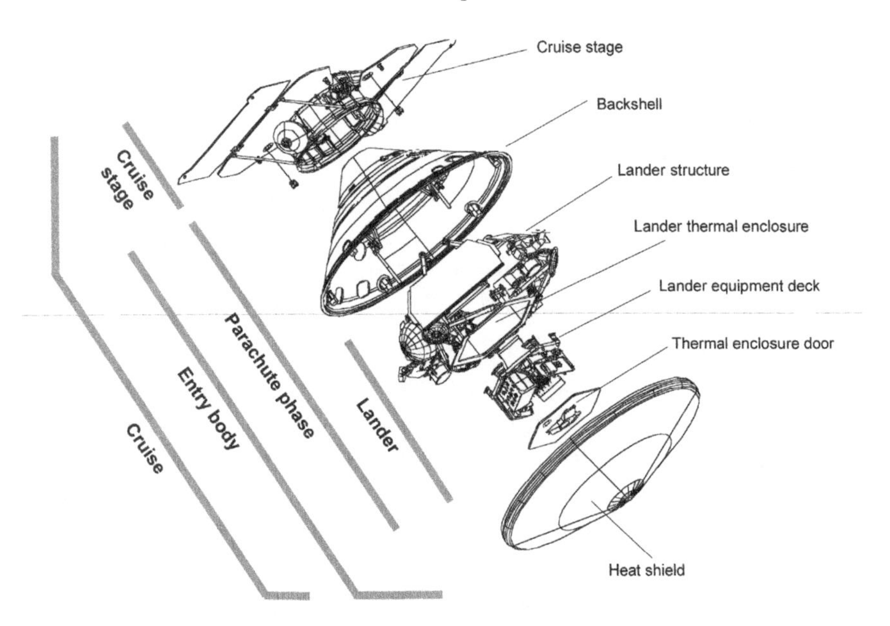

Mars Polar Lander (MPL), in an exploded cruise configuration. Unlike Mars Pathfinder, MPL's cruise stage did not contain fuel tanks. The lander's fuel tanks provided fuel for altitude control and trajectory correction maneuvers during the trip to Mars.

difference between the frequencies they transmit and those they receive from surface echoes; just as a listener on Earth hears the tone of a whistle on a moving train change, a Doppler radar will measure a change in frequency if the radar has moved relative to the surface. The Viking project had spent a large amount of money developing a spaceworthy (and Mars-worthy) Doppler radar, but Polar Lander would not have to. Stafford's engineers intended to use one developed for cruise missiles.

At the end of January 1996, McNamee had lost his struggle to keep the Russian dust lidar off the lander (see chapter 3). In February, the negotiations over whether to include the Deep Space 2 microprobes had concluded with the deal to decide to fly them after Polar Lander had passed its thermal vacuum test successfully.[1]

These late changes to the payload meant that the project's status going into the preliminary design review was more "preliminary" than was desirable. This was particularly true for the electrical interfaces between the instruments and the spacecraft. One of the basic tenets of faster-better-cheaper design was to reduce the need for detailed interface design and control through the use of already-developed, and preferably standardized, interfaces, but since both of the

Mars Surveyor 1998 project's spacecraft were new designs, and many of the instruments were as well, getting the interface designs finalized took up most of the rest of 1996.

The project's reviewers were also unhappy that the payload did not seem to be stable yet. NASA kept adding more. In addition to the Russian lidar and NASA's New Millennium Program's Deep Space 2 microprobe addition in February, NASA had the project considering further additions. But neither spacecraft had sufficient mass margins to accommodate more payload. The orbiter's mass margin was already only 11 percent, while the board members believed that a new design at this point should have at least a 20 percent margin. The lander was in somewhat better shape, with 16 percent margin, but JPL's recent experience with Pathfinder had seen its mass double in development. So the board did not find this a comfortable figure either.[2] The board wanted NASA to quit adding to the payload and McNamee to create additional mass margin through redesign or descoping.[3]

The board members were also critical of the project's financial outlook. Since the contract start the preceding year, both Lockheed Martin and JPL had increased the overhead rates charged against the project. McNamee and his Lockheed counterpart, Stafford, had responded by cutting the projected number of work-hours on the project—but without reducing the project's scope. The board did not find that particularly credible. Board member Glenn Cunningham objected to the financial estimates on different grounds; Lockheed's cost performance to date on his Global Surveyor contract suggested to him that the 1998 project's estimates were optimistic. It would almost certainly overrun.[4]

The review board's final major concern was the project's software plan. The flight software development effort did not seem to be well integrated into the project planning effort. The Surveyor 1998 missions were following the Mars Observer, Global Surveyor, and Pathfinder computing architectures, using a single general-purpose computer to control all of the spacecraft's activities. Since the software controlled everything, it also interacted with everything, and its development had to be tightly integrated with all of the electronic subsystems. This software-intensive nature of 1990s spacecraft was new to most of the project team, and the board members didn't think the project team really understood or had planned effectively for it.

The board's criticism resulted in a few revisions to the project's designs. For the orbiter's mass margin challenge, there were few available options. Throwing one of the two instruments off was not acceptable to NASA. Another possibility, starting aerobraking in a higher Mars orbit and extending the aerobraking period, would save fuel and its associated mass. But NASA wanted the opposite, a shorter aerobraking period to ensure that the orbiter was already in its final orbit when Polar Lander arrived three months later. That way, if Mars Global Surveyor failed, Climate Orbiter could provide relay service to the lander. So the higher orbit option was not an option at all. Nor were there any credible ways to reduce the orbiter's mass significantly without spending quite a bit of money to, for example, miniaturize the flight electronics (which were already the product of separate miniaturization effort, in the Small Spacecraft Technology Initiative program).

Instead, McNamee got NASA officials to relent on the requirement to use the "Med-Lite" version of the Delta 7325 to launch the orbiter. The rocket originally specified for Climate Orbiter was to use a new upper stage, a Star 37, instead of the larger Star 48. The Star 37 had never flown on a Delta before, so its certification would cost money and impose additional risk. McNamee was able to convince his superiors at NASA that using the more powerful Star 48 was lower risk and wouldn't cost more. It would also get him about 100 kilograms additional mass. The rest of it the solution came from adding a fourth solid rocket motor to the Delta, for about $1.6 million. This would add another 50 kg mass capability and make the rocket, in the manufacturer's terminology, a Delta 7425. The upgrade would improve both the mass margin situation and permit enough fuel to be carried to enable the accelerated aerobraking that NASA and the project scientists sought. McNamee did not get NASA to pay for it, however; NASA simply allowed him to use his already-approved budget to pay for the upgrade.[5] This seemed the least expensive, and lowest risk, way out of the orbiter's mass margin problem.

A similar launch vehicle increase would not help Polar Lander's mass margin challenge, however. This was because, as for Mars Pathfinder, the maximum entry mass was controlled primarily by the relationship between the entry angle and heat shield capabilities, not by the launch vehicle. Only the cruise stage's mass margin could be improved by a more powerful Delta, and neither the review board nor the project staff regarded this as a problem in need of solution. But with NASA willing to upgrade the orbiter's rocket, the New Millennium program office saw an opportunity to ensure that the Deep Space 2 microprobes were guaranteed a ride to Mars. The two probes were to be about 15 kg mass total, and the addition of a fourth solid motor to the lander's launch vehicle would provide more than enough capability to fly them.

So Sarah Gavit, the Deep Space 2 microprobe project manager, approached McNamee about re-opening negotiations over their memorandum of agreement of February. Their new deal resulted in the New Millennium program paying for the fourth solid motor, with the specific caveat that McNamee couldn't claim the excess mass capability for his mass margin calculations.[6] By early April, the microprobe project had a guaranteed ride to Mars—assuming its team could meet the demanding launch schedule.

The review board's concern with the project's instability was well founded, as NASA had still not finished adding to the payloads. In May, the project got another request from NASA headquarters, to add a piece of art to the Polar Lander. This was a 10-centimeter-diameter sphere constructed of an alloy that would change shape with the diurnal temperature cycle on Mars. The sphere would not fit within the already tightly packed aeroshell, at least without a great deal of redesign effort, and project leaders succeeded in rejecting it. The same month, the Planetary Society's Lou Friedman proposed adding a microphone to the Russian lidar instrument that would record wind noise on Mars. Friedman planned to use it as part of a Planetary Society education effort. McNamee put this proposal forward for discussion at the June meeting of the project science group, where the consensus was that the microphone "cannot be justified scientifically for the 1998 lander mission."[7] In a June 17 letter, McNamee recommended its rejection. In August, however, NASA accepted the microphone anyway. Its inclusion was a political decision, as Friedman acknowledged in an apologetic e-mail to the very unhappy Donna Shirley.[8] But the deal worked out required that the microphone's integration to the lidar be paid for by the Russian principal investigator, Sergei Pershin, and not out of the project's funds. This effectively made Pershin responsible for the microphone, too.

In part due to the changing payload, project engineer Sam Thurman did not get all of the interface design agreements finalized and signed until December. The late payload additions contributed to the delay, but Thurman reflects that the main problem was that it took time for all the instrument teams to gel. The individuals in the different partner institutions had to develop communications and trust with each other and with the Lockheed Martin and JPL teams, too. That very human process could not be hurried, and it had nothing to do with the slim manpower allocations they all had. The result was that the team moved their critical design review from August 1996 to January 1997.

One of the project's innovations that the review board praised was a mechanism for funding changes to the spacecraft design or to the testing program without having to renegotiate the primary spacecraft contract with Lockheed Martin. Significant changes to either Lockheed's proposed designs or to its proposed test program could be considered changes in the contract's "scope," and thus allow the company to demand additional payment. Yet changes were inevitable given that both spacecraft were new designs. Mars Pathfinder, of course, had seen many design changes and many revisions to its test program. But as an in-house JPL project, only some of its subcontracts with suppliers had to be renegotiated for changes in scope. The situation for McNamee's team was quite different. As prime contractor, Lockheed could use one change in scope to "bundle in" many other costly changes; this practice was common in aerospace and was one of the main reasons large cost overruns were prevalent in the industry.

McNamee's solution had been to use a unilateral contract modification to establish a "risk reduction pool" of funds to pay for additional testing and design changes. McNamee's spacecraft manager, Paul Sutton, and his counterparts at Lockheed Martin had to agree on the tasks to be funded this way. From McNamee's perspective, the risk pool reduced the possibility of overrunning the project's budget while reducing both the project's technical risk and management burdens.[9] The risk pool was funded by the difference in cost between JPL's estimate of how much the spacecraft contract was likely to cost ultimately, about $94 million, and Lockheed's cost proposal of $84 million. In essence, it served as a $10 million contingency fund, in addition to the contingency funds held by both Lockheed and McNamee.

One example of risk pool use Thurman remembers was Polar Lander's radar. Having struggled with Pathfinder's altimeter during 1995, Thurman came onto the Surveyor 1998 project in 1996 thinking that Lockheed's radar subcontractor, Honeywell, had underestimated the need to characterize the radar's performance through testing. He recalls that their test plan was essentially to fly the radar around the manufacturing facility's local area, outside Minneapolis, Minnesota. The Minneapolis area, with its low, rolling hills, lots of lakes, and few surface rocks didn't "sound much like Mars," Thurman remembers thinking.[10] Further, Honeywell's plan did not include independent measurements to check against the radar's results, so there was no "ground truth" against which to evaluate its performance.

Thurman convinced McNamee that the project needed to send the radar out to the China Lake testing range Pathfinder had used and to expand the testing to include drop tests. The radar's actual motion in its descent to Mars would be primarily vertical, while airplane flights are primarily horizontal. So merely flying the radar around in an airplane was not a realistic test. Dropping it out of a helicopter or from a balloon would at least get the direction vector right, even if the velocity would not be. This change would cost a few hundred thousand dollars but would certainly improve confidence in the radar. The risk reduction contract enabled the change to take place without reopening the spacecraft contract for negotiation.

The Impact of Mission Choice on Design

With the payload finally stabilized a few months after the preliminary design review, more attention turned to mission design. The scientific activities the two spacecraft were to perform served as one set of constraints on the details of their missions, but there were many others: capabilities of the spacecraft in terms of power, thermal management, amount of available on-board computer memory, availability of the Deep Space Network and orbiter relay (for the lander), and so forth.

Over the project's lifetime, most of the emphasis had been on understanding the lander's mission. Pathfinder and its Viking predecessors had all landed in the equatorial zone, where there was a strong daily cycle to power and heat that forced a strong structure to surface operations: work during the martian day, hibernate at night. Polar Lander, however, would land in late south polar "spring," when the Sun would be up continuously, until sometime in the polar "autumn." It was a much different operations scenario.

The earliest effort to define the lander mission had focused on understanding the polar surface environment using data from the two Viking orbiters. The martian poles were already known to have two "ice caps," a "permanent" ice cap composed of water ice and a "seasonal" ice cap of frozen carbon dioxide that forms during late fall and grew throughout the hemispheric winter. Polar Lander had to arrive after the seasonal cap had retreated from its desired landing zone.[11] The seasonal cycle also determined the mission's end. Martian southern fall would descend in January 1999, and the Viking data indicated that ambient temperature would be back below the lander's thermal capabilities again by the beginning of February 2000. By March, it was likely to be frozen into a block of carbon dioxide. There was very little chance of an "extended mission" for Polar Lander.

Figuring out the lander's power needs as they related to Sun availability drove much of the early mission design effort. While the Sun would be "up" through the entire mission, its angle to the lander's solar panels would vary throughout each sol. Array output would drop to almost zero around local midnight, when the Sun was very low on the horizon. Of course, as spring turned into summer, average daily power available would increase then decrease again as summer waned into fall. This meant that the mission's first few days on the surface drew the designers' attention.[12] If the lander had enough power for its first week, it was likely to meet all of its scientific goals. Ultimately, the team added small auxiliary arrays on the north-facing edge of the lander deck to generate a little more power from the Sun at low angles.

Primarily for thermal reasons, the lander would only be able to use its medium-gain antenna to transmit data to Earth directly for about an hour per day.[13] Its Cassini-derived transmitter had been optimized for deep space operation, not the martian surface. This window was not enough time to transmit all the data the lander could collect in a day's operation, so the mission design favored use of the UHF relay available from either Mars Global Surveyor, if the orbiter was still "alive" when the lander arrived, or from Climate Orbiter. In turn, the lander's need for the UHF relay immediately after reaching the surface to ensure that the first few days' data all made it back to Earth had strongly influenced the decision to shorten the orbiter's aerobraking period instead of extending it.

Staying within Mars Climate Orbiter's aerothermodynamic limits but being in a relay orbit 90 days after arriving meant capturing into a much lower orbit than Mars Global Surveyor's 48-hour initial orbit. The mission design group found a capture orbit of 29 hours duration that would get them down to the desired two-hour mapping orbit, although with respect to local martian time, the result would put the orbit's overpass time a little past 4:00 p.m.[14] The orbiter had to launch in the first 14 days of the launch window to be capable of capturing into the desired 29-hour initial orbit. Launching after the first week of the window also reduced the likelihood of being on-station for the landing, simply because the reduced number of days available for aerobraking started eating into schedule margins. So launching in the first eight days was a key mission goal.[15]

Mars Surveyor 2001

With John McNamee's project finally moving forward, and its major decisions made, the staff members at JPL's Mars Exploration office could focus on getting their next set of missions defined. As of early 1997, NASA still sought an orbiter-lander pair for the 2001 launch window, with the orbiter to carry the last of the Mars Observer instruments: Bill Boynton's gamma ray spectrometer (GRS). The lander payload was far less clear, however. Two different NASA offices wanted the lander to carry their priorities. The Space Science Enterprise, which was paying for the mission, wanted a payload aimed at some aspect of the scientific "follow the water" strategy worked out during the previous year. The Human Exploration and Development of Space Enterprise (HEDS, to insiders), emboldened by NASA administrator Dan Goldin's desire to gain White House approval in

2004 for a humans-to-Mars decision, wanted the lander to carry instruments that could be used to study the Mars surface for human biological compatibility. This conflict was not resolved gracefully. In its first year of life, the Mars Surveyor 2001 project would go through four major design changes as conflict over these payloads collided with engineering and financial realities.

Norm Haynes and Donna Shirley had hired George Pace as the Surveyor 2001 project manager after he'd finished his duties as spacecraft manager for Mars Global Surveyor. When Pace first came on to the project in late 1996, the basic mission concept for the 2001 lander was to deliver a large, autonomous rover to Mars. Unlike Sojourner, which had been tethered to the Pathfinder lander by short-range radio and needed the lander to send its data back to Earth, this new rover would be self-contained. It would not require lander services of any kind after debarking, so the lander itself would not have to carry a power system or telecommunications gear.[16] Around JPL this concept became known as the "dead on arrival lander." The lander would operate only as long as necessary to deliver the rover to the surface. After that, the rover would be on its own. In this form, the predicted launch mass was about 550 kg, and the rover mass about 45 kg.

In early January 1997, members of the Mars Exploration office at JPL met with counterparts in NASA's HEDS enterprise in Houston in response to a November directive from NASA requiring them to "fully integrate robotic and human Mars exploration study and planning into a coherent overall approach."[17] This meeting revealed that the HEDS leaders primarily wanted the 2001 missions to perform demonstration of a small set of technologies and to provide data on in-orbit and surface radiation levels. Human mission planners considered aerocapture, precision landing, and in-situ propellant production crucial technologies for human expeditions to Mars.

Aerocapture was simply an advancement over aerobraking, using the martian atmosphere to achieve all of the velocity reduction necessary for orbit insertion. In essence, a single, deep pass into and through the atmosphere would remove enough velocity to achieve the desired orbit; the vehicle would not need a propulsion system for orbit insertion. The implementation was not as simple as the concept, however. The orbiter would have to be packaged into an aeroshell–heat shield combination just as if it were a lander, and it would need a cruise stage to provide power during the voyage to Mars, too. The vehicle would have to maneuver itself during the aerocapture pass to control the operation and achieve the proper orbit. Just as the short landing sequence and high atmospheric variability had required the Viking and Pathfinder landers to be autonomously responsive to their descent environments, the aerocapture-configured orbiter would have to be the same. The precision landing demand carried with it essentially the same requirement. Instead of using a simple ballistic trajectory during landing, the lander would have to steer itself during entry to obtain a much smaller landing ellipse. The desired precision was plus or minus 10 km, a substantial improvement over Pathfinder's 100-by-300 km landing ellipse.

The in-situ propellant production demonstration experiment was intended to show that human flights to Mars would not have to carry fuel for their return journey to Earth with them. If possible, it would be an enormous mass, and cost, savings. The basic architecture the demonstration was to support derived from a Martin-Marietta effort in early 1990 called Mars Direct. This approach would land automated fuel factories on Mars to make propellant using the carbon dioxide in the atmosphere. A couple of years later, once the fuel had been produced, the human crew vehicle would be dispatched to Mars. The astronauts would have to land near the refineries in order to get back to Earth, making precision landing a requirement.[18] At this point, it was not clear how much mass and power, or how many other lander resources, would be required to support the propellant production demonstration.

The JPL participants in the January 1997 meeting returned to Pasadena knowing that their "dead lander" idea was probably dead. The 2001 lander would have to be able to provide power, telecommunications, thermal control, and more to whatever the human exploration payload experiments turned out to be. Because JPL had entered into a long-term contract with Lockheed Martin, the lander design would have to be a variant of the 1998 lander. And almost certainly, it would have to be scaled up to carry both the science directorate's rover and the HEDS experiments.

For the next couple of months, the Mars office worked with counterparts in the HEDS enterprise and at Lockheed Martin to determine what the payload details would be. In late March, the affected NASA offices reached a tentative agreement to partner on the 2001 missions. The HEDS enterprise would develop the instrumentation with its own funds and under this agreement would support a $40.5 million increase in the Mars Surveyor 2001 project's budget, to pay for the increased cost of the "alive" lander.[19]

For the orbiter, the negotiated experiments were the aerocapture demonstration, a radiation monitor, Boynton's GRS, and a "mineralogy mapping investigation."[20] For the lander, Mike Malin's Mars Surveyor descent imager (MARDI) was part of the negotiated baseline payload, along with a surface radiation monitor, the Mars in-situ propellant (MIP) production experiment, a soil and dust experiment, and the rover. The rover's payload, estimated to be about 15 kg, would be chosen via announcement of opportunity. NASA also wanted the rover to be able to collect and cache soil and rock samples on the surface to be recovered by the 2003/2005 sample return mission.[21]

The Mars Surveyor 2001 missions' announcement of opportunity went out to the public in draft form in May. JPL and NASA received the scientific community's proposals for the mission the first week of September. The selection process ran in two tracks, with the Office of Space Science choosing the mineralogy and rover payloads, and the Office of Life and Microgravity Sciences choosing the radiation and soil experiments. In early November, the Office of Space Science announced the rover payload.[22] It went to Cornell astronomer Steven Squyres, who had proposed a suite of mineralogical instruments called Athena. Athena, in its original Surveyor 2001 form, included eight experiments: a "mini-corer," to take samples from rocks; a microscopic imager; an alpha-proton X-ray spectrometer like Sojourner's; Mossbauer and Raman spectrometers; a sample container; a miniature version of Phil Christensen's thermal emission spectrometer (named "mini-TES"); and a high-resolution panorama camera similar to Peter Smith's on Mars Pathfinder.

Christensen was also awarded the mineralogy mapper slot on the orbiter. He had proposed an experiment called a thermal emission imaging system (THEMIS). This instrument allowed production of finer-scale mineralogical maps of the surface.[23] Among other goals, Christensen hoped to be able to identify thermal "hot spots" on the surface that might indicate the presence of liquid water near the surface and thus be potential abodes of martian microbes. These oases would also likely be geologically interesting places for future landings.

The remaining experiments were not chosen until February 1998, by which time a great deal had changed. During summer 1997, NASA's flagship International Space Station project unveiled a substantial cost overrun; in September, the Office of Management and Budget informed NASA that the agency budget would be cut $500 million in the following year. The funds for the HEDS initiative on Pace's Mars Surveyor 2001 missions were redirected to pay for part of the space station's overrun. But the HEDS experiments were not dropped. Instead, the Mars Surveyor program had to take them on as an "unfunded mandate" that was estimated in November to be about $56 million. Lockheed Martin's September bid for the two spacecraft, $208.5 million (compared to its $84 million bid for the 1998 spacecraft), was an additional "sticker shock" that put the overall project budget at about $300 million.[24] Already outside the project's cost envelope, even this estimate proved optimistic.

Pace's project office, which was formally constituted at the Phase B start in August 1997, had allocated about $30 million for the rover, along with the 45 kg mass. Robotics engineer Jake Matijevic's rover team did a "bottoms up" review of the rover concept design that put it well outside the constraints of mass, money, and schedule. Matijevic recalls that initially the team had not "properly anticipated what the electronics builds would be like for this, as well as the rest of the equipment that had to now be part of the vehicle system in order for it to act like [an] independent vehicle."[25] JPL's Mars Exploration Directorate had originally planned the rover mission to last two weeks, like that of Sojourner, but Squyres had proposed one Mars year (two Earth years), and NASA's selection of his mission proposal drove rover cost and complexity still higher. By February, Matijevic could offer a best estimate mass of 57 kg. But the lander did not have that much mass margin to give the rover; it was already over its entry limit.[26]

The increasing size of the rover imposed further challenges for the lander. The initial idea for debarking the rover was to have the lander's robotic arm pick it up and deposit it on the surface. But the robot arm developed for Mars Polar Lander was not powerful enough, and a larger arm had to be designed. It also created a "tip over" problem. As the arm lifted and swung the rover over the lander's side, the lander would experience a center-of-gravity shift that could tip it over if it were not on a flat surface. The same would happen even if the team used ramps, instead of the arm, because the lander deck would be about a meter above the surface. Finally, accommodation of the entire Athena experiment package, which had oversubscribed the initial rover design's resources, further drove up the rover's size and cost.

At the project's first formal quarterly review for NASA officials on February 12, 1998, Pace presented an updated budget estimate of $377 million.[27] It was well beyond what the Mars program could afford, or what NASA headquarters was willing to give it. Funding a 2001 project of this financial scope left nothing with which to fund the 2003/2005 sample return initiative that was still planned as the next mission for the Mars program. So JPL and NASA each initiated an independent review of the project. JPL's review was chaired by Pathfinder's Tony Spear. NASA assigned its review to the Independent Program Assessment Office at Langley Research Center. What NASA officials hoped the two panels would find was a means of implementing the two missions for at most $300 million—preferably much less.

Spear's review was completed in late April. In his draft report, he was quite blunt: "There is no approach to project implementation of all requirements for $300 million. There is no magic."[28] This was due to "unchecked escalation of requirements," including the demand for longer mission lifetimes and the HEDS experiments and demonstrations. One of the demands of the faster-better-cheaper management method was maintenance of tightly focused scientific objectives and minimal technical requirements. As had also been true for Mars Observer, NASA itself was unable to resist piling on demands.

Spear's team concluded that the 2001 lander design should be dropped entirely. It was a great departure from the 1998 lander design in order to support the big Athena rover and HEDS experiments, which was driving the cost up, but the money spent on it would not provide any cost or risk reduction for the sample return mission.[29] The investment in it would not generate the technologies that the 2003 sample return mission needed. So they recommended switching to a rebuild of Polar Lander, with the minimal modifications necessary to fly to the martian mid-latitudes instead of to the pole. The Polar Lander design could only support part of the HEDS payload, part of the Athena payload, and probably no rover.

The Mars Exploration Directorate at JPL came in for criticism of its own. Spear's panel did not think it had done an acceptable job of planning across all of its projects. In their view, the program office should have ensured that the individual missions preceding sample return contributed to the technical base needed for sample return.[30] Of course, Shirley had argued the same thing for two years. But her office had had its planning authority removed by headquarters. Thus the mission sequence that had evolved was not optimized for early achievement of sample return—although perhaps it suited Dan Goldin's desire to gain a 2004 announcement supporting human expeditions to Mars.

Spear further criticized Pace, Shirley, and Norm Haynes—and implicitly their superiors at JPL—for not "pushing back" at NASA headquarters strongly enough over the increasing requirements and chaotic planning.[31] This was a difficult subject around JPL. As a contractor, JPL was not in a particularly strong position to do this in the first place. And its primary in-house mission at the time, Cassini-Huygens, had been threatened repeatedly with cancellation, which would be disastrous for both JPL and Caltech. While Pace, Shirley, Haynes, and Dan McCleese had all tried various ways of getting across to their counterparts at headquarters that the program could not afford everything being asked of it, director Ed Stone had not taken the subject up with Goldin.

In Washington, the dissent within JPL over the scope of the program and the rapidly increasing costs produced an image of a schizophrenic organization. An Office of Management and Budget examiner had visited the Lab and told

McCleese and others that he had the sense that the "old" (expensive, slow, gold-plated) JPL was reasserting itself over the "new," faster-better-cheaper one.[32] Spear's panel report in April lent credence to this view, as it called for getting back to faster-better-cheaper methods by reducing requirements and stabilizing them very early in the project. Spear and his Pathfinder gang had been able to do that, with Associate Administrator Wes Huntress preventing major scope changes after Phase A. Neither McNamee nor Pace had experienced that kind of high-level protection. Another tenet of "faster, better, cheaper" was being responsive to one's customer, which meant acceding to NASA's demands.

During the first week of May, Huntress accepted Spear's recommendations and told the Mars Exploration Directorate at JPL to replan the 2001 missions around rebuilds of the 1998 spacecraft. The orbiter would still be required to carry the GRS and THEMIS, but the radiation monitor, an instrument from Johnson Space Center called MARIE, would be reduced to a "goal."[33] The lander payload would include the three HEDS experiments and elements of the Athena payload to be chosen at a later science meeting. The new project budget would be capped at $220 million. The long-life Athena rover was kicked to the 2003 mission, tentatively, while the entire longer-range program was re-replanned to support the sample return mission.

At the beginning of June, a very unhappy Mars Expeditions Strategy Group met to decide which elements of the Athena payload should remain aboard the 2001 lander. This group, assembled by McCleese in mid-1996 to provide a scientific strategy for the revived search for life on Mars, had replaced MELTSWG as the chief advisory group to NASA for Mars science. At this meeting, George Pace made clear that his engineering team only wanted to fly the HEDS experiments. This was the lowest-risk, least-cost, approach. Squyres argued, without much enthusiasm, that the lander should carry the two remote sensing instruments that were part of the Athena package, the panorama camera and mini-TES instrument. Squyres remembers that he was not very persuasive, either. He and the majority of the collected scientists did not think the mission was worth flying with this payload. At the end of the first day, though, the NASA headquarters official at the meeting stipulated that the lander had to include the mini-TES. Because the Office of Space Science was paying for the mission, a HEDS-only payload was unacceptable. Because mini-TES and the panorama camera were tightly integrated with a lot of common hardware, that stipulation essentially meant they would both fly. On the second day, Squyres pitched including the Mossbauer spectrometer as it was to be donated by Germany and thus cost nothing to

NASA.[34] This payload became the strategy group's recommendation to NASA, with the caveat that they understood that it did not fit the budget allocated to the project.[35] The group also advocated for retaining the precision landing requirement, on the grounds that it was necessary for sample return. Both NASA and JPL agreed to the payload. The shrunken version of Athena Squyres named "Athena Precursor Experiment," or APEX.

The turmoil surrounding the project was hugely frustrating for its staff. Both Pace and his spacecraft manager, Roger Gibbs, understood that they faced a hostile environment at JPL. With many members of Spear's committee also involved in the sample return project, they were not exactly neutral parties: sample return needed the 2001 project's funds. So Gibbs and Pace saw the review not as a means to help their project succeed but as a way to hobble it and to divert its budget to the sample return effort. Their project engineer, Lynn Lowry, found the constant changes demoralizing. Pace remembered that she would threaten to quit every few Fridays, leaving him to wonder whether he'd have to find somebody new the following week. She always came back, though, and eventually he asked her (half-jokingly) to start quitting on Mondays instead. That way they could work through the troubles during the week and he wouldn't have to spend his weekends worrying.[36]

Lynn Lowry was not the only person frustrated by the turmoil. One evening in August 1997, Wes Huntress was having dinner with his family when his son reached over and hit him on the shoulder. "Dad, what are you doing?" he asked. "I'm eating my spaghetti." But he wasn't. Huntress was explaining his budget conflicts with Goldin to his plate of spaghetti. "I knew right then that I had lost it, and that's when I decided to leave."[37] He spent a few months looking for a successor in private, and announced his resignation in February 1998. But he remained in the job until September to see the Space Science Enterprise through that budget cycle; his successor, astronomer Edward J. Weiler, came on board at the end of the month.[38]

The changes to Surveyor 2001 didn't stop after Spear's review. It didn't take very long for the idea of simply building a copy of the 1998 lander to fall apart. By the beginning of July, Lockheed's power consumption analysis showed that the 1998 lander design would not survive its first night on the surface. The project science group wanted the lander to go to a mid-latitude site that would not have the benefit of perpetual sunshine, and accommodating the short days and very cold nights in the mid-latitudes forced changes to the power system, batteries, and the provision of larger solar arrays. The need to tailor the spacecraft to a specific location made reuse of designs very difficult.

The payload lasted almost two months. It was, as Squyres and many others knew from the outset, a boring payload. It had little to interest a public that clearly loved Mars rovers. It was not terribly interesting scientifically, either. Mars Pathfinder had not been highly regarded by Squyres's Mars Science Working Group of the mid-1990s because it was rather light on science, and the 2001 lander was, if anything, even lighter. It lacked even the excuse of being primarily a technology demonstration mission. The removal of the rover and, perhaps more to the point, the imposition of an unfunded mandate by the human spaceflight side of the agency on the Space Science Enterprise drew a rebuke from the Office of Management and Budget. Shirley reported to JPL director Stone that OMB was "unhappy with the delivery of HEDS requirements to Code S [the Space Science Enterprise] without money."[39] Apparently, so was Congress. In June, Congress gave the Surveyor 2001 project the additional $56 million to cover the HEDS instruments.[40]

It isn't clear who initiated the idea of adding a microrover to the payload, but by July, Pace's team was studying the possibility of adding Marie Curie, Sojourner's Earthbound twin, to the payload. Like Sojourner, it would carry an alpha-proton

Mars Surveyor 2001 payloads

Spacecraft	Instrument	Investigators (institution)
Orbiter	Gamma ray spectrometer (GRS)	William Boynton (University of Arizona)
	Thermal emission imaging system (TES)	Philip Christensen (Arizona State University)
	Martian radiation environment sensor (MRES)	Gautam Badwar (Johnson Space Center) Cary Zeitlin (National Space Biomedical Research Institute)
Lander*	Mars in-situ propellant production precursor	David Kaplan (Johnson Space Center)
	Athena integrated rover payload**	Steven W. Squyres (Cornell University)
	Martian radiation environment sensor (MRES)	Gautam Badwar (Johnson Space Center) Cary Zeitlin (National Space Biomedical Research Institute)
	Mars descent imager	Michael Malin (Malin Space Science Systems)
	Microscopy, electrochemistry, and conductivity analyzer	Thomas Meloy (West Virginia University) Michael Hecht (JPL)

Source: Adapted from 2001 Mars Odyssey Launch press kit, April 2001; Mars 2001 Minutes and Viewgraphs, November 4, 1997, EDS D-15672, PSG #1; and Squyres, *Roving Mars*, chapter 3.

*The Marie Curie rover, the flight spare of the Sojourner rover on Mars Pathfinder, was part of the lander payload for the later part of 1998 and 1999 but was descoped in January 2000.

**The Athena payload originally consisted of the panorama camera, the mini-thermal emission spectrometer, the miniature corer, the microscopic imager, the alpha-proton x-ray spectrometer, the Mossbauer spectrometer, the Raman spectrometer, and a sample container. It was descoped during the spring of 1998 into the APEX package, which included the panorama camera, the mini-thermal emission spectrometer, a copy of the Mars 1998 robot arm and robot arm camera, and the Mossbauer spectrometer.

X-ray spectrometer for mineralogical studies. In August, the Mars program office formally adopted the plan.[41] Marie Curie would be deployed by a clone of the 1998 lander's robotic arm; Squyres then got the project to accept putting the Mossbauer spectrometer out on the arm too, so it would have a greater range of samples available to it.

International Collaboration: Sample Return, Mars Express, Mars Micromissions

While the Mars Exploration Directorate at JPL was struggling with the formulation phase of the Surveyor 2001 missions, it was also trying to define three other initiatives. First was the long-desired sample return mission. Second was an upcoming possibility to place one or more U.S. instruments on a future European Space Agency mission to Mars called Mars Express. Third was the possibility that one or more "Mars micromissions" might be flown on a European "carrier" spacecraft derived from the Ariane 5 launch vehicle and the Rosetta mission spacecraft bus. All of these initiatives were designed around international collaboration, to permit doing more science with fewer U.S. dollars.

The largest by far was the sample return initiative. Central to Spear's critique of the Mars program office, the sample return project office had been set up late in 1997. JPL veteran Bill O'Neil was the project manager, and Roger Bourke handled much of the international relations. The principal partner was to be the French space agency. The basic architecture worked out between JPL and the Centre National d'Études Spatiale was to have a lander arrive on Mars in 2003, collect a sample with a "grab sampler" arm immediately, and to deploy a clone of Squyres's Athena rover to collect about 60 more samples. The rover would return the sample container to the ascent vehicle on the lander, and the lander would fire the sample into Mars orbit. In 2005, a French-built orbiter would arrive carrying a U.S.–supplied Earth return vehicle and sample capture apparatus. A second 2005 lander would collect more samples and place them in orbit too. Then the orbiter would collect the sample containers and shoot them back to Earth. The Earth return vehicle would plummet into the Utah desert in 2008.

The estimated cost to the United States for this effort was $684 million. But this estimate had little credibility within the project itself. The Pathfinder and Surveyor 1998 and 2001 landers had deliberately been kept smaller than the Viking landers to enable reuse of the Viking parachute and heat shield designs. These projects had depended on the components' flight heritage. The sample return landers were much heavier than the Viking landers had been, invalidating this "heritage" argument. They would probably need new designs and require a substantial testing program to qualify them. This would be expensive. But the lander budget for 2003 was to be $135 million—essentially the same as the 2001 lander, unlikely to be sufficient given that it was three times the 2001 lander's launch mass.[42]

The second international effort explored during 1997 and 1998 was Mars Express. This spacecraft would be the European Space Agency's first planetary orbiter. The payload selected for Mars Express included a "subsurface sounding radar/altimeter" that was of great interest to the Mars Expeditions Strategy Group, due to the radar's uniqueness.[43] While other Mars Express instruments were similar to those already in the U.S. mission roster, the radar was not. It would be able to look for water ice in the top few hundred meters of the surface, very substantially improving on the one meter or so of water detection capability of Boynton's GRS. This capacity would allow a much better estimate of the total amount of water on Mars. The radar's data would also permit analysis of the structure of the martian subsurface. It might, for example, permit identification of buried craters or of underground magma conduits.

The final international effort was to make use of multi-payload "bus," developed by the French company Arianespace, that would be part of the Ariane 5 launch vehicle program. It would be able to deploy several spacecraft toward Mars simultaneously, each as much as 150 kg in mass. Several concepts for these spacecraft were in the early phases of study during 1998. At Langley and Ames Research Centers, teams were studying Mars airplane concepts. These would be released after the supersonic phase of a Mars entry and would be able to fly a few hundred kilometers, providing a regional-scale view of surface conditions.

Deep Space 2

The goal of the New Millennium program's Deep Space 2 project was the development of small, inexpensive penetration probes for doing various kinds of in situ scientific studies, including seismology and meteorology. While the basic concept of simple, inexpensive microprobes came from JPL's Mike Hecht (also an experimenter on the Surveyor 2001 orbiter), they had been strongly supported by Huntress prior to his retirement from NASA. He had wanted to see landers under $50 million. "JPL needs to know that expensive penetrators are unacceptable," Huntress had told Dan McCleese in October 1995.[44] The concept had been turned over to the New Millennium program, founded specifically to develop new technologies for space science, for development into flight-worthy machines.

The basic concept behind Deep Space 2 was to design, build, and demonstrate the ability to "land" small (under 10 kg) probes on Mars using an entirely passive entry system. They would simply plunge to the surface, impacting at several hundred kilometers per hour. The probes would be contained inside an aeroshell made of a ceramic glass. When the vehicle hit the surface, the aeroshell would shatter and the probe would plunge into the surface like a lawn dart. The rapid deceleration would cause the probe to separate into two pieces, with a dart-like forebody penetrating the surface to a depth of a meter or so, while its afterbody stayed on the surface. A cable would connect them, and the afterbody would have a low-power transmitter able to relay data up to either Mars Global Surveyor or Climate Orbiter.

Deep Space 2 got going in the fall of 1995. Initially the small engineering team led by Sarah Gavit took two approaches to finalizing their design concepts, dropping models out of an airplane to try to identify aeroshell shapes that would be aerodynamically self-righting and canvassing the weapons industry for ideas on how to package electronics to withstand impact accelerations in excess of 20,000 times Earth's gravity. Some artillery shells are designed to penetrate a target before exploding, so weapons laboratories were an obvious place to look for expertise.

Saverio d'Agostino, the project's packaging engineer, reflected later that Sandia National Laboratory provided the team with valuable expertise in testing their designs. And Air Force Research Laboratory personnel told him that electronics actually rarely failed, or even changed characteristics, from high impacts as long as the circuit boards were not allowed to bend. So they had to be well-supported mechanically. In artillery shells this was normally achieved through "potting." Essentially, the circuitry for an artillery round was inserted into a hollow space in the casing, and then the empty space around the board was filled up with a resin, usually epoxy, which hardened, producing a rigid structure that protected the circuitry from bending.

But potting prevents testing the circuitry, as the resin can't be removed without damaging the board. D'Agostino knew from his previous work on the Voyager and Galileo projects that this would never fly given JPL's way of designing and testing spacecraft. The armed services accept that some proportion of the artillery rounds fired will not work and plan for this, but JPL could not accept the risk that some of its spacecraft might fail from untested electronics. It was well known in the aerospace industry that electronics have a high rate of what's called "infant mortality," parts that fail within a few dozen hours of first being energized. Such parts are weeded out in the life-test programs carried out prior to launch. Mars Pathfinder's leaders had imposed a requirement of 2,000 operating hours on electronics to insure themselves against such a failure. The Deep Space 2 probes, intended to have lifetimes of only a week, did not need so many test hours. But they did need a few hundred, so the design had to allow for bad components to be removed and replaced.

D'Agostino realized that he did not need to use potting if he could devise a mechanical design for the probe that absorbed all the impact loads for the circuit boards. As long as the probe structure provided rigidity to the boards, the boards themselves could be designed and manufactured just like any other spacecraft circuit boards. The board fabricators would not have to meet any special rigidity standards. This was particularly important given that the project had many business partners developing various bits of its electronics.

Tom Rivellini became Deep Space 2's lead mechanical engineer in early 1996, initially working part time while waiting for Mars Pathfinder's flight airbags to be manufactured and delivered to Cape Canaveral. He and D'Agostino worked out the basic probe design together. One principle they adopted early on was that the probe had to be short to attain the necessary rigidity without being far too heavy for their mass limits. Short structures are inherently more rigid than long ones composed of the same materials, but they have the disadvantage of being aerodynamically less stable.

The final forebody design came to D'Agostino while riding his lawn tractor around his yard. The interior structure would be a three-sided prism, a shape that is structurally strong and would provide the mounting surfaces for the circuit boards. The exterior would be cylindrical, which would give the best penetration performance. A blunt, half-spherical, solid nose would complete the mechanical structure.

The final afterbody design took a good deal more study. It had to "stick" to a fairly wide variety of surface materials when impacting at a variety of possible angles. Testing done using a portable airgun at Sandia National Lab led the designers toward a "top hat" configuration: a squared-off top surface containing the probe's antenna and a rounded lower surface that would help right the vehicle if it struck at an off angle.

While Sandia's airgun was instrumental to the project, it offered no way to do a full system test of the complete aeroshell and probe assembly. In fact, there was no conceivable way for the small project to do such a test. Tom Rivellini did make one attempt at an airgun test of the aeroshell, but the aeroshell shattered while being loaded into the gun—at 0.02 inches thick, it was extremely delicate. But the test attempt did prove that the aeroshell was mechanically untestable. They could not test the probes as they would actually fly, a violation of JPL's technical norms. But this was the faster-better-cheaper era, and that untestability was an acceptable risk.

About seven months before they had to deliver their microprobes to Kennedy Space Center for integration to Polar Lander, the project suffered its biggest crisis. The telecommunications team could not complete the "telecom-on-a-chip" that they had promised. Gavit formed a new telecom team to generate a new design that would still fit in the probe's afterbody and could be delivered in time. The new, larger circuit board design forced Rivellini and D'Agostino to redesign the interior of the afterbody.

On May 28, 1998, the airgun team fired an afterbody with the new telecom design, verifying that it would survive the expected impact, and the project was able to move ahead with assembly of the two flight probes. The team started on these on June 1. The two flight probes had to be at the Cape by early November. Even if they made this schedule, the team would miss the lander's thermal vacuum test and would have to have to undergo it separately.

The small team went from working long hours to working very long hours; by late September, they were putting in 16 hours a day, 7 days a week to make the launch.[45] Sarah Gavit remembers that when the probes finally arrived at Kennedy Space Center on November 8, McNamee's lander integration team had already given up on them. They were three days late and had been removed from the assembly schedule. But the single day allocated for the probe installation hadn't been reassigned to other work yet, so the two little spacecraft and their release mechanisms were installed on the lander's cruise ring on November 10.

The rush to make it to the launch pad left several things not done that the team made up for as well as they could after launch, using their qualification and engineering models. The probes had not undergone a complete thermal vacuum test sequence because the telecommunications components were still in their own subsystems testing. That meant, among other things, that the flight probes were not tested in the very low pressure Mars environment. An engineering model, the so-called qualification probe, was tested, but only after launch. The team also continued performing airgun tests. In many of these cases, either due to low impact angles or very hard, very soft, or very icy surfaces, the probes skipped.[46]

Getting Surveyor 1998 to the Pad

John McNamee's Mars Surveyor 1998 project had held its preshipment review for Climate Orbiter in Denver in mid-August. The orbiter had come together very well, and significantly underweight, a testament to Lockheed Martin's mass-reduction efforts. The major concerns for the review board, instead, were with the solar array and with software.

The solar array questions descended from the broken yoke on Mars Global Surveyor. The board wanted to know what Lockheed had done to ensure that this break would not recur. Despite being fairly complex in nature, the problem was relatively simple to fix. The solar array structures were strengthened, and more powerful dampers installed. Lockheed also mandated testing the array deployments in vacuum, a change to its normal testing policy.[47]

Initially, McNamee's team had thought that they would be able to save money by inheriting a good deal of software from Mars Global Surveyor and from Mars Pathfinder for the functions that all spacecraft shared: attitude control, power management, and so forth. However, much of the software, especially Pathfinder's (also a faster-better-cheaper project), was not well documented. So the software task had been underscoped significantly. Surveyor 1998 had also, in retrospect at least, put too much of its early software effort into the lander software. Lockheed had intended to develop a new, modular approach to the entry and landing software, but this development failed technically, putting the lander project well behind and causing McNamee and his Lockheed counterpart, Ed Euler, to start again on a more Pathfinder and Viking-like software system.

But the software problems got even worse in the last few months before launch, after discovery that some very common electronics components used in many of the flight avionics had developed faults. Diodes are small electronic components used primarily as solid-state switches. Spacecraft can have hundreds of them. The team for JPL's Stardust mission, which was also being built by Lockheed in Denver and used the same flight avionics as Climate Orbiter and Polar Lander, had discovered that 33 of the diodes on its flight boards had cracked in testing, under circumstances that should not have caused any damage. The cracks caused some of the diodes to perform incorrectly, but not all of them.

The reality that Stardust and Surveyor 1998 had hundreds of these diodes put both projects into a crisis. Tearing the spacecraft down, removing the boards with the diodes, replacing the diodes with new ones from an unaffected batch, and reassembling everything would be time-consuming and expensive. It also imposed a great deal of risk—risk that the teams might not meet the launch schedule, that they were introducing untested diodes into already-tested electronics. Pulling all the affected avionics would also delay the already-behind software schedule because it would prevent testing the software on the spacecraft themselves.[48]

After a great deal of discussion that extended all the way to the JPL director and Lockheed Martin's president, the Surveyor 1998 project leaders decided that the risk posed by the bad diodes to the two vehicles was great enough to require replacing all the diodes. In both cases, the refurbished boards would have to be installed on the vehicles after they had been shipped to Kennedy Space Center. This affected Polar Lander more severely than it did Climate Orbiter because there was a great deal more work involved in removing the lander from its aeroshell to get at the assemblies in question and then reassembling it. It was "a major program impact and technical risk," costing about $1 million to repair.[49] The decision also meant that the final builds of flight software could not be tested on the flight vehicles.

At the time of the orbiter's preshipment review, more than three-quarters of the vehicle's fault protection software still had to be tested and integrated. The orbiter would be shipped to Kennedy Space Center without it installed; once the software had been tested in Denver, it would be loaded into the spacecraft down in Florida. They would not have time in Florida to complete an end-to-end test the way Pathfinder had. In effect, it meant that the spacecraft would launch with a software load that had not been thoroughly tested. The project staff and review board saw this as a significant risk. But the team had little choice beyond a 26 month delay, which was unaffordable.

One month after the orbiter's preshipment review, it was time to evaluate Polar Lander. Many of the issues at the lander's review were the same as had been addressed with the orbiter. One unique challenge that the project had confronted back in May drew the reviewers' attention, however. Polar Lander had failed its cruise configuration thermal vacuum test, which was designed to ensure that the vehicle's internal temperatures remained within safe limits during the voyage to Mars. Its "capillary pumped loop" thermal control system simply didn't work.[50] It relied on the difference in temperature between the avionics inside the vehicle and the outside world to generate circulation of its coolant. The problem had been, Sam Thurman recalled years later, that there was not enough difference between the electronics' temperature and the temperature inside the aeroshell during cruise to generate circulation.[51] The problem was traced to an inadequate thermal analysis done early in the lander design phase. This was a "show stopper" of a failure. Fixing it meant providing a "sink," a very cold spot somewhere on the outside of the spacecraft, that the fluid could circulate to, but there was no way to do that without redesigning the interface between the lander and cruise stage. And it was far too late for that.

The failure was not a surprise to some members of the project, or to the JPL thermal engineers McNamee had employed as part-time consultants. Duncan MacPherson at JPL had been particularly vocal about the likely uselessness of the system. He had doubted that the lander needed an active thermal control system at all. The thermal loads had seemed manageable with traditional passive measures—thermal blankets, heaters, and radiators. Left with little choice this late in the project, McNamee challenged this group of critics to prove that the craft could survive without an active thermal control system. They drained the working fluid, removed the exterior radiator from the lander and replaced it with a passive radiator, sealed up the remaining piping, and reworked some of the internal electric heaters and thermal control software, and thermal blankets. Their quick rework passed retesting in August. McNamee then presented the junked radiator to MacPherson as a trophy, along with a red "remove before flight" tag.

Conclusion

Kennedy Space Center launched Climate Orbiter toward Mars one day after the beginning of its launch window, December 11, 1998. This small delay was due to uncertainty over on-board software. The final software build containing all of the missing fault protection elements had been loaded into the spacecraft while it was sitting on the launch pad. This was a risky operation because it meant the spacecraft had never operated on that software—forbidden under traditional JPL practice but accepted by the project and its review board as legitimate under the looser constraints of faster, better, cheaper. Some new, last-minute results out of Denver's software testbed, however, suggested that the spacecraft might not "boot up" with the new software, and John McNamee delayed the launch while this possibility was investigated. This apparent problem came to naught, and the launch went forward the second day without trouble.

After launch, however, Climate Orbiter's inactive propulsion system started to overheat. Sam Thurman remembers that the operations team started getting high temperature readings on some of the valves as soon as the attitude control system locked on the Sun.[52] Quick analysis pointed to a simple design problem.

Sunlight was reflecting off the thruster nozzle's inner surface into the propulsion system's interior, overheating the valves. The valve seats could decompose at the temperatures they were reaching, and the only solution was to turn the spacecraft off its preprogrammed attitude control profile so that the Sun was not reflecting off the nozzle. This adjustment could be done manually, although maintaining the spacecraft in that orientation manually meant more work for the flight team until they could develop and upload a new attitude control profile for the flight to Mars. With only two weeks until the launch of Polar Lander, however, the reprogramming task had to be put off until after their second vehicle had reached space and undergone its initial checkout.

The project's luck with the inconstant Florida weather held, and Polar Lander launched on the first day of its window, January 3, 1999. It too had an immediate post-launch difficulty—its star cameras could not find their reference stars.[53] Attached to the cruise stage, these cameras fed position information to both the attitude control system and to the navigation system. They turned out to be blinded by sunlight reflecting off the lander backshell. Here again, the solution was to turn the spacecraft away from its intended attitude control profile manually and then program a new set of sequences so that the vehicle could maintain it automatically all the way to Mars.

For Polar Lander, the new attitude control profile posed some secondary challenges that it had not for the orbiter. The lander's solar arrays and the medium-gain antenna used to communicate with Earth were fixed to the cruise stage body and could not be moved to compensate for the new off-Sun alignment. The solar arrays would generate less power than anticipated, which would be a management challenge for the flight team as the lander got further and further from the Sun. The antenna had been designed always to have Earth in its "view" while the spacecraft was on its Sun alignment so that the flight team would not have to command it to turn toward Earth; under the new attitude profile, it would not see Earth for portions of the flight. The result of the two spacecraft's problems was more work for the Mars Surveyor Operations Project.

Right after the end of launch operations, the Surveyor 1998 team held a "lessons learned" review with their Lockheed Martin counterparts. Part of the review examined the project's overall development budget in the context of the Pathfinder and Mars Global Surveyor costs.[54] On a percentage basis, the project had spent relatively less on the spacecraft, project management, and operations costs than its predecessors had, and more on the scientific payload. This seemed all to the good. Once NASA had finished adding to the payload, the project also rated itself successful at preventing cost increases from the incorporation of "nice-to-haves"—desirements instead of requirements—by ruthless vetting of proposed changes.

The development had not been without problems, however. The spacecraft contract had exceeded its $83 million budget by $38 million. The additional funds had come from the reserve and contingency funds established at the beginning of the project, and from Lockheed's fee. The company had effectively sacrificed the $12 million fund set up as a performance reward pool. Ed Euler gave several reasons for the overrun. The assumption that they could reuse software from Mars Global Surveyor and Pathfinder had turned out not to be true, leading to a significant portion of the overrun. Most of the major subsystems were delivered late, too, leading to continuing schedule problems. "Hands-off" management of subcontractors—one of the basic principles of faster, better, cheaper—had resulted in the tardiness. Relying on subcontractors to deliver flight-quality hardware that met the design specifications, on schedule, with little oversight, had not worked out well.

Lockheed had also counted on gaining some cost efficiency from the belief that parallel assembly and testing of two vehicles would speed the test and integration schedule. But this hypothesis proved largely untrue because the vehicles were not similar enough. Instead, in Euler's words, "parallel integration and test activities on two unique vehicles cause[d] resource conflicts instead of efficiencies."[55] The assembly and test staff in Denver had also made mistakes that had resulted in equipment damage and additional delays—the team had been too small and lacked experienced supervision.[56] The two spacecraft test laboratories in Denver had been oversubscribed trying to fulfill all of the project demands, leading to delays and increased costs. The higher-than-expected costs had caused Lockheed to increase its proposed price for the Surveyor 2001 missions, which

Mars Surveyor 1998 and Deep Space 2 costs (millions of U.S.$)

Project	Development	Operations	Total
Surveyor 1998	193.1	42.8	235.9
Deep Space 2	28.0	1.2	29.2

Source: NASA 1998 Mars Missions press kit, December 1998, www.jpl.nasa.gov/news/presskits.cfm (accessed 7 February 2012).

Note: As the Mars Surveyor 1998 project was accounted for as a single project, it is not possible to accurately cost Climate Orbiter and Polar Lander separately. Total includes science funding but not the cost of launch vehicles.

played out in the inflation, and resulting descoping, of those two ambitious vehicles.

Yet John McNamee, Ed Euler, now-retired Parker Stafford, and their team were proud that they had delivered two working spacecraft for less than Mars Pathfinder had spent on one. It had been a difficult achievement.

Revenge of the Great Galactic Ghoul

If 1997 had been a great year for JPL, 1999 would be a banner year for JPL's bane, the Great Galactic Ghoul—and an annus horribilis for NASA, in the view of *The Economist*.[1] In September and December, the Ghoul would dine happily on the Mars Climate Orbiter, Polar Lander, and the Deep Space 2 microprobes. Their loss would effectively end the Mars Surveyor program and its two-launches-per-launch-period edict. The 2003/2005 Mars sample return project would die too.

It was not obvious that the Surveyor 1998 project's outcome would be so terrible. The initial difficulties Climate Orbiter and Polar Lander experienced right after launch were not severe; in fact, they were less traumatic than the post-launch problems experienced by both Mars Global Surveyor and Pathfinder. Donna Shirley, former JPL Mars exploration director, followed Associate Administrator Wes Huntress into retirement late in 1998, frustrated with the Mars Exploration Directorate's problems with NASA headquarters but thinking that project manager John McNamee's two Surveyor 1998 missions were going to succeed. Years later, she would recollect going to see David Baltimore, Caltech's president, and telling him that George Pace's Surveyor 2001 missions would probably fail, not the 1998 missions.[2] The losses were shocking to the project team, and to JPL.

Site Selection for Mars Polar Lander

Polar Lander had been sent off to Mars without a final decision about where it was to land. With Mars Pathfinder, Matt Golombek had done the site selection long before launch, giving the project's engineers the opportunity to design the vehicle for its likely landing zone. Polar Lander's principal investigator, Dave Paige, and his former mentor at JPL, Richard Zurek, though, had wanted to wait for the high-resolution imagery and altitude data that Mars Global Surveyor would return before making their final choice. The Viking datasets from a generation ago gave little information about the south polar region, and no direct measurement of altitudes. So the laser altimeter on Global Surveyor would provide valuable new information. Assuming that Global Surveyor would get into its Mars orbit without too much trouble, Paige and Zurek had decided to wait until that data became available partway through Polar Lander's flight. So Polar Lander was launched generally toward the martian south pole region, and its navigation plan included a site selection maneuver that would enable targeting a specific site once it was chosen.

Paige's hope in proposing the mission had been to land poleward of 60° south latitude, within what showed in the Viking imagery as intricately layered terrain. As the mission design effort had progressed during 1995 and 1996, the lander engineers gradually had been convinced that higher latitudes actually would be an easier design problem. This was because the closer to the pole the lander went, the longer the weak martian daylight would last. Both the power and thermal environments became more manageable. So during the design phase, the target latitude moved southward, reaching 75° south by early 1997.[3] Using the Viking orbiter imagery from the late 1970s, the project scientists had looked for interesting sites within the polar layered terrain and that appeared to be within the lander's capabilities. They developed a set of ten candidates and asked Mike Malin to image the region with his Mars orbiter camera on Global Surveyor once it was in a suitable orbit.

In January 1998, the project science team had met to look at the first images Malin's camera had obtained. These convinced the Polar Lander team that they didn't yet have enough information to make a site selection choice. So they had the mission designer plan the initial trajectory to Mars so that it would target the general region they wanted and could be adjusted to a specific site sometime after the second trajectory correction maneuver (TCM) in March 1999. At the end of June 1999, the science team held a landing site workshop to shrink the list of potential sites to four. Global Surveyor's laser altimeter had gotten five "stripes" of data across the desired landing ellipse. Rich Zurek remembers that these data disabused the group of a naive belief that the south polar layered terrain would be relatively flat, with little topography. Instead, "there was lots of variation," including a kilometer-deep hole inside their landing ellipse but near its edge.[4] Their landing ellipse was about 250 km long and 20 km at its maximum width, far too large an area for the Mars orbiter camera to photograph at its highest resolution. Instead, they imaged a handful of sites and assumed the rest of the region was similar. After some discussion, the group chose a primary landing site centered at 76° south, 195° west, and a backup site at 75° south, 180°

west. NASA headquarters approved of the choices after a review on August 24. On September 1, 1999, the Surveyor Operations Project team carried out the lander's site adjustment maneuver to target the primary landing site.[5]

Flights to Mars

About a year before the launches of Climate Orbiter and Polar Lander, Sam Thurman had started to transition over to the Mars Surveyor Operations Project, which would operate the two vehicles after launch. Engineer Pete Theisinger was responsible for developing and deploying the ground systems necessary to operate the two spacecraft. Theisinger had a budget of about $5 million from the Surveyor 1998 project to do this; it had not been enough, and about $2 million more from the Mars Global Surveyor project's underrun had also been allocated to complete the operations system.[6]

The Surveyor Operations Project had maintained the same basic operations structure for John McNamee's two spacecraft as it had built for Global Surveyor. Out in Denver, Lockheed Martin engineers monitored the spacecraft engineering data and were responsible for building and testing the command sequences that told the vehicles what to do. JPL maintained the navigation and tracking functions. A daily phone conference served to keep all the parts of the operations effort working in concert, and the JPL and Lockheed segments exchanged in-progress command sequences electronically.

The one substantial difference for the 1998 missions was the Polar Lander's surface operations, which were to be run from the University of California, Los Angeles, where Dave Paige intended to gather the project's other scientists and a considerable student workforce. UCLA's mission operations facility was not set up until after launch, and would not be ready until July, when the first entry, descent, and landing (EDL) and surface operations tests would be held. One significant personnel change happened, too. Glenn Cunningham retired as manager of the Surveyor Operations Project in mid-June 1999, and Richard Cook, who had been the mission operations manager for Mars Pathfinder, replaced him.

The Surveyor Operations Project's place within JPL's hierarchy had changed, too. After Donna Shirley retired late in 1998, Mars Exploration Directorate chief Norm Haynes had also announced his intention to retire, so JPL director Ed Stone had decided to merge the Mars Exploration Directorate with Charles Elachi's Space and Earth Science Programs Directorate (SESPD). Cook, as head of the Surveyor Operations Project, now reported to Elachi, not to the Mars Exploration Directorate, which became responsible only for future project planning.

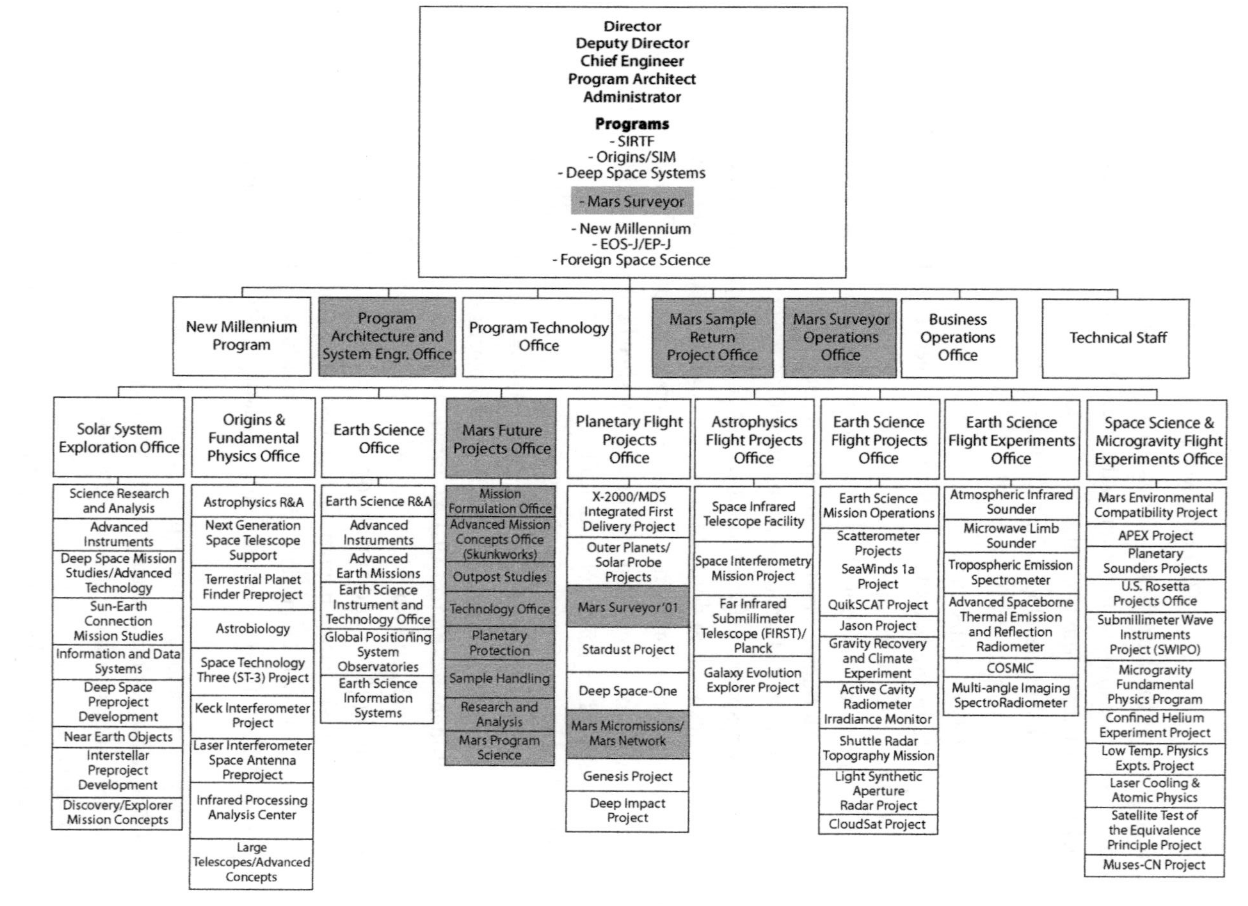

Organization chart for the Space and Earth Science Programs Directorate from the Mars Program Independent Assessment Team report. Note that the Mars flight projects, highlighted in gray, no longer reported to the Mars Surveyor

George Pace no longer reported to the Mars office either but instead reported to Tom Gavin, head of Elachi's Planetary Flight Projects Office.[7] The Mars sample return project was its own organizational entity, separate from the Mars office, the Surveyor Operations Project, and the Planetary Flight Projects Office. In its early 1999 configuration, Elachi's SESPD had 15 offices, responsible for seven different programs, reporting to him. Pace felt buried under the new organization. He remarked later that the reorganization had "negated any synergy that the Mars projects had benefited from by at least having a common manager."[8]

Shortly after launch, the Surveyor Operations Project staff was surprised by the orbiter's behavior. Like most spacecraft built after the mid-1970s, Climate Orbiter used reaction wheels as part of its attitude control system. The wheels were really just gyroscopes aligned along the vehicle's three axes, and, like a child's top spinning, they could absorb a certain amount of torque (a "push" that causes rotation) before they would saturate. When that happened, engineers on the ground had to use the spacecraft's thrusters to remove the stored energy. Mars Global Surveyor had required these "desaturation burns," as they are known, once a week. But Climate Orbiter needed them twice per day—a substantial workload for the small team, and one that few of the operators had expected. This was not a flaw in the vehicle, however; it was actually operating exactly as Lockheed's engineers had intended it to.

Global Surveyor had been slowly spinning around its Sun axis during cruise to save a little fuel and in that operational mode had not needed to carry out these firings very often. But Lockheed's engineers had decided to fly Climate Orbiter in its three-axis stabilized mode, and without the slow spin, it needed more frequent firings. Solar pressure on the large, off-center single solar array was what caused the need for the firings. The array's location to one side of the vehicle caused it to try to "spin" the spacecraft around its north-south aligned axis (also known as its z-axis), and one of the three reaction wheels was constantly resisting this torque. This Sun-induced torque was causing that one reaction wheel to saturate every half-day or so, necessitating the frequent desaturation burns.

The flight team at JPL was surprised by this behavior due to a miscommunication. The original navigation plan for the orbiter had been based on a Global Surveyor–like spin mode, but after the decision to switch the cruise to three-axis stabilized mode, it had not been revised. So while Thurman had expected Climate Orbiter to be doing these frequent desaturation burns, the majority of the flight team had not.

The orbiter's first trajectory correction maneuver (TCM-1) had gone off without a hitch on December 21. The second, initially scheduled for January 25, 1999, was delayed after the Polar Lander's post-launch difficulties began to consume the operations project's limited manpower. The orbiter's TCM-2 was rescheduled to March 4, to give the project team the time to troubleshoot and correct Polar Lander's star camera problem—the cameras were unable to find their reference stars (see chapter 6)—while also keeping their commitment to provide ground operations for Stardust, another simultaneous JPL mission, after its launch in February.[9] A week after TCM-2, Thurman's operators got the orbiter's new attitude control profile uploaded, permanently correcting for the thermal problems with the main thruster. On March 15, they performed the lander's second TCM, again without troubles.[10] By mid-April, they had resolved their remaining known problems by software updates, including an odd problem that the attitude control thruster data being sent back by the spacecraft was arriving at JPL in an incorrect format. Once the data became readable, Climate Orbiter's navigator began to realize that the thruster data did not match the Doppler tracking data from the Deep Space Network. On April 14, the navigator visited Lockheed's Denver-based spacecraft team, and on April 26, he e-mailed to ask them to look into the discrepancy between the thruster and Doppler data sets. During May, Lockheed's attitude control group began to work on the problem but didn't resolve it before a reorganization replaced the group leader in early June.[11]

The Mars Pathfinder team had spent a large part of their flight to Mars running EDL simulations on their testbed in between operational readiness tests (ORTs) in order to uncover previously undiscovered errors. The Surveyor Operations Project had not had money to do this for Polar Lander at first. But McNamee was able to convince officials at NASA headquarters to provide about $4 million to pay for this "stress testing," as he called it. The testing started in late May and ran through the end of August. Once Polar Lander's surface operation center at UCLA was functioning, staff there also began rehearsing the first few days of lander surface operations.

On July 23, the operations team sent Climate Orbiter the command sequence for its third TCM. The vehicle carried it out on the morning of the 25th. But afterward, the spacecraft reported a problem in the solar array's gimbal mechanism, which allowed the array to track the Sun. During the maneuver, the spacecraft automatically put the array into the mechanical restraint to protect the drive mechanism. Apparently after the thruster firing, as the spacecraft moved the array back out of the restraint, the sensor reporting the array's position had malfunctioned. Fault protection software switched to a backup sensor and completed repositioning the array.

The apparent fault caused great concern on the project, despite the lack of immediate consequences. The array had to be moved in and out of the restraint for the remaining TCMs, for the orbit insert burn, and for every aerobraking pass. But it wasn't clear from the telemetry whether the sensor was at fault, the gimbal was malfunctioning, or whether anything was wrong at all. A failure of one of the two sensors was tolerable, though hardly desirable. But the gimbal mechanism was mission critical. If it were failing, the mission would be over shortly. The gimbal had no backup. It had to be proven to be working correctly. For the next few weeks, Lockheed orbiter manager Steve Jolly assigned his Denver-based team to work on this issue—to identify diagnostic tests that could be performed on the spacecraft to check the gimbal's operation, to carry them out, and to understand the results of those tests.[12] This all had to be done in time for TCM-4, which was scheduled for September 15.

After this maneuver, the Surveyor Operations Project was confronted with another challenge. The navigation solutions for the vehicle were diverging. JPL navigators used three methods to estimate a spacecraft's probable location. Normally these methods produced position solutions that were in close proximity to each other, and as more tracking data came in after a TCM, the solutions would converge. But for Climate Orbiter the solutions remained dozens of kilometers apart.[13] Something seemed to be wrong with the data, and the orbiter's navigator—Climate Orbiter had only one full-time navigator with a couple of part-time assistants—started to be concerned about it.

The worst-case solution, the one putting the spacecraft closest to Mars at orbit insert, was at about 110 km altitude, and the lead navigator raised the issue with the project in early August. He was concerned both with the apparently low projected periapsis for the first orbit and with the magnitude of the divergence in the solutions. These seemed to him to indicate a significant problem with either the spacecraft or the way it was being operated.

A series of meetings between the lead navigator, Sam Thurman, Richard Cook, and other members of the Surveyor Operations Project later in the month did not resolve the issue. The navigation team came away frustrated at not having gained a higher priority for their investigation of the trouble, but they thought they had an agreement to use the final TCM in the schedule, which was formally a "contingency" maneuver, to raise the projected orbit. The project management came out of the meetings thinking that while they might well find themselves in a lower initial orbit than they sought, even the worst-case solutions the navigators presented were still within the spacecraft's thermal and structural capabilities. So the problematic solar array remained their highest priority.

By early September, Steve Jolly's team in Denver thought they understood the solar array incident and had prepared command sequences to test it. These were transmitted to Climate Orbiter on September 2. The next day, the tests showed that the array was hitting the spacecraft during the unstowing procedure, so they had to alter the software that controlled the array's movements. Jolly put his spare people on the array problem again, as it was still the biggest risk to the mission. Over that weekend, they reworked the software and tested it in the spacecraft test lab. On September 8, they ran the new sequence on Climate Orbiter, fixing the problem.[14]

But the navigation problem returned after the fourth TCM on September 15. The "quick look" navigation solutions put the projected periapsis about at 138 km, instead of the 210 km the maneuver had been designed for.[15] The lower periapsis altitude was tolerable but not desirable; given that the navigation solutions had shown a consistently large spread, it left a possibility that the spacecraft's second orbit after insertion would be dangerously close to the planet. The project's risk management plan specified a minimum 150 km altitude, so Thurman and Jolly had built and tested a command sequence that would use a contingency TCM-5, scheduled for September 20, to raise the orbit about 44 km. The sequence was tested in Denver satisfactorily on September 17.

On the next morning, John McNamee called a meeting of the Mars Surveyor Operations Standing Review Board at JPL to review the desirability of carrying out TCM-5. Three of the eight members were able to attend; he didn't invite the project's navigator. At the meeting, Thurman argued that the current periapsis estimates of 150–170 km did not warrant carrying out the final TCM. The sequence that commanded the orbit insertion burn was already running aboard the spacecraft, and he was concerned that the TCM sequence would interfere with the orbit insertion sequence. The two sequences had been run together in the simulator, but one could never be certain that the simulator was a perfect representation of the spacecraft. He was also concerned that the project had never practiced TCM-5 via an operational readiness test, so the flight team was not well prepared for it and whatever might go wrong during it.

After some discussion, the board members present agreed with the decision not to carry out TCM-5, with the caveat that if the orbit solutions generated that afternoon grew worse, the decision would be revisited. Those solutions stayed within the 150–170 km range, and on the September 19, at the go/no-go decision meeting, they agreed that the trajectory correction maneuver would not be carried out.[16]

Orbit insertions are generally big events at JPL, and Climate Orbiter's was no different. Because the orbit insertion would take place a little after 2:00 a.m. Pacific time, mission scientists had explained their goals to assembled reporters at a press conference in the Lab's Von Karman Auditorium the previous morning. Cameras started providing a live video feed from the mission support area and from Lockheed Martin's similar facility in Denver at 1:30 a.m. Climate Orbiter's orbit insertion events started with a final desaturation burn of the maneuvering thrusters, followed by a final telemetry downlink. At 16 minutes prior to start of the main engine firing, telemetry shut down, leaving only the carrier signal broadcasting via its medium-gain antenna. The burn was expected to last about 16 minutes—the engine would fire until it ran out of oxidizer, so the timing could vary by a few seconds. Five and a half minutes after the burn started, the spacecraft should disappear behind Mars, its radio signal blocked by the planet itself. About 27 minutes later, it should reappear, having finished the burn, redeployed the solar array, and activated the high-gain antenna.[17]

The navigation solutions available to the flight team the night before projected the orbit periapsis at 150 km, lower than they'd wanted but well within the spacecraft's capabilities. But two hours before the insertion burn, a new solution put the periapsis at 110 km. While acceptable for the first orbit, it would result in the second orbit passing dangerously low in the atmosphere, so the team started preparing a command sequence to raise the orbit periapsis immediately after the insertion burn.[18] But they were concerned that they might not have enough time to get the command uplinked before Climate Orbiter disappeared behind Mars again on its second orbit the next day. This was their chief worry.

All of the preprogrammed events of the orbit insert went nearly perfectly prior to occultation. The Deep Space Network picked up the Doppler shift in the carrier signal precisely when the engine should have started firing, proving that everything to that point had gone well. The next event, occultation by Mars, was the first indication that not all was well. They lost the signal 39 seconds early. No one in JPL's mission support area noticed it at the time, though. The Mars atmosphere determined when the last moment of signal reception would be, and its high variability made this prediction uncertain. Interviewed a few minutes later, the mission manager told the television audience that everything was "nominal."[19]

Climate Orbiter never emerged from occultation. Ten minutes or so after the predicted time of emergence, Lockheed's mission operators started working through their preplannned loss of signal procedure, which was based on the premise that the spacecraft had gone into its safe mode. In this mode the vehicle would maneuver automatically to align the solar array with the Sun and then listen for commands from Earth. The flight team sent up a command to turn on the low- and medium-gain antennae, which the Deep Space Network could detect through a large range of possible spacecraft orientations. But they still heard nothing.

After 30 minutes without a signal, Cook also had other recovery procedures started. This included a review of all the navigation data from the last 24 hours. That provided the first tentative answer. As the spacecraft had neared Mars, the planet's gravity and the presence of Mars Global Surveyor within the Deep Space Network's field of view permitted a much more accurate trajectory determination. And a shocking change in the solutions appeared. The vehicle seemed to be about 100 km closer to Mars than predictions made the day before.

By 8:00 a.m. Pasadena time, the Surveyor Operations Project knew what had happened, though not yet why. Instead of entering an orbit with a 110 km altitude periapsis, the craft had entered an orbit with a 60 km periapsis. Lockheed's analysis showed that at about 98 km altitude, the orbiter would have exceeded its thermal limits; at 85 km altitude, the attitude control system would no longer have been able to counter aerodynamic forces imposed by the atmosphere.[20] At right around 60 km altitude, the spacecraft's solar array would probably have been torn off.

Richard Cook, John McNamee, and Carl Pilcher, NASA's solar system exploration director, answered questions from the press that morning. Cook did most of the talking, explaining in his opening statement that they appeared to have had a serious navigation error and that the spacecraft had probably entered the martian atmosphere and been destroyed. The more experienced reporters in the room were in disbelief; it was the first time JPL had ever lost a spacecraft due to a navigation error. Cook told them he was as shocked as they were—a 100 km error was so far outside anyone's experience that it simply hadn't been conceivable.[21] Typical navigation errors were 2 to 3 km.

The team continued trying to contact Climate Orbiter for the next day or so, sending up various commands relating to transmitters, orientation, and finally, in Steve Jolly's words, made a "last-ditch attempt to uplink and execute a large periapsis-raise maneuver" before the second orbit. Uncorrected, this second orbit would have had a periapsis of only 40 km, "a passage in which the spacecraft would enter and surely not survive."[22] But they had little hope. McNamee understood they'd lost the vehicle as soon as the navigation error had been revealed, but due diligence required that they try to reach it until it passed behind Mars again. No one believed it would survive that second orbit.

That evening, Ed Stone sent out an all-hands e-mail message announcing the probable loss of the orbiter.[23] On the September 24, Steve Jolly sent an e-mail to all of the mission personnel explaining in detail the efforts to revive the spacecraft and concluding, "The Flight Team personnel at both LMA and JPL would like to express our deepest regret and disappointment in the loss of MCO. This goes out to all of the Flight Systems and JPL personnel that have dedicated so much to see that this mission was successful."[24]

On September 29, Sam Thurman sent out the formal end of mission report.[25]

Navigating the Investigations

Sometime during the early morning between the orbiter's disappearance and the 8:00 a.m. press conference, John McNamee had called Tom Gavin to tell him he thought they'd flown Climate Orbiter into Mars. Gavin had recently been the Cassini spacecraft manager and had just become deputy to SESPD director Charles Elachi at JPL. Gavin, an electronics engineer by training, called Frank Jordan, who had recently joined the Mars program office after a long career in Lab's navigation section, and told him to show up at the press conference. Sitting near the back in Von Karman, Jordan heard Richard Cook admit to the 100 km navigation error in disbelief, and then tears. The next day, Jordan was formally appointed chairman of the Mars Climate Orbiter Navigation Peer Review committee, one of three investigations started after the loss.[26]

It did not take very long to figure out what had happened. The large divergence in the navigation solutions in the weeks prior to the orbit insert strongly suggested to Jordan that some sort of modeling error was involved. There were only two things that the Deep Space Network could measure very accurately: the range to a spacecraft and the Doppler shift imposed on its radio signal by accelerations. These allow determination of the spacecraft's distance from Earth much more accurately than they do its angular position in the sky (what navigators call its plane-of-sky location). To produce more accurate estimates of a spacecraft's angular position, navigators have to estimate the magnitudes of many other forces on a spacecraft, including solar pressure, the gravitational accelerations imposed by the other planets, and the effects of the accelerations imposed by the spacecraft's own activities. Because JPL operated many other spacecraft, the solar pressure and gravitational models used on the orbiter could be checked by applying them to another spacecraft and seeing if they produced that vehicle's known location. They did. So something was wrong with the modeling of the spacecraft's self-imposed accelerations.

There were only a couple of possibilities for error in those calculations. If the spacecraft had sprung a leak in one of its pressurized tanks, it would have produced a large, unmodeled acceleration. But it would also have showed up in the Deep Space Network's Doppler data as an unexpected, large acceleration, and no such event appeared in the data. The other possibility was that the spacecraft's attitude control thruster model was incorrect. These small thrusters kept the spacecraft in its programmed orientation and enabled the desaturation of its reaction wheels. Each time one of them fired, it imposed a small acceleration on the spacecraft in a desired direction. But the firings were never perfect—they also always produced a small acceleration in an undesired direction.

JPL-built spacecraft, like Mars Pathfinder, generally used "coupled" thrusters. Coupled thrusters were thruster pairs on opposite sides of the spacecraft that fired in unison to produce a pure rotation around the spacecraft's center of mass. In other words, they would not produce any translation, or linear motion. Mars Pathfinder chief engineer Rob Manning commented years later that they employed this method because precision navigation was considered important to the Pathfinder project. "I wanted a quiet spacecraft that followed Kepler's laws and did not have a mind of its own."[27] They went to the trouble of designing and calibrating a coupled-thruster system to ensure their spacecraft would be easy to fly. But Steve Jolly's engineers had not really understood how important these small forces, as they were known, were in deep space navigation and had done what was common in Earth orbiters: used uncoupled thrusters.[28] These were single thrusters that would produce both rotation and linear motion. The undesired linear motion was then calculated and, if necessary, removed in subsequent firings. For Climate Orbiter, the linear component of motion was to be calculated on the ground and compensated for during the TCMs. The Global Surveyor flight team had made these adjustments during their maneuvers without problems, so what had gone wrong when the same people had performed the calculations for Climate Orbiter?

Interviews with the navigators and operations staff led Jordan to understand that the navigators had never discovered that Climate Orbiter was not being operated in its "barbeque" spin mode as specified in the navigation documents.[29] This was a key miscommunication. If Climate Orbiter had been spinning, the errors, whatever their magnitude, in its thruster calculations would have been distributed "spherically," located as if inscribed at random points on the surface of a sphere. But in its three-axis stabilized mode, the errors would all build in specific directions. Worse, since Climate Orbiter's thrusters fired so often to desaturate only one of the reaction wheels, due to the high solar torque produced by its solar panel, most of the error was in a single direction: toward Mars.

Still, when the thruster calculations were applied to a nonspinning Climate Orbiter, they did not put the vehicle anywhere near its actual Mars orbit. There was still a source of more error. In fact, the amount of error necessary seemed to be about 4.5 times as much as the model showed. Jordan's panel was set up on a Friday. Over the weekend, Jordan realized that the amount of error also happened to be the same as the unit conversion necessary to turn a metric force estimate into one in English engineering units (Newtons to pounds-force). On Monday he requested a teleconference with Lockheed's attitude control team. It took place on Tuesday; on Wednesday, Steve Jolly called Gavin to report that this had, in fact, been the error.[30] Ground software used by Lockheed's operations group to calculate the thruster forces lacked this crucial conversion factor. While the requirements document specified that the company provide the thruster data in Newton-seconds, they had actually been providing it in pounds-force-seconds.

Jordan had the navigation section run the trajectory analysis from TCM-4 to Mars using the corrected thruster data. This "yielded a trajectory 160 km closer to Mars."[31] The root cause, a simple software error, had been found.

Two other investigations started in parallel. Ed Stone had John R. Casani, recently retired after completing his assignment as Cassini-Huygens project manager, run an internal investigation. Casani focused on process. How had such an egregious error gotten through all of the checks and balances that were supposed to find and fix inevitable human errors? The second committee, appointed by NASA, was led by Arthur Stephenson of Marshall Space Flight Center. Stephenson's committee, which got going more slowly, was rendered partly moot by the rapid discovery of the root cause. It ultimately reported on project management failures in NASA more broadly.[32]

Casani's panel found there had been many opportunities to discover the problem, but they were missed and drew no corrective actions. The initial software error had been made by a "freshout" at Lockheed Martin. He had been assigned the task of revising a piece of Global Surveyor's software, called "sm_forces," to make it compatible with Climate Orbiter. Climate Orbiter had different thrusters, so the programmer had deleted the Global Surveyor thruster-specific code and replaced it with code corresponding to the orbiter's thrusters. He had not realized he needed to apply the Newtons-to-pounds-force correction; while experienced aerospace engineers know that the U.S. industry is caught in a weird hybrid of English and metric systems and probably would have checked this, the fresh-out did not. And, as had also been noted in the final development report back in January, supervision at Lockheed's plant had been weak. Nobody had told the programmer this key bit of information.

The software flaw also could have been caught during the early operational readiness tests. If these ORTs had been carried out properly, the data table being provided by the software would have been examined for correctness.[33] Then the examiners would have discovered that the software was creating the table in an unreadable output format. But no one read the data table during or after the ORT. Discovered in December after launch, this format error was not fixed until April. So the erroneous data was not even being used to fly Climate Orbiter until then. The navigator, who knew about the output format error, made his own estimates of the thruster firing impulses instead.[34] Once the format error was fixed, he started using the software-generated data, as it saved him work. That was why the navigation solutions only started to diverge after the third TCM. The navigator's estimates were actually more accurate than the software's.

The Deep Space Network's Doppler tracking data also could have revealed the problem. The Doppler shift produced by the thruster firings was consistently about four and a half times the magnitude expected, reflecting the miscalculation. The navigation team had noticed the divergence between the Doppler data and the calculated firings in late April and asked Lockheed's spacecraft team to investigate.[35] An attitude control engineer was assigned to the problem, but at the beginning of June, Lockheed had reorganized the operations staff. That engineer went to the Stardust mission and his replacement never knew about the Doppler discrepancy. JPL and Lockheed Martin each maintain a formal problem reporting system known as incident, surprise, and anomaly (ISA), and no one filed an ISA either. ISAs must be tracked and responded to; without the formal requirements imposed by the ISA system, the Doppler issue got lost in the press of other work during the busy period preparing for Climate Orbiter's orbit insert and Polar Lander's site selection maneuver.

Instead, the orbiter navigator became concerned about the large "spread" among the three position solutions without connecting this issue with the anomalous Doppler results. This was likely the single most fatal disconnect among the many revealed in the investigation. Because the navigation team itself did not "connect the dots," as it were, into a consistent story that exposed much larger uncertainties than were nominal for Mars missions, they could not convince Thurman, McNamee, and Cook that they needed to raise the orbit periapsis. McNamee, Thurman, and Cook all came from the navigation and operations sections of

JPL, and they knew that 2–3 km errors were typical. Their own experience gave them a false sense of security that could only have been broken by a solid case for larger-than-expected uncertainties. And the navigation staff didn't make one. Finally, there had been no leader for JPL's navigation section to whom the project's navigator could appeal after being rebuffed by the project staff during the flight to Mars; the section manager had left, and a new one wasn't chosen until the week of the orbiter's arrival.[36] Even Cunningham's retirement in June was ill-timed, in retrospect. He had known the navigator from many years' service on previous missions and might have provided him with a more sympathetic ear. Climate Orbiter had suffered a perfect storm of errors, oversights, and understaffing.

Casani's report made 14 recommendations. Central to them were inducing better communications within projects and between projects and JPL's technical specialists (known as the "line organizations" at JPL) and restoring formalized processes and procedures that had been abandoned in the effort to do things less expensively—especially problem investigations. Arthur Stephenson's report put essentially the same recommendations in terms of reinvigorating the discipline of systems engineering at JPL. Systems engineering was created to exert management control over complex development programs. At its core is interface control, a process to ensure all of a spacecraft's systems would work together as intended. At least at JPL, this was understood to be more than a technical process; it's humans that build things, and systems engineering had to ensure that adequate, and accurate, human communication occurred and kept occurring throughout a project's life cycle.

Demise of the Polar Lander

Orbital mechanics dictated that Polar Lander would reach Mars on December 3, regardless of the outcome of the three Climate Orbiter investigations. Charles Elachi reorganized SESPD and immediately put the Surveyor Operations Project under Tom Gavin. The operations project's navigation staff was also reinforced immediately, to ensure no other navigation problems recurred. Polar Lander did not have the same ground software as Climate Orbiter, so at least on the surface, it did not share that same risk. It also did not have the asymmetric architecture that had caused all of Climate Orbiter's thruster firing errors to build up in the same direction. Nonetheless, Polar Lander's flight team moved their scheduled October 10 TCM back ten days to allow a thorough scrubbing of their navigation data and process.

The increased navigation attention and staffing also led the team to discover that they had not fully understood the ramifications of the post-launch discovery of the star cameras' stray light problem. The solution had been to keep the lander oriented slightly away from its designed Sun line, resulting in the lander's cruise stage antenna being pointed away from Earth. So the operators had to command the spacecraft to turn toward Earth in order to transmit engineering telemetry every day. The additional firings of the uncoupled thrusters had, as they had with Climate Orbiter, built up an error that had to be removed at TCM-4. So TCM-4 had to provide a larger velocity change than originally expected.[37] This didn't seem problematic; the lander had plenty of fuel.

But TCM-4 then had an unexpectedly large execution error. The team traced it to the firing being long enough to allow a pointing error (an error in the direction of the firing's velocity vector) to build up but not long enough to enable the attitude control system to correct it. This was another consequence of having not designed the vehicle for high-precision navigation. And TCM-5 could not be used to correct the trajectory either. So the lander's final course would take it to the extreme western edge of the planned landing zone, which contained a crater near the northernmost extent of the ellipse.[38] Extensive analysis by the navigation team, with the help of Richard Zurek, showed that the lander was not in significantly more danger than it had been. It just would not land quite where they had wanted it.

Ed Stone had also asked John Casani to set up a "red team" review of Polar Lander. This was headed by Chris Jones of JPL, who had developed the fault protection software for the Voyager spacecraft. Jones's team looked at several potential issues in detail, one of which seemed a potential killer. Arthur Stephenson's review committee had commented that it was not clear that Polar Lander's descent thrusters would fire properly, and this uncertainty drew a great deal of attention.[39]

The terminal descent thrusters needed certain thermal conditions to exist for them to operate properly. Polar Lander used hydrazine thrusters in a monopropellant mode. The hydrazine passed over catalyst beds, inducing a reaction that converted the liquid to a gas. The catalyst beds lost efficiency below 7°C, and hydrazine froze at 1°C, so the issue was whether the project's thermal analysis and testing had been adequate to ensure that the catalyst beds were above that temperature when the spacecraft computer ordered the thrusters fired. Because the thrusters had to turn on and off very rapidly to control the descent, even a short "warm up" delay was intolerable. And, of course, the lander's thermal analysis had already been impugned by the failure of the capillary pumped loop system during thermal vacuum testing. Stephenson's group wanted this potential problem addressed.

It turned out to be a significant issue. During TCM-3 and TCM-4, the fuel outlet temperature from one tank (the other tank had no temperature sensor) did drop very close to hydrazine's freezing temperature. So local freezing within a tank was possible, despite the operation of tank heaters. The propulsion review team also unearthed data from the cruise thermal vacuum tests done before launch that showed the catalyst beds to be at –30°C. The thrusters would not have fired at that temperature. JPL and Lockheed had the thruster contractor perform tests to determine what could be done in flight to bring the beds up to a useful temperature; these showed that activating electric heaters attached to the fuel valves several hours before landing could bring the beds up to an acceptable temperature.[40] The operations team made changes to their procedures in the days before landing, resolving the problem.

Polar Lander's descent to Mars was scheduled for the afternoon of Friday, December 3. It was a long day for the flight team because TCM-5, their contingency maneuver, had to be carried out at 4:00 a.m. Pasadena time. After examining their projected entry corridor vis-à-vis the desired landing site, they had decided to use TCM-5 to shift the projected landing spot southward slightly to improve their chances of missing the crater that was sitting on the north edge of the landing ellipse.

The landing was scheduled for 12:15 p.m.; then the team would have to wait to hear from the lander. Unlike Pathfinder, Polar Lander had no communications with Earth during its descent, so JPL wouldn't know it had landed successfully until about a half hour had passed. Aboard the vehicle, the first 20 minutes after touchdown were devoted to deploying the solar panels and performing "gyrocompassing." The lander needed to know what its precise orientation on the surface was in order to aim the medium-gain antenna at Earth. It figured this out by analyzing data collected from its inertial measurement unit during the descent and first few minutes on the surface.

The gyrocompassing period was critical because the antenna had to be aimed within 6° of the Earth's true location in the sky to be detectable by the Deep Space Network's antennae. The Monte Carlo simulations run out in Denver had shown there was a quite significant chance that it would not find Earth on its first try, and both Zurek and Sam Thurman cautioned during a press conference that they should not consider the mission lost if there was no signal. Instead, the lander had been programmed to hibernate for about seven hours if it didn't find Earth the first time, charge its batteries, and then carry out a search pattern with the medium-gain antenna. So it was very possible that the lander could descend successfully but no one would know it until the next day.

The Deep Space 2 microprobes would also not be heard from immediately. They depended on the Mars relay aboard Global Surveyor to communicate with Earth; specifically, the relay would take the probe data, encode it as if it were a photograph, and store it in a memory area in Mike Malin's Mars orbiter camera. Global Surveyor would then send it back to Earth, where the Deep Space Network would transmit it to Malin's facility for decoding. Microprobe project manager Sarah Gavit's team at JPL would first find out from Malin whether they had any data. The first pass was scheduled for 7:50 p.m., but this pass was at a low sky angle, and not hearing from either probe could simply be a product of the low overpass angle. Every two hours, Global Surveyor would fly over again, climbing higher in the sky for four passes and then descending again. So it could take nearly a day to determine whether the probes had survived, and Gavit's small team would have no information at all when the first post-landing press conference was scheduled, at 1:30 p.m. on December 3.

JPL was transmitting Polar Lander's descent to Mars live on NASA's own TV channel. Perhaps due to Climate Orbiter's failure, the mission support area in building 264 was crowded with dignitaries, too. NASA administrator Dan Goldin was there, with Ed Stone, Charles Elachi, Tom Gavin, and Chris Jones. Planetary Society executive director Lou Friedman was also in attendance, along with California congressman David Dreier. Cameras outside the mission support area at JPL, at Lockheed's operations facility in Denver, and at Dave Paige's lander operations room at UCLA recorded an excruciating four days for the Polar Lander and Deep Space 2 teams, and for the assembled higher-ups.

Outside JPL's fence, the Planetary Society had organized another PlanetFest around Polar Lander's descent. Held at the Pasadena Convention Center, the three-day event's theme was "A New Millennium of Exploration." Lou Friedman's crew had gotten scientist Chris McKay to speak about the now-famous "Rock from Mars." Ed Stone spoke on solar system exploration, and astronaut Story Musgrave presented pictures of the Earth from space. There were children's activities, artists displaying space and Mars art, and lectures by science fiction writers.[41] There were also giant television screens linked to JPL so that, as with Mars Pathfinder, Polar Lander's arrival could be witnessed almost firsthand by the 20,000 or so attendees.

The December 3, 12:39 p.m. opening of the first communications window for Polar Lander after its descent came and went; after a few minutes without a signal, Rob Manning snuck away from the mission support area to be sick. He already believed the lander had failed and felt horrible for the team. McNamee,

Mars Polar Lander mission support area on landing day, December 3, 1999. In the foreground are Sam Thurman (*right*) and mission systems lead Philip Knocke (*left*). Just over Knocke's left shoulder are John McNamee (*left*) and NASA administrator Dan Goldin (*right*), sitting. Standing behind McNamee to his left is Charles Elachi, Matthew Landano of JPL's mission assurance organization, and Richard Zurek (*arms crossed*). Standing above Goldin's right shoulder is NASA associate administrator Ed Weiler (*arms crossed*). JPL director Ed Stone, Tom Gavin, and John Casani are visible in the back row behind Elachi. Charles Whetsel is standing under the TV monitor in the upper left. JPL D-120399B7, courtesy NASA/JPL-Caltech.

Thurman, Cook, and Gavit were more confident, and it was Richard Cook, the eternal optimist of the bunch, who appeared at the 1:30 press conference and told the assembled reporters that they had not heard from it.[42] Cook also told them the next opportunity was a little after 8:00 p.m. that night and would last about an hour, while the medium-gain antenna went through its expanded search for Earth.

JPL held another press conference at 10:40 that night, originally expecting that there would be data from Polar Lander and the Deep Space 2 probes, but there were none. Gavit's project engineer had bounded into the mission support area an hour before the 7:50 Global Surveyor pass all smiles, but as she and everyone else there heard Mike Malin over the speaker announcing no probe data from the Mars relay's first pass, her mood turned tense, too.[43] Gavit joined Cook on the stage at Von Karman this time, to explain that it wasn't really surprising that this low angle pass did not hear anything from the probes.[44]

But by the evening of the next day Gavit's team largely understood they had not succeeded; while they had to spend the night of the third in the mission support area to hear the null result of each successive communications pass, for the Polar Lander team the next opportunity was not until the evening of December 4. The Polar Lander team reassembled for this opportunity, with many of the dignitaries still in attendance. Chris Jones spent a lot of time explaining to the television audience what was happening and what Polar Lander might be doing at its south pole perch, but the communications window came and went with nothing received. Cook, Gavit, and Sam Thurman faced the Von Karman audience together at a 9:40 p.m. press conference.[45]

Cook explained that there were several more opportunities to communicate with Polar Lander over the next two days, but the last pre-programmed communications window was on December 7. By that time, the lander would have attempted to communicate via the Mars relay on Global Surveyor, as well as by its own medium-gain antenna, and the operations team would have tried each variation of search command available to them. He and his team reassembled for each communications window, with a diminishing group of spectators. They didn't hear anything.

On December 7, Ed Stone began the painful process of notifying all the Mars program's stakeholders, to use a popular bit of management jargon, of the probable loss, with an address to Caltech's Board of Trustees. He explained what had happened to Climate Orbiter in some detail, and the corrective actions taken, apparently in vain, to ensure Polar Lander didn't suffer a similar fate.[46] On the December 9 he went to Washington to take responsibility for the two mission's failures.

On the flight back to Pasadena the next day, he composed an address for a town hall meeting in JPL's Von Karman auditorium. His audience greeted him with a standing ovation that took a few minutes to quiet down. Then he told them that his chief concern was that JPL, and NASA, would learn the right lessons from the losses. "Scrutiny and criticism can be good things. We are accustomed to this as an internal exercise," he said. "We excel at 'finding the flaw, and then fixing the problem' because this activity helps insure mission success."[47] The magnitude of the Lab's failure meant that this normal internal process would be applied from the outside, too, by NASA. There would be many critical reviews of all sorts of things in the coming months.

Recriminations and More Investigations

Public reaction to the Polar Lander's loss was harsh. Television comedian David Letterman had a mockup of the vehicle crash onto his stage during one show. Myriad political cartoons about it appeared in newspapers and magazines, and on the Internet. Some were sympathetic to the team, but many were harsh condemnations of NASA or faster, better, cheaper, or both.

Long before JPL finally gave up looking for Polar Lander, recriminations about the Surveyor 1998 project had started. On December 8, already tired of attacks from inside JPL about the lack of telemetry during Polar Lander's descent, John McNamee had written a lengthy e-mail to Ed Stone and many others explaining the decision. From its beginning, the project's financial and mass constraints mitigated against anything that was not directly concerned with "getting a spacecraft safely to its' [*sic*] destination and conducting science operations with that spacecraft."[48] Telemetry during entry, descent, and landing did nothing to improve the chances of the lander succeeding—it would only improve the chances of future landers succeeding. So "absent a Program level requirement to have a downlink during EDL," his team had not pursued one. That decision had been explained at many reviews, to both JPL and NASA management. It should not have been a surprise.

The lack of telemetry left the inevitable investigation with little data to work with. Ed Stone gave the unenviable task of investigation to John Casani, who assembled a diverse technical committee to review all aspects of the design. There were many potential points of failure to address: if the cruise stage had never separated from the lander, none of the vehicles would have survived entry. If the lander heat shield, parachute, radar, or descent thrusters had failed, Polar Lander would never have been heard from (although none of these would have affected the probes). The lander could have landed safely, but if the solar panels had not deployed, they would have prevented communication via both medium-gain and UHF antennae—again, this wouldn't have affected the microprobes, so Casani's team had to postulate independent failures for those. Because they had so many potential failure modes, they didn't get much traction on the problem for several weeks. Their breakthrough came from George Pace's Surveyor 2001 project.

In February, during testing of the 2001 lander's leg extension process out in Denver, technicians noticed that the legs rebounded slightly after reaching full extension.[49] That rebound caused the sensors that sensed touchdown on the surface to send a false touchdown signal to the flight computer, which was stored in memory. The legs deployed while the lander was still on the parachute, so nothing would have happened immediately. But in the actual mission, when the computer released the parachute and started the descent thrusters, it would start looking for that touchdown signal, find it already in memory, and shut the engines off again—while still around 40 meters above the surface.

Roger Gibbs, George Pace's spacecraft manager, had the test run 47 times. In 32 cases, the leg sensors generated the spurious signal.[50] Tests run on the engineering models of the Polar Lander's legs also generated spurious signals. So it looked like in the majority of cases, the 2001 lander would have crashed. Since it was a rebuild of Polar Lander's design, this finding implied that it was also likely to have been the fatal moment for Polar Lander.

Lockheed's Parker Stafford reflected years later that he had realized this possibility might exist but that it was easy enough to solve by wiping the memory address for this "touchdown flag," as it was called, immediately after the legs deployed.[51] There was plenty of time to do this, relatively speaking. The wipe would take a microsecond, and the lander rode the parachute for more than a minute after leg deployment. The memory wipe never became part of the requirements that specified how the software was to work, however, so it hadn't been implemented.

Casani's committee gave leg extension triggering premature engine shutdown as the likely cause of Polar Lander's loss, but they also specified six other potential failure modes as "plausible," based on examination of the lander design and the project's testing documents. Further, they found areas where the lander test program had been inadequate to ensure the proper operation of subsystems, such as the heat shield ejection mechanisms, and criticized Lockheed for use of excessive overtime. "Records show[ed] that much of the development staff worked 60 hours per week, and a few worked 80 hours per week, for extended periods of time." The manpower shortage also meant that some technical areas had only one person working them, leading to inadequate peer interaction. So "there was insufficient time to reflect on what may be the unintended consequences of day-to-day decisions."[52]

The landing leg fault did not explain the silence of the Deep Space 2 microprobes, and the lack of data there was even more daunting for the investigation. With Polar Lander, the committee at least knew the lander's status up to the moment at which its telecommunications had shut down prior to cruise stage separation. The probes had no communication with anyone from the moment they had been attached to the cruise stage back in Florida, so it was impossible to know whether they had even survived the flight to Mars. But based on the possibility that the surface conditions were not as expected, the committee postulated that they had either bounced or landed on their sides, preventing communications. They could also have suffered electronic or battery failure on impact, as the flight lot of batteries had not been impact tested. Finally, the probe electronics had not been tested in the low-pressure Mars atmosphere, and very low pressures could induce electronic failure.[53] The board also was not satisfied by the lack of an end-to-end test of a complete probe, although they did not list this is a potential cause of the loss.

Years later, Sarah Gavit would point out that Casani's report was unusual for an aerospace investigation. Typically investigators look for a single point of failure that could explain all of the evidence. In the context of Polar Lander and the Deep Space 2 probes, the only single fault that could explain the total lack of communications with any of the three vehicles was failure of the cruise stage to separate from the lander's aeroshell.[54] Because the mechanical movement of the aeroshell away from the cruise stage was the trigger for probe deployment, the probes would not have deployed either. All four of the vehicles would then have been destroyed in the martian atmosphere. Casani had rejected this scenario because the pyrotechnics that trigger the release are extraordinarily reliable, and the electronics that commanded the pyrotechnics had been properly designed and tested. To her, the question of what had actually happened remained an open one. Several years later, the Phoenix project would discover that she had probably been correct.

In addition to the JPL special review board led by Casani, Dan Goldin asked A. Thomas Young, recently retired as executive vice president of Lockheed Martin and once the operations manager for the Viking project, to chair a broader review of the Mars program. This panel came to three major conclusions. First, McNamee's project had been substantially underfinanced. Using Mars Pathfinder as their standard for a minimum successful project, they figured the Surveyor 1998 project had been underfunded by at least 30 percent.[55] This had led to inadequate staffing, testing, and training.

Second, poor communications had hobbled the project and the Mars program overall. The Mars Exploration Directorate at JPL had seen NASA's requirements regarding costs, schedule, and launch vehicles as nonnegotiable. Fear of losing the planetary science business, JPL's primary role in NASA, to competitors at Ames Research Center and the Applied Physics Lab caused JPL's senior management to downplay the risk the project faced. Thus, "what NASA Headquarters heard was JPL agreeing with and accepting objectives, requirements, and constraints."[56] The requirements creep on the Mars Surveyor 1998 and 2001 projects was a product of JPL senior management not "pushing back" against additional requirements until Tony Spear's intervention. This had been further exacerbated by a complex reporting relationship between the Mars office and NASA headquarters, where Norm Haynes had to deal with four different officials within the Space Science Enterprise, plus others in the Human Exploration and Development of Space Enterprise, for the experiments on the Surveyor 2001 missions.

JPL had made the communications channels worse still in its own 1999 reorganization. The Mars program director no longer reported to the JPL director, and the Mars Surveyor project managers no longer reported to the Mars program office. The resulting lack of senior management attention was inadequate. Young's team concluded that "the current organization is not appropriate to successfully implement the Mars Surveyor Program in combination with other commitments."[57]

Finally, the panel reinforced what Tony Spear's 1998 review had concluded. The program had not succeeded in creating a "Mars architecture" that would permit continuous evolution of appropriate technologies, with each project contributing to capabilities needed for the longer-term goal of sample return. This, of course, had a lot to do with poor communications between and among NASA, JPL, and the individual projects. JPL had not succeeded in communicating to headquarters officials the substantial technical differences between the Mars projects, leaving headquarters officials confused as to why Surveyor 2001 and the sample return project had such high cost projections.

But the panel was also careful to conclude that faster, better, cheaper itself had not failed (although they admitted no one could find a definition of faster, better, cheaper, so they made up their own).[58] Mars Pathfinder and Global Surveyor had succeeded, as had the New Millennium program's Deep Space 1, launched late in 1998. So JPL *could* implement such projects. But not for the aggressively low cost of the Surveyor 1998 missions, and not given the managerial relations that existed between and within NASA and JPL. In particular, JPL senior management had to take a stronger role in project management and in dealing with NASA headquarters. So the Young panel's view was that faster, better, cheaper could continue to be policy, with around one-third higher budgets and significantly increased management attention.

A *Nature* editorial did not let NASA off quite as lightly. "The political realities of the Mars programme were not misinterpreted by the JPL engineers and scientists in the trenches. They understood that they could not ask for more money, nor could they radically 'descope' their missions. Their only choice was to sigh and accept more risk. That, or resign."[59] Perhaps understanding that this was the reality of the faster-better-cheaper drive he'd created, on March 28 Goldin took McNamee, Cook, Gavit, Thurman, and their teams out to dinner at Monty's restaurant in Old Town Pasadena to apologize. The next day, he made his apology public in a speech at JPL that NASA broadcast on its national TV channel: "I told them that in my effort to empower people, I pushed too hard, and in so doing, stretched the system too thin. It wasn't intentional. It wasn't malicious. I believed in the vision . . . but it may have made failure inevitable."[60]

Conclusion

Down at the annual Lunar and Planetary Science Conference in Houston, held each March since 1970, Carl Pilcher discussed the ramifications of the demise of Climate Orbiter and Polar Lander with the assembled scientists. As many expected, he essentially admitted that Mars sample return was no longer likely anytime soon. And he confirmed that the Surveyor 2001 lander was likely to be cancelled. But the 2001 orbiter would still fly on schedule, probably. His meeting was held in sight of a giant new topographic map of Mars being shown off by a beaming David Smith, created from a little over an Earth year's worth of his laser altimeter data.[61]

The twin losses forced what Rob Manning called a "reality check" on the Mars program, and on the cost-cutting fervor fostered by faster, better, cheaper. While feeling horrible for John McNamee, Sam Thurman, Sarah Gavit, and their teams, he knew that he could never deliver the giant-sized sample return landers for anything like the $130 million budget he'd been given. Polar Lander's failure had left NASA and JPL management with no confidence in the basic Lockheed Martin design that formed the basis of the Mars landers, and Wes Huntress's successor as associate administrator for space science, Edward J. Weiler, said as much to *Aviation Week* reporters.[62] Manning's conscience was relieved, although it meant he, along with the rest of the 2005 sample return project, no longer had a job to do.

George Pace also understood the Lockheed lander design no longer had any credibility. But he saw his job as project manager with a nearly finished lander—by January 2000 nearly all the 2001 lander hardware had been delivered, and the spacecraft team was well along in putting it together—as restoring the design's credibility. So while sample return was evaporating, Pace, Roger Gibbs, and their counterparts at Lockheed were absorbing the recommendations of the failure investigations and trying to figure out how to implement them. By March they had to have a recovery plan to present to whoever wound up in charge of the reorganized Mars program.

Faster, better, cheaper had been predicated on accepting more risk, including risk of occasional mission failure. Goldin had thought that as long as most missions succeeded, Congress and the public would accept the failures as a normal cost of doing business. That seemed to be true in the defense world, where the major media routinely ignored the occasional loss of military satellites, even those costing many times what the Surveyor 1998 missions had.[63] David Letterman didn't pillory the Pentagon for a billion-dollar spy satellite lost in 1998; at a tenth the cost, Polar Lander drew far more public derision. NASA had also lost a small astronomy explorer, WIRE, in 1999, without much attention. So the public eye was less trained on NASA in general than on Mars in particular.

Goldin had realized that Mars was popular. He hadn't really understood, though, how the word "Mars" would focus national attention on both success and failure. Risk turned out not to be acceptable for the Red Planet.

Recovery and Reform

The loss of Climate Orbiter, Polar Lander, and the Deep Space 2 probes was the worst series of failures JPL had experienced since the 1960s, when its Rangers had suffered six sequential failures. Then, the Lab had been saved by the success of two Mariner missions, to Venus in 1962 and to Mars in 1965. Speaking at the town hall meeting at JPL a week after Polar Lander's disappearance, director Ed Stone reminded everyone that not everything JPL had touched recently had failed. Stardust and another New Millennium mission, Deep Space 1, had both launched at almost the same time as Climate Orbiter and Polar Lander and were operating successfully.[1]

So all that seemed necessary to Stone was a small adjustment to the Lab's management processes, and perhaps a little more money. "We can't turn back the clock to another era," he said. "I am convinced that the intelligent application of technology and process is the key to breakthroughs in exploration. I am personally committed to reshaping the way we do things."[2] In short, Stone wanted faster, better, cheaper done, well, better—not thrown out. In a later interview, he put it a little differently: JPL needed to learn to "manage in the middle," somewhere within the spectrum of possibilities lying between excessively high risk but low cost and excessive conservatism with attendant high cost.[3]

Relatively quickly, Stone set out to stop the Lab's downsizing. He had agreed to reduce the Lab's "full time equivalents" to 5,000 in fiscal year 2000, down from 7,600 in FY 1992.[4] By the end of 1999, the Lab had been shrunk to 5,344 full-time employees, and in March, Stone asked Ed Weiler, the associate administrator for space science, to raise the goal to 5,500. Stone thought another 150 full-time employees committed to the Mars program's missions would be needed to "assure mission success."[5] Weiler agreed, and for the next few years, the shrinking stopped.

Stone also decided that whatever the historical baggage carried by JPL's old Flight Projects Office—that bastion of systems engineering and mission assurance, and lightning rod for attacks by faster-better-cheaper advocates—the Lab really did need it to successfully implement flight projects. He appointed Tom Gavin to resurrect it. Gavin, who had been at JPL since 1960, in mission assurance until becoming Cassini spacecraft manager, then hired John Casani and another specialist in mission assurance, Matt Landano, to help him reestablish the division.

Gavin's goal was to reinstitute uniform design and project management standards at JPL. He thought the faster-better-cheaper advocates had unwisely thrown away many of the tried-and-true practices of the spacecraft business, and he planned to bring them back. Central to his criticism of the era was inadequate testing. He had thought it unconscionable to have loaded Climate Orbiter's final software build into the spacecraft while it was already sitting atop its launch vehicle. That guaranteed inadequate testing. It also bothered Gavin that different projects adopted different test standards for things they shared in common.[6] So he set Casani and Landano to devising what became known as the design principles and flight project practices, electronic versions of the Flight Project Office's old 600 series publications. They were to become the standard that every JPL flight project had to work to.

There were other institutional changes, too, most notably the abandonment of the Surveyor Operations Project. Its founding idea had been that a single team could operate several spacecraft at once, and no one was willing to argue that this was still a good idea after the twin losses. The development-phase project managers got "cradle to grave" responsibility for their vehicles back again. A Multimission Support Office was established to provide non-mission-specific services that all planetary spacecraft needed, such as telecommunications.

At NASA headquarters, NASA administrator Dan Goldin and associate administrator for space science Ed Weiler set out to resolve the problem of poor communications by recruiting a new Mars program manager. Shortly before the release of the Young committee report, Goldin called Scott Hubbard of the Ames Research Center, who had led the early conceptual work for the Mars Environmental Survey (MESUR) network and then been project manager for the Lunar Prospector mission. Goldin told him to fly to Los Angeles for a meeting, then proceeded to tell him how badly he thought JPL had screwed up. He wanted Hubbard to take over the Mars program at NASA headquarters to fix it. Hubbard, in turn, wanted complete authority over the program. Though he hadn't yet seen A. Thomas Young's panel report on the Mars program and its criticism of communications between NASA and JPL, he was already pretty sure that communications problems existed from his service on various review boards.[7] There had been too many people at NASA headquarters who thought they could tell JPL what to do; he wanted it clear that he would be the only person giving JPL's Mars Exploration Directorate direction.

JPL's Mars Exploration Directorate was removed from the Space and Earth Science Programs Directorate and made an independent directorate, reporting directly to Ed Stone once again. After the loss of Polar Lander, Chris Jones had replaced Norm Haynes as head of the Mars Exploration Directorate, but Jones wound up hating the job. It involved too much politics for his taste.[8] So Stone appointed Firouz Naderi in April 2000. Naderi, born in Iran, had completed a Ph.D. in electrical engineering at University of Southern California in 1976, returned to Tehran for mandatory military service, and then fled the Iranian Revolution in July 1979. He landed at JPL that September; by 1999, he had been manager of JPL's Origins Program Office for five years.[9]

All these administrative changes occurred behind a steady drumbeat of findings streaming from Mars Global Surveyor. The two most programmatically relevant were the discovery of a large region of the mineral hematite in the Meridiani region of Mars by Phil Christensen's thermal emissions spectrometer and the announcement that newly formed "gullies" had been found in a crater rim by Mike Malin's Mars orbiter camera (MOC) team. Both were considered signs of water. They provided political cover—the Mars program was doing something right, if not everything—while also suggesting some exciting new targets for exploration.

What Do We Do Now?

In the first few months of 2000, the principal challenge facing NASA and JPL was figuring out what they had to do to recover from the Surveyor 1998 failures. The institutional and leadership reforms were one component of recovery, but leaders at both institutions knew they had to get back their reputations in scientific circles and in the public arena. Flying the 2001 orbiter successfully would not really serve to reignite public interest in Mars, as orbiters were "boring" and didn't attract media attention. But the 2001 lander had no technical credibility outside George Pace's Surveyor 2001 project staff. And it was also too late to salvage the 2001 launch opportunity, even if Weiler had been willing to support flight of the 2001 lander. Surveyor 2001 project manager George Pace's team would need a good deal more test time to meet the new standards that were coming, and the inexorable planetary launch calendar had none to give them. So if the lander were to be launched, it would not be until 2003.

Beginning in February 2000, NASA and JPL leaders held what amounted to a very abbreviated competition for the 2003 Mars opportunity, without the formalized announcement of opportunities process. It started with a big "ideas" conference held in Pasadena at the end of the month. There, at a dinner meeting in the Athenaeum, Charles Elachi posited four options for 2003. First, fix Pace's 2001 lander and fly it in 2003, with Steve Squyres's APEX payload (see chapter 6). Option 2 was to fix the 2001 lander but replace the payload with a rebuild of Dave Paige's Polar Lander payload suite. This could not be done prior to 2002, so the spacecraft would have to be launched in 2002 using a Venus gravitational assist to reach the martian south pole. Elachi's third option was to fly a Mars "micromission" in 2003, sending six very small landers to Mars equipped for meteorology—a derivative of a MESUR-like Mars network study done in collaboration with France. And, finally, the fourth option was to send Squyres's Athena rover (see chapter 6) to Mars, using a Pathfinder-like airbag lander to deliver it to the surface.[10]

Option 1, flying the 2001 lander in 2003, was undesirable to everyone but the 2001 lander team for two primary reasons. A new requirement to provide guaranteed telemetry back to Earth during landing meant the removal of some of the scientific payload so mass and power could be reallocated to telecommunications and a new "flight-recorder" black box, and a lot of scientists, including Steve Squyres, didn't think the payload had been particularly interesting to begin with. A further-reduced payload was even less attractive. And second, no amount of rework on the 2001 lander would overcome the basic problem that many of JPL's engineers—particularly those associated with the sample return and Pathfinder missions—did not believe that legged landers could be made robust enough for Mars. The place was simply too rocky, and the fragility of any reasonable set of landing legs to rock damage restricted the choice of landing site to only the most geologically dull spots on Mars. It didn't help this option's chances, of course, that Ed Weiler had already announced he didn't think the vehicle had any credibility anyway.

Option 2 was intended to recover the lost Polar Lander scientific objectives before the orbits of Mars and Earth made the south pole inaccessible for the next decade or so. However, it was technically difficult. The 2001 lander had been altered from the original Polar Lander design for a mid-latitude landing site and would have to be altered again for the polar site, and doing a rebuild of the lander's instrument suite for a 2002 launch was risky, schedule-wise. Finally, the sunward trajectory required by the Venus gravitational assist would pose thermal problems for the lander, sealed as it had to be in its insulated aeroshell. In short, option 2 was scientifically and possibly politically viable but not technically credible.

The same was true for option 3. The microlanders it postulated were conceptual, not completed designs. And so were their payloads. With no extant hardware, whatever the scientific merit of the idea, the probability of a wholly new development effort making the 2003 launch window was very low, and it would be very expensive. So option 3 lacked technical and financial credibility.

The final option was Squyres's favorite. His payload engineer at JPL, a Brooklyn-born math major named Barry Goldstein who'd started at JPL in 1982 working on the Galileo spacecraft attitude control system, had put together an idea he called "Athena in Bags." It would use a "dead on arrival" version of Pathfinder's airbag lander to deliver the Athena rover, complete with Squyres's original payload from the 2001 Athena proposal (see chapter 6). With Pathfinder's successful 1997 landing, this methodology was seen as technically credible. Most of the payload had already been built for the 2001 lander's APEX suite, so new instruments did not have to be selected and built. In fact, the Athena rover itself already largely existed; a nonflight model had been built at JPL as a proof-of-concept effort during the late 1990s.

By the end of the night, Squyres thought Athena in Bags was the top contender. But his initial enthusiasm lasted barely two weeks. Goldstein couldn't make the Athena rover fit into the Pathfinder lander structure without a complete redesign; the redesign, of course, removed the all-important (at least in the environment of early 2000) "flight heritage."[11] An all-new design had no more credibility than the 2001 lander—and possibly would have less before a selection committee, as the cost and schedule needs for the 2001 lander could be defined more credibly given the large amount of already-completed hardware. So Athena in Bags went away too.

In mid-March, JPL set up an internal panel of senior engineers—Tom Gavin, John Casani, Duncan MacPherson, and some others, to review still more proposals that were percolating around various corners of the Lab. One finalist was a new, large orbiter that Squyres dubbed GavSat because it had been Gavin's suggestion.[12] The basic idea was to fly a comprehensively instrumented orbiter to Mars in 2003, instead of a lander. Orbiters performed more science than

The Athena rover in JPL's Mars Yard in 2007. Behind it is the "Scarecrow" rover, a chassis built to resemble the Mars Science Laboratory rover's mobility system.
JPL D2007-0619_D569, courtesy NASA/JPL-Caltech.

Mars Surveyor 2001 costs, as built (millions of U.S.$)

Development	Operations	Launch
165	79	53

Source: NASA 2001 Mars Odyssey Launch press kit, April 2001, www.jpl.nasa.gov/news/presskits.cfm (accessed 7 February 2012).
Note: Because the 2001 Mars Odyssey orbiter and the 2001 Mars Lander were accounted for as a single project, their costs cannot be separated accurately. The 2001 lander was eventually flown as the Mars Phoenix mission (see chapter 12). The operations cost for 2001 Odyssey's primary mission includes the science costs.

landers could, at least in the sense of comprehensiveness, and at far less technical risk. Gavin had given the idea to Pete Theisinger, who had moved from the Surveyor Operations Project to a brief stint on the sample return project, to develop into a formal proposal.

Independently, Mark Adler had also been thinking about getting a rover to Mars. Adler had come to JPL in 1992 after earning a Ph.D. in particle physics at Caltech, first working on Cassini and then as Mars program architect and on Mars sample return. Knowing that Barry Goldstein's brief effort to squeeze the Athena rover into a Pathfinder lander hadn't worked out, Adler turned the problem around and asked if there was any kind of rover that could be stuffed into the Pathfinder aeroshell. And, preferably, could also fit at least part of the Athena instrument payload. Because those instruments had already been built for the 2001 project, using them would reduce the putative project's overall risk. (Later, he would discover others had proposed this idea in 1997 and had been shot down due to the Lockheed contract awarded for the Climate Orbiter and Polar Lander missions.)

Adler approached engineer Rob Manning to learn more about the Pathfinder lander and came away convinced that the basic lander design had a lot of overlapping margins. So while many others at JPL thought the Pathfinder crew had merely gotten lucky with a not-so-good design, Adler thought it had been relatively robust. His chief concern was whether he could get a rover into the tight confines of a tetrahedron. He asked Dan McCleese about it, wondering why nobody else was taking another look at a rover mission, and remembers being told, "The only reason is that there isn't a champion. You should do it."[13] So Adler became the champion. He turned to Frank Jordan, the Mars office's advanced projects manager, for funding. Jordan gave him $100,000 for the study, and Adler recruited Manning, mechanical guru Howard Eisen, rover specialist Jake Matijevic, and geologist Joy Crisp to help. They had a little under three weeks to prepare for their first encounter with Gavin's risk review board.

Adler wanted the vehicle to carry at least the mast-mounted panorama camera and mini-TES (thermal emission spectrometer), and like Sojourner, have a mechanical arm that could place an alpha-proton X-ray spectrometer (APXS) against rocks. The arm built for the 2001 lander already carried an APXS and the Mossbauer spectrometer, so those instruments existed, too. The 2001 lander arm also had a camera, and Joy Crisp advocated changing its optics to make it a microscopic camera for imaging mineral grains. They also included one new device. Adler called it "the surface preparation device, which would clean off the dust and grind it a little bit and knock off some of that couple billion years of weathering so you could look inside the rock."[14] On Earth, geologists carry hammers to break open rocks so they can see the unweathered insides; the rover couldn't do that, but it might be able to grind off the weathered surface to reveal the unaltered material underneath.

Adler and Rob Manning first presented the concept to JPL's risk review board on April 26, 2000. The board's charter was to provide an independent evaluation of the technical and schedule risk of all the mission concepts that were supposed to be presented to Scott Hubbard and Ed Weiler on May 3 and 4. He initially wasn't going to be allowed to present the idea, though, because Firouz Naderi had already finalized the May meeting's agenda and sent it to headquarters—and Adler's idea was not on it. Naderi thought adding a new concept at the last minute would appear underhanded, particularly since Adler was proposing it as a JPL "in-house" mission. And it was well-known that JPL needed the business. So Gavin told Adler and Manning to go away. Manning remembers John Casani interrupting and saying, "Wait a minute, Tom. How do you know this isn't the safest option, and lowest developmental risk?"[15] They decided that they might as well review Adler's idea too.

Adler had named the idea Mars Geologist Pathfinder 2003, and he opened the presentation with the words "As it happens, we have a robust, proven landing system called Mars Pathfinder. Let's use it."[16] He then argued that the availability of "as built" documentation for Pathfinder would allow essentially rebuilding Pathfinder from the lander tetrahedron out—lander structure, airbags, associated support and pyrotechnic gear, and cruise stage—while developing a new rover. So the developmental risk for a landing system was already retired. The development risk for the rover, he also pointed out, had been reduced, though not entirely eliminated, for the rover by the Athena rover development and field-testing of another JPL rover technology project called FIDO.

As presented, the project would provide a 30 sol surface mission, arriving at Mars in January 2004 after an early June 2003 launch. The 2003 launch opportunity enabled a Delta II rocket to put a 1080 kg payload onto a Mars trajectory, and the team's estimate was that their spacecraft concept would be about 890 kg. Of that mass, only 100 kg belonged to newly designed hardware. The rest was in "build to print" equipment, so the risk of exceeding the launch vehicle's capacity seemed low, and Adler had allocated half the excess to the launch vehicle (essentially as an additional mass margin "bank") and the rest to the spacecraft.[17]

Adler saw his principal risk as getting a project start decision made by June and the project established in time for a September preliminary design review.[18]

That would be an extraordinarily rapid decision by NASA, but there was no room for delay. A June 2000 project start only left 36 months until the July 2003 launch period, and, as Pathfinder had, Adler wanted an 18-month assembly, test, and launch period (ATLO) starting about February 2002. All of the major parts had to be built by then, which would not be easy. He also foresaw some risk in not being able to get the key Pathfinder personnel away from their current commitments, and in not being able to replicate the Pathfinder project's experiment in co-location. In the years since Pathfinder, the Cassini-Huygens project's mission operations team had occupied Pathfinder's old office suite, and as they were on a ten-year primary mission ending in 2007, they were not going to be kicked out. And JPL didn't have another similar space.

Amazingly enough, Gavin's board essentially agreed with Adler that his idea was low risk. On a scale of one to four, with one representing the highest risk, they gave the risk of missing the launch period the proposal's only two. They considered mass, software, science, and technology readiness risks low (all fours). They also considered cost risk low. And probably most importantly, they assigned mission success the lowest risk.[19] The board did have some criticisms. Adler had assumed that the data return would be via relay from an orbiter, but that was not considered a safe assumption. The 2001 orbiter and Mars Global Surveyor could both have failed by the rover's January 2004 arrival, so direct-to-Earth communication was necessary. They also thought the volume assigned to the rover's electronics might be optimistically low and wanted to see a better scrubbing of its design.[20]

After this first meeting, Adler and Manning got approval to move forward to the May 3 presentations, and Adler finally called Steve Squyres, whose instruments he needed for the concept, and told him what he was proposing.

The following week, Adler and Manning presented the concept to Hubbard and Weiler, in a preliminary competition with George Pace's lander and with Pete Theisinger's 2003 orbiter. At this meeting, Pace tried to emphasize the reality that while Adler's newly renamed Mars Mobile Pathfinder was a design concept, with all the attendant development risks, his 2001 lander was already basically built. Pace embellished his PowerPoint presentation with photographs of the built-up lander, cruise stage, and aeroshell in Lockheed's clean room. The lander's ATLO was 25 percent complete, the flight software was already integrated, and he had 400 hours of system-level avionics testing already completed. His team had passed a return to flight review for the lander on February 28, after fixing the EDL software's leg extension problem, removing Marie Curie from the payload in order to add a flight data recorder, and switching to a larger Delta

7925.[21] So he could make a strong case that his development risk had already largely been retired. The 2001 lander, unlike Mars Mobile Pathfinder, had very little risk of missing the 2003 launch window.

Adler's advantage in this competition with Pace was twofold. He knew NASA leadership had no confidence in the Lockheed lander, regardless of Pace's efforts to convince them it really would work. And he knew that the scientific community didn't really support Pace's mission either. Squyres, the principal investigator for the payload on Pace's lander, was much more interested in the Mobile Pathfinder. A geological rover was what he'd designed his payload for in the first place, so that's what he really wanted.

Richard Zurek and Pete Theisinger presented the orbiter concept. Their Surveyor Orbiter's primary goal was to recover the Mars Climate Orbiter payload's science and to provide relay services for future landers. It would not be a simple "build to print" of Climate Orbiter, though, as Theisinger, Zurek, and even Dan McCleese sensed little interest in such an unambitious mission. McCleese had proposed a miniaturized version of his pressure modulated infrared radiometer (PMIRR)—only about a quarter the original size, freeing space, mass, and power for other instruments. Lockheed, which would provide the spacecraft bus, had also chosen not to propose the Climate Orbiter bus but a slightly modified Mars Global Surveyor bus. This eliminated the navigation difficulties and also provided more power and space.

The availability of more vehicle resources led to questions about what to do with them, both from the review committee and among scientists serving as informal advisors. There was no obvious payload for their Surveyor Orbiter. Zurek had made an informal poll of the planetary science community to see what instruments might be available on the three-year schedule the project had and found only a few things that were already mostly built, or were spares left over from other missions. There was an ultraviolet spectrometer for atmospheric science at the University of Colorado, developed for the Contour mission, and the spare of a Cassini instrument very much like the visual and infrared mapping spectrometer that had been deleted from Mars Observer. But not much else.

Zurek also remembers everyone wanted a more powerful camera than Malin's Mars color imager (MARCI) on Climate Orbiter had been, but they couldn't agree on how much more powerful. Many scientists wanted a medium-resolution camera that could provide context for the higher-resolution images from the MOC on Mars Global Surveyor. Other scientists, and most engineers, wanted an even higher-resolution camera than the Mars orbiter camera for both stratigraphy (study of geologic layers) and for landing site selection and certification. But a bus similar to the Mars Global Surveyor could not support both cameras. So while Adler was able to present a payload for a pretty specific mission, Theisinger and Zurek presented a set of options that could be flown in 2003.

A key pitch for both proposals was their scientific possibilities. Joy Crisp and Rich Zurek presented the two "science stories," focusing on how each contributed to the "follow the water" strategy to identify sources of water on Mars. The orbiter's key advantages were its global coverage, its recovery of the final Mars Observer instrument, PMIRR, and its high-resolution imager's ability to look at the recently discovered martian "gullies" in greater detail. It might also have the ability to detect surface deposits of hydrates and carbonates remotely, a capability deleted from Mars Observer back in the late 1980s.[22] Scientifically, it was a pretty compelling mission.

The rover mission's scientific advantage was "ground truth." Crisp concentrated on the discovery of hematite at the surface in the Meridiani region of Mars. Phil Christensen's TES instrument on Mars Global Surveyor had mapped a huge region where the mineral was right at the surface; part of that region was low-lying enough that the Pathfinder-type entry, descent, and landing system could land on it. Hematite on Earth forms in water, so its detection suggested that this region had been wet at one time. For that reason, it had been the primary landing site choice for the 2001 lander, and it was also well suited for the Mobile Pathfinder—it was relatively flat and seemed to have low rock abundance. The Athena payload had been designed to look at the mineralogy of rock types that tended to form in water, so the rover could confirm what the TES data seemed to say.

Crisp also had to explain to the audience that the rover mission couldn't go to the recently discovered "gullies" site for a couple of reasons. One was power; the gullies had been found in the southern mid-latitudes, where the vehicle would receive much less power from the Sun. Another was that the rover probably wouldn't be able to land close enough to where the gullies had been spotted to drive there in the short mission lifetime.[23]

The next morning, at a meeting in Ed Stone's office at JPL, Ed Weiler asked the Mars program's new chief scientist, James Garvin, which of the mission concepts he'd heard the previous day was the most scientifically desirable for the 2003 opportunity. Garvin, who had been part of David Smith's Mars orbiter laser altimeter team on Mars Global Surveyor, told him, "Well, it's pretty apparent. The MGR [Mars Geological Rover] [is] the most ambitious and highest science-value mission we could fly." And Garvin remembers Weiler responding,

"Well, that means you're willing to risk the Mars program, your program, on probably the riskiest of the things we could do."[24] The group discussed the subject a while longer and concluded that a bit more study was in order before making a decision. On May 13, Weiler and Hubbard approved two-month studies of both the rover and the Surveyor Orbiter.[25] Pace's lander was clearly dead in the minds of NASA leadership, and while NASA never sent him a formal stop work order, Pace eventually negotiated an agreement with Lockheed to store the lander hardware. It might be used on a later mission, when the Washington environment improved. His lander engineers were transferred to the orbiter, to help that team with the greater workload imposed by the need to respond to increasing test standards and reviews.

Even though a formal decision by NASA officials hadn't been made yet, JPL's leadership decided in favor of the rover immediately. The Mars Geological Rover was obviously the best choice for JPL because much of the work would be in-house. It also gave the Lab more control over its own fate. It would not have the risk involved in systems contracting. So Tom Gavin had approached Mark Adler after the May 4 meeting and asked if he thought Theisinger would be a good choice as project manager for it. Adler himself was not qualified for a project manager's spot, having never delivered flight hardware, but he thought Theisinger was a good choice. So did JPL's executive council. Theisinger remembers that at noon on May 5, Gavin told him, "You've got the Rover project. [Richard] Brace is your deputy. [Richard] Cook is the flight system manager. Goldstein is the Rover manager."[26] Theisinger was elated. A baseball fan, he remembered it was like being asked to pitch the last game of a World Series. "No" was never an option for him.[27] It promised to be the hardest thing he, or JPL, was likely to do any time soon. David Lehman, who had recently completed the Deep Space 1 mission, took over the orbiter study.

NASA headquarters scheduled the final "shootout," as participants called it, between the two proposals for July 13 and 14, in Washington, DC. Before that, both missions went through a preliminary mission system design and cost review at the Pasadena Hilton. By this time, NASA had selected the orbiter's instruments, choosing a "high resolution imager" proposal that offered a resolution better than 0.9 meters/pixel; the University of Colorado's ultraviolet spectrometer; and a near-infrared spectrometer from the Applied Physics Laboratory in addition to the reflight of the two Climate Orbiter instruments, MARCI and PMIRR.[28]

Each drew the board's concurrence that the project's risk was low and that they had adequate preliminary designs and been properly costed (around $397 million for the orbiter and $411 million for the rover mission). So when the two groups of advocates assembled in Washington, they had already been vetted.

The key discriminating factors in the shootout were public appeal, orbital mechanics, and perceived risk. The original Pathfinder lander had been a tremendous public success, while the public had largely ignored the highly scientifically successful Mars Global Surveyor. That mattered, as NASA was trying to recover its public image. Orbital mechanics also strongly favored the rover proposal. Because of the relative positions of Earth and Mars in 2003 and 2005, the rover could go to Mars in 2003 but probably not in 2005. It would almost certainly be too heavy for a Delta rocket in 2005. Against that, the orbiter seemed a lower-risk effort. As Squyres put it later, the orbiter was "safe and boring."[29] The Mars Geological Rover was the opposite: risky and exciting.

The shootout took place in front of a committee of senior NASA and aerospace specialists, including Jim Martin and Glenn Cunningham. The committee was split about which to support, and Cunningham recalls that Martin made an impassioned plea for the rover, the last chance for a high-profile landed mission for many years. It was "absolutely imperative that we go back to the surface."[30] The committee voted support for the rover after all, with the caveat that they thought the project was underbudgeted. Weiler then told Dan Goldin, who initially questioned the choice. Goldin had assumed that the orbiter, with its big, high-resolution camera, would be chosen to investigate the gullies that Mars Global Surveyor had found. So the need to convince Goldin they'd made the right choice delayed the selection a few days.

Goldin then delayed the selection further by asking if JPL could build two rovers. Goldin understood that the biggest risk to the mission was in landing. If the vehicle worked perfectly but landed on a sharp rock or at the bottom of a steep crater, it wouldn't matter how good a job JPL had done engineering and testing it. The mission would still fail. The only way to mitigate this risk was to build and launch two of them. Scott Hubbard called Firouz Naderi on Friday, July 21, and told him that JPL had to give him a cost estimate for including a second rover by the end of the day in Washington, giving, Squyres, Goldstein, and Adler less than an hour to come up with it: $665 million.[31] It took more time for NASA to "find" the money necessary to pay for this. After a rebuff from the White House, Goldin "passed the hat" around to NASA's other operating directorates to accumulate the additional funds.[32] Most of it ultimately came from the agency's Human Exploration and Development of Space Enterprise. Meanwhile, headquarters abruptly cancelled a July 24 press conference intended to announce the decision, sparking media interest.[33]

The two-rover option also got a review by JPL's Governing Program Management Council. There were some obvious questions about the wisdom of trying to build and launch two Mars spacecraft in the scant 34 months left before the launch period started. Did JPL have adequate manpower and facilities necessary for simultaneous assembly and testing? Could the Deep Space Network handle the communications needs of two spacecraft in the same field of view? Would the surface operations teams be able to manage two semiautonomous surface vehicles in different martian time zones, having never operated a single one?

Theisinger found a strong advocate for the two-rover option in an unlikely person, Tom Gavin. Gavin, who was widely known as a rigorous conservative in engineering practice, supported the idea based on his experience in the 1960s and 1970s, when it was routine for JPL to build two identical spacecraft for a given mission.[34] While the second spacecraft required additional manpower to build, the reality that the second vehicle was essentially a complete spare for the first substantially reduced risk during the test program. If a piece of electronics failed during testing, its twin could be substituted and the test could resume while the hardware was repaired. The entire test program wouldn't be delayed by a single failure. Further, several "tiers" of testing are done on spacecraft. System-level tests required the entire vehicle for testing, while assembly-level testing needed parts of it. With two vehicles, these two testing tiers could proceed in parallel. And some of the system level testing did not have to be repeated on the second spacecraft. The net result would be a test program that was shorter in calendar terms—parallel testing could get the test program finished more quickly. The more refined cost estimate at this review was raised to $685 million.[35]

At the review on August 7, considerable opposition to the two-rover option arose from members of JPL's executive council. Not every member believed the argument that two vehicles could be tested more quickly than one. Further, not every member thought JPL had enough people available to support the test program.[36] Here the issue was simply that JPL no longer operated in an environment in which it had only a single project going to which it could devote the entire Laboratory. It had other flight projects in progress that also required manpower and facilities to stay on their own schedules. And the Lab had been shrinking for the past decade. It was a smaller institution than it had been. For all these reasons, it was not clear to some of the council members that JPL could handle two rovers. Finally, while Theisinger's team had been able to put together a coherent plan for building and testing two vehicles in the time available, they had not assembled a credible operations plan for two.

Mars Exploration Rover payload

Instrument	Providing institution
Panorama camera	JPL
	Ball Aerospace
Mini-TES (thermal emission spectrometer)	Arizona State University
	Santa Barbara Remote Sensing
Microscopic imager	JPL
Alpha-proton X-ray spectrometer (APXS)	Max Planck Institut für Chemie
Mossbauer spectrometer	Johannes Gutenberg-Universität
Rock abrasion tool (RAT)	Honeybee Robotics

Source: Adapted from Squyres et al., "Athena Mars Rover Science Investigation," and "Athena Science Team Members," http://athena.cornell.edu/the_mission/scientists.html.

Barry Goldstein remembers that after the review committee's recommendations had been made, JPL's deputy director, Larry Dumas, gave a little speech attacking all the negativity in the room. Dumas reminded his staff that JPL had once seen itself as a place where everything was possible. Where did the "can't do" attitude come from? JPL's business was in doing things that were hard. Dumas was a little embarrassed about the vehemence of his speech years later, but Ed Stone agreed with him.[37] The risk to the Lab of taking on two was worth it.

The logic behind the two-rover option was persuasive in headquarters, too, and Scott Hubbard recommended to Weiler going ahead with it. Goldin approved it on August 8.[38] On August 10, the mission was announced at a press conference at NASA headquarters under the name Mars Exploration Rover 2003.[39]

Is There a Better Way to Land on Mars?

While the Mars program office was trying to finalize the 2003 mission, it also initiated an effort to figure out if there was a better way to land on Mars. Beyond the lack of credibility that Lockheed's Polar Lander design had after its loss, the sample return project's effort to scale the Lockheed legged lander up to the 1,200 kg needed to carry out the mission had left many of the project's engineers cold. The legs were fragile, and the need to squeeze them into an aeroshell left them too short to provide much ground clearance, making the lander inherently vulnerable to rocks. The old Viking lander design that Lockheed used as a basis for Polar Lander had itself been derived from the old Lunar Surveyor landers, and the Moon wasn't Mars. By early 2000, JPL engineers were convinced that the legged lander concept was a longstanding case of "inappropriate heritage."

The effort to conceive a better Mars landing was embedded within a Mars Exploration Directorate–initiated sample return architecture review. Set up in February 2000, the review's purpose was actually to protect the sample return architecture from being reviewed externally—perhaps by the Johnson Space Center, which everyone knew wanted to be involved in any Mars sample return but which had little role in the sample return effort as it was being implemented in 2000. So the review was structured to prevent major changes to the architecture, which might then be used as justification for reassessing institutional roles. Separate teams were to review each component of the mission—a team for the Earth return vehicle, a team for the lander, a team for the Mars ascent vehicle, and so on—so that they couldn't change anything major. This became known around JPL as the "bubblehead study."[40]

Brian Muirhead was chosen to lead the bubblehead team looking at the lander architecture, along with Tom Rivellini and Pathfinder veteran engineer Dara Sabahi. Muirhead decided that his tasking was dumb, though. Like Manning and Sabahi, he didn't think the Lockheed legged lander made any sense when scaled up to the size necessary to carry out sample return. So he ignored the charter and had his team re-envision the lander entirely.

This group devised some new options for the lander. Back in the 1960s, there had been some investigation of what was called a "pallet lander." The pallet lander concept replaced the legs with a structure that would absorb the lander's impact energy by crushing, exactly like the "crumple zones" on passenger cars. One advantage of the pallet approach was that the lander deck would be closer to the ground than it would be on the legged lander. This would make disembarking the large sample return rover easier, and with less potential for a "tip-over" disaster.[41] There were also two clear disadvantages. The lander still had pressurized propellant tanks underneath its top deck, and these couldn't be punctured or crushed, so the structure had to be designed to protect them too. And because the structure was destroyed on impact, the test program necessary to verify its performance under a wide variety of Mars-like environments was going to be expensive.[42] But the team still thought it better than the legged lander.

The other option became known as the 4pi airbag lander. This concept was derived from the Mars Pathfinder approach of having a rocket-equipped backshell zero out the vertical velocity before dropping the lander on the surface. Like Pathfinder, the lander would be protected by a set of airbags on impact. Unlike Pathfinder, the airbags would also provide the self-righting system. (Pathfinder's self-righting was performed by the metal lander petals, not by the airbags). And instead of using solid rockets in the backshell, the backshell would be equipped with a throttleable liquid-fueled rocket. A throttleable rocket would enable more precise control over the touchdown velocity, so the airbags would not have to be as robust (and heavy) as Pathfinder's had been.

Muirhead's team also decided to remove the Mars ascent vehicle from the lander and put it on the sample collection rover—a big change, as it meant a much larger rover. They did this because this team, which was composed of mechanical engineers sensitive to things like allowable mechanical tolerances, thought that the task of getting the rover back onto the sample return lander, in the proper mechanical alignment to deposit the samples into the ascent vehicle's receiving mechanism, was much too hard to do from 150 million miles away. If the rover simply carried the sampling apparatus and ascent vehicle around bolted to its own deck, this problem went away entirely. It could all be aligned while still sitting in JPL's clean room. In an odd echo of the past, they wound up with a rover plus ascent vehicle concept with a mass of around 800 kg, the same as the sample return rover from Donna Shirley's studies in the late 1980s.

Their efforts never came to anything final, though. Due to the changing leadership in the Mars program during March and April, the bubblehead study was redirected into a different set of studies. Barry Goldstein got an assignment to look at establishing a new line of competed Mars missions while James Graf took charge of a "large lander study" that absorbed the work (and membership) of Muirhead's bubblehead team. With sample return no longer in the cards, the large lander study's purpose was to figure out how to land a large enough payload to carry out sample return without actually doing the sample return. On the recommendation of John Casani, Graf recruited Lockheed's Steve Jolly to the lander team so that Lockheed's point of view was also represented.[43]

This large lander study team further developed the pallet and airbag concepts from the bubblehead effort while a subgroup argued for a third approach. Jolly, Rob Manning, and Dara Sabahi thought the team was not taking a radical enough approach. Sabahi explained later, "You have a rover with a very capable mobility system. We should just suspend the rover on the backshell and land with the wheels."[44] If there's no lander or airbags to egress from, there's no egress problem. The rover's wheels will roll off most rocks. Because the rover's center of gravity would be low, it wasn't as likely to tip over on a high slope as a higher-standing legged lander carrying a rover. To this little troika, the most obvious solution was just to get rid of all the impediments to the rover.

They didn't win the argument with the larger group, though. There were two good questions that they couldn't resolve within the study period. The first was whether the backshell's propulsion system could control the terminal velocity precisely enough to assure a touchdown speed of no more than about 2 meters/second, the most a reasonable rover suspension system could handle. This was principally a question of how good Doppler radars really were. Rivellini explained later that Polar Lander's problematic radar had raised a lot of questions about the performance of radars on Mars, questions that simply couldn't be addressed without much more research.[45] The second issue was one of controllability. How would the backshell keep control of a rover dangling on a cable 10 or 20 meters below it? If the rover started to swing on the end of its cable, could a reasonable control system stop it before touchdown?

Without answers to those questions, the team shelved the wheel-landing idea and kept the pallet and 4pi airbag lander concepts alive. The Mars program office ultimately funded a technology-development task to build and test proof-of-concept models of the two.[46]

Road Map for the Near Future

Separately from the short-range effort to figure out what to do in 2003, Hubbard and Naderi had also sponsored an effort to make a "road map" for Mars exploration over the next decade. The roadmap was designed to overcome the technological incoherency of the Surveyor program's mission set—the program's inability to select missions that would build capabilities needed for the long-sought sample return. The road map swept in the "Surveyor Orbiter," baselined for a 2005 launch. The pallet lander became the basis of a "Mars Smart Lander" project that would launch in 2007, along with a "Telecommunications Orbiter" developed jointly with the Italian national space agency. Telecom Orbiter was justified by the high data rates that Smart Lander was supposed to be capable of, and by the program's intent to carry out sample return in 2011. The pallet lander would be the basis for the large sample return lander, and part of Smart Lander's purpose was to demonstrate new precision landing technology. The new program also envisioned starting a series of competed missions called "Mars Scout," and selecting the first one in 2003, for a 2007 launch.[47]

Mars Surveyor Orbiter was the first of these road map projects to be established. Like the other orbiters in the Mars program, it was to be done as a systems contract, but not necessarily with Lockheed Martin's Denver operation. NASA leaders wanted the contract rebid. And since the launch date of August 2005 was now comfortably far in the future, they also planned to solicit instruments via the normal announcement of opportunity process. When the orbiter project was established late in 2000, JPL assigned Jim Graf as project manager, and Graf approached Rich Zurek and asked what he thought scientists' biggest desire for future Mars exploration was likely to be. Zurek, he remembers, told him higher data rates.[48] During the 1990s there had been an enormous improvement in both computing technology and digital imaging technology; it was possible by 2000 to build a camera that would simply overwhelm the ability of Mars Global Surveyor's telecommunication system to send its imagery back to Earth. The amount of data that could be returned from Mars was limited by the power available in Mars orbit, the size of the spacecraft's antenna, the data compression scheme in use, and the availability of time on the Deep Space Network. So in drafting the request for proposals, Graf had his project team specify a dramatic increase in data rate for bidders' prospective spacecraft.

From the standpoint of sample return strategy, the orbiter's function was to enable safer landings through very high-resolution imaging. MOC's resolution was not high enough to image rocks of sufficient size to endanger a lander, but by the early 2000s it was possible to build such a camera. It simply needed a big spacecraft to carry it (and a high data rate to get its images back to Earth). The alternative, designing landers to survey their own landing sites during the descent phase and maneuvering to avoid rocks, was possible but difficult. Merely testing a spacecraft that would be able to decide on its own where to land would be challenging, because under JPL's rules, every option available to the spacecraft, and every combination of options, had to be tested. The test program would expand exponentially as the spacecraft's flexibility increased. It would be far simpler, and less expensive, to have humans on Earth look at high-resolution images of landing sites and verify ahead of time that there were no hazards to avoid. So the orbiter's strategic purpose was less scientific than engineering support. The mission gained the name Mars Reconnaissance Orbiter, reflecting its strategic function as an intelligence-gathering asset.

But the orbiter also had some scientific purposes. One was simply to try to recover, again, Dan McCleese's atmospheric temperature measurement through a third reflight. For this attempt, McCleese and Tim Schofield proposed a revamped instrument that was smaller than its predecessors and consumed less of the spacecraft's resources. This new version was named Mars climate sounder.

NASA also owed Mike Malin another flight for his MARCI camera, lost on Climate Orbiter. Another purpose was to complete the mineralogical survey that had been descoped when Bill Purdy had removed the visual imaging and mapping spectrometer (VIMS) from Mars Observer's payload. VIMS had promised the ability to identify minerals whose spectra could not be identified by instruments like TES or the thermal emission imaging system (THEMIS), including many carbonates, a class of minerals that generally form in water. A third purpose was to carry to Mars a radar for higher resolution subsurface mapping of Mars than the radar already selected for the European orbiter, Mars Express.

The roadmap also included a "Mars Scout" mission, pegged for a 2007 launch. The Mars Scout missions were to be competed the same way Discovery program missions were, but they would be restricted to Mars and administered by NASA's Mars program. The selection process for the mission was scheduled to start in 2001, with selection planned for mid-2003. The program's goal was to have a Mars Scout launch every other launch period, about every fourth Earth year.

Mars Surveyor 2001

For the first half of 2000, George Pace's team had focused their efforts on their orbiter, to keep it on schedule. They were already fairly sure that the lander would not fly in 2001—Ed Weiler had said as much in a presentation in March. But the 2003 launch opportunity was still available for the lander, and Pace and his spacecraft manager, Roger Gibbs, had a small group of people start working on figuring out how to fix the lander. Since the leg extension problem had not yet been identified when they'd begun reevaluating the lander's possible problems, they'd done, in essence, what John Casani's failure review panel was doing: work their way through fault trees. Gibbs explained later that he had provided a few of his engineers to support the investigation and had been surprised that within a couple of days' work, they reported back, "We have six or eight possible causes, and we have a leading candidate that looks pretty strong."[49]

To Gibbs, that was terrible. "I thought, here is a mission that has gone through the development process which is supposed to deliver you a product, a flight system, a ground system, that are, through a rigorous process of verification and validation, assured to have a high probability of success. And yet a handful of smart people ask questions, and within 10 to 20 working hours they came up with a set of failure mechanisms that were credible. And it struck me that those two things should not have gone together."[50] The problem should have been far more difficult to unearth.

So Gibbs had had the lander team start building and reviewing fault trees for the vehicle even before Polar Lander had vanished. The fault tree analysis uncovered gaps in the project's test coverage that were still correctible. One example was in the catalyst beds for the vehicles' hydrazine thrusters. The Surveyor 1998 project had accepted an industry standard lifetime for cold starts on these thrusters without doing their own cold-start testing; given the post-loss scrutiny the thrusters received, that approach was no longer acceptable. Gibbs had a batch of thrusters tested to ensure they really did meet the project's requirements.

For the orbiter, both the project staff and the program officials believed that since Climate Orbiter had operated well throughout its flight—Gibbs remembered it "went right where we sent it, too deep into the atmosphere"—the design didn't need much rework for the 2001 flight.[51] The navigation failure had been on the ground, not on board the spacecraft, and the navigation process had already been fixed for Polar Lander. The program level recommendation was to fly the 2001 orbiter essentially unchanged from the 1998 design except for the instrumentation but with an improved radio beacon to allow more accurate tracking. As a consequence of Climate Orbiter's loss, the Deep Space Network restored a capability called Delta differential one-way ranging (Delta DOR) that provided an independent check on spacecraft "plane of sky" locations, and Delta DOR needed modifications to the vehicle transponder.

The Mars program imposed a red team on the Surveyor 2001 project to "aggressively review" the spacecraft's development history and upcoming plans.[52] A major goal of faster, better, cheaper had been reduction of formal reviews—Tony Spear had tried to fight these off on Pathfinder, with unexpected help from Jim Martin—because reviews were expensive and time consuming. The project brought in JPL's technical specialists review the orbiter's design, development, and testing program. Pace had already thought John McNamee had gone too far in minimizing reviews on his Surveyor 1998 missions, so he wasn't in principle averse to the increased oversight. But the review burdens imposed on the small project team after the loss of Climate Orbiter strained their ability to keep to their schedule, Gibbs remembered years later. The project had been staffed for the reduced-review era of faster, better, cheaper and didn't get many more people to deal with the increased demand for reviews. Because the lander was being put in storage, Pace put his lander engineers to work responding to the red team review. Pace ultimately made a bit of a joke out of the review onslaught by inserting a slide titled "Review Your Way to Mars" into the mission readiness review package.[53]

All the reviews resulted in one significant change to the orbiter. The 2001 orbiter was designed without the check valves that had been suspect in the Mars Observer design. The 1998 and 2001 orbiters were designed to operate in "blow-down mode," with the oxidizer and pressurant tanks isolated by pyrotechnic valves from the fuel lines until their Mars orbit insertion firing. These valves had no history of leakage, so no one on the two projects believed they needed check valves in addition to the pyrotechnic valves. They would serve no purpose in the system design. And, of course, by this time the team knew that Climate Orbiter's propulsion system had worked properly—it was clearly not the cause of the loss. The new propulsion review panel wanted the project to add check valves anyway. The propulsion system was already built, though, and cutting open the system piping to add the valves presented considerable technical and schedule risk in itself.

Gibbs didn't want to make the change. So he went to Tom Gavin to complain and got a lesson in NASA politics. He remembers Gavin telling him, "Roger, from an engineering point of view, you may be right. But you cannot win this argument. There is no way that JPL, nor NASA, will let this mission go when there are technical specialists, including some non-JPL people who are propulsion people, who say this is a credible failure."[54] The recent failures had made both institutions risk-averse, so a design that had been acceptable in 1998 no longer was. Gibbs and Steve Jolly had the check valves installed.

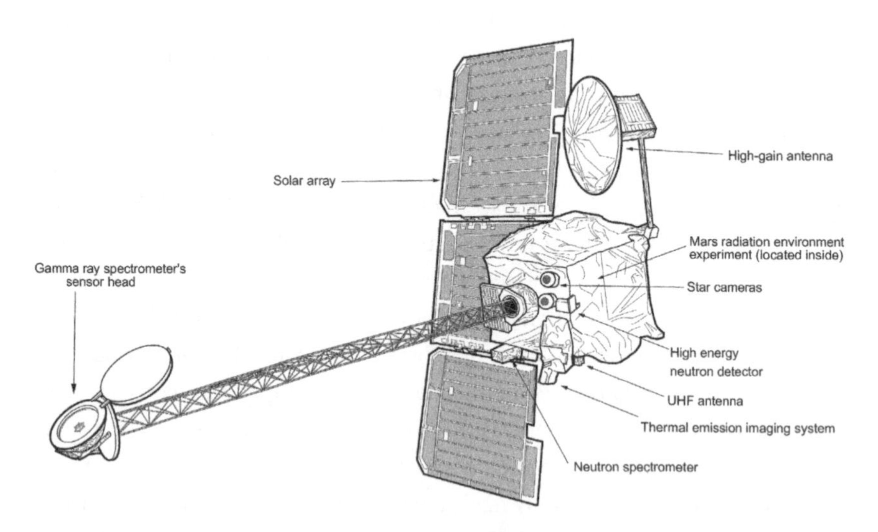

2001 Mars Odyssey spacecraft. Note the basic similarity to the Mars Climate Orbiter (see chapter 6) due to its derivation from that earlier design.

The removal of the 2001 lander mission provided one benefit to the project, too. Originally, the project team had intended to launch the orbiter from Vandenberg Air Force Base in California and the lander from Cape Canaveral. This was due to a more favorable trajectory available from the Vandenberg site during the launch window using the Delta 7425. But Vandenberg was used exclusively by Earth orbit missions, not for planetary launches. It needed upgrades to handle the launch, which of course cost money. When the lander was mothballed, the team decided to use the lander's launch vehicle, a higher-capacity Delta 7925, to send the orbiter from Cape Canaveral instead. The vehicle had to fly an unusual "dog-leg" trajectory to achieve the inclination necessary for the desired Mars orbit, but the additional thrust from the more powerful rocket made it possible.[55]

Phil Christensen's THEMIS instrument caused the project some trouble late in its development. Christensen had pushed the contractor to let him bring the instrument outside to take an image of the team framed against the mountains. The company was pressed for time on the project and didn't really want to do it. But they relented, and the resulting image revealed a substantial problem. Christensen remembers: "So we take this beautiful infrared image, but when we looked at it, the scene was fine but the people were black, just like ghosts. You could see the silhouette, but there was no information at all. The people were just black. And we thought, oh my God, what's this? We played around with it for a day. We called it 'the people problem,' because what's wrong with the people? Why can't THEMIS image people? Everything else literally was fine."[56] Humans are much warmer than their surroundings and should have been glowing in the image—not black.

It took them five months to track down the problem. The heat radiating from people was saturating the instrument and shouldn't have been. On one circuit board they found some bad solder joints, which were producing processing errors. The joints had been cracked during vibration testing of the instrument intended to simulate the launch environment for the spacecraft, revealing that the board had not been stiff enough. The joints might have failed completely during launch if they hadn't been discovered during the unscheduled imaging test, rendering THEMIS useless. The redesign and repair meant THEMIS was delivered to Cape Kennedy barely two weeks before launch.

Bill Boynton's gamma ray spectrometer narrowly avoided its own calamity. When his team delivered the instrument to Kennedy Space Center it still needed some of its thermal vacuum testing, so they had brought along a portable testing chamber cooled by liquid helium. The helium compressor was itself water-cooled,

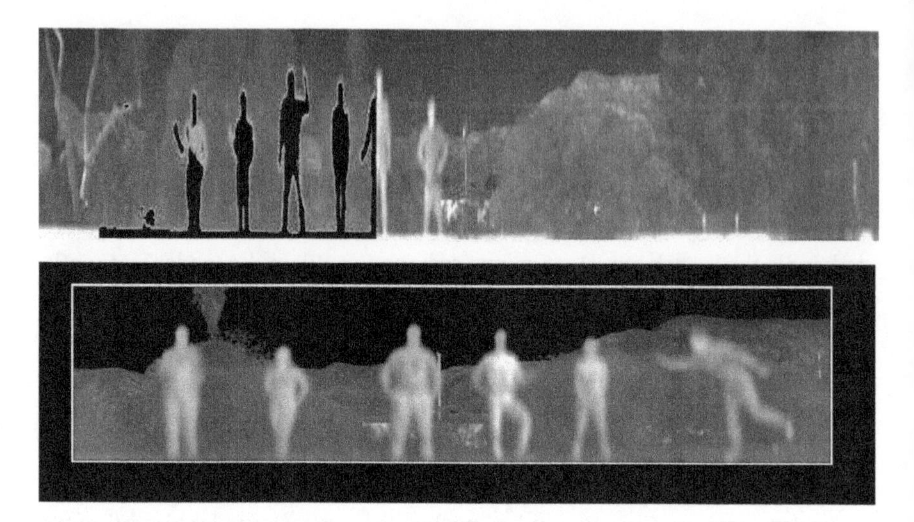

The "people problem" image from the thermal emission imaging system (THEMIS). The upper panel shows what Phil Christensen's team saw when they first tested their instrument, while the lower panel shows what they should have seen and did see once the instrument was repaired.

Top, courtesy Phil Christensen / Arizona State University; *bottom*, JPL image P50873Ac, courtesy Arizona State University and NASA/JPL-Caltech.

with water provided through connections to the building water supply. When Boynton's crew tried to hook the water supply up, they were told they had to use fittings provided by Kennedy Space Center, and they agreed. One evening while the team was off celebrating, a member of the project's outreach staff who had set up a Webcam to stream the preparation work noticed on the video feed a brown stain spreading out on the floor beneath the spectrometer. He called George Pace and told him about it. One of the fittings had failed, and the cooling water was leaking out. So a couple of Boynton's team members spent the weekend coordinating a cleanup. Fortunately, the instrument hadn't warmed up enough to be damaged.[57]

Throughout the last year of the orbiter's development, the pressure to succeed was enormous. The orbiter had to work. Pace recalls that he was asked "at every press conference about the feet to meters problem. Was the mission going to work? The highlight came when I was taking a shuttle to catch a plane for the pre-ship review. I struck up a conversation with the driver and when he asked what I did I told him I was working on the next mission to Mars. His response: 'Oh, that's got to work.'"[58]

Conclusion

George Pace's team launched their orbiter on April 7, 2001. Pace retired not long after, and Roger Gibbs became project manager. The spacecraft was named 2001 Mars Odyssey in honor of Arthur C. Clarke's famous Jupiter-bound vehicle. It had an uneventful flight to Mars and completed aerobraking into its initial mapping orbit on February 19, 2002. It didn't take very long before it had produced spectacular results. The first two months of data from the gamma ray spectrometer revealed that there was water ice bound into the martian regolith nearly everywhere poleward of about 45° latitude, confirming a nearly generation-old hypothesis from the Viking mission.[59] Much of Mars was covered by kind of dirty red permafrost, a finding that validated the Mars program's focus on following the water.

By the time Odyssey got to Mars, Ed Stone had stepped down as JPL director, having reached his 65th birthday at the end of 2000.[60] Caltech's board of trustees chose Charles Elachi, formerly head of the Lab's Space and Earth Science Projects Directorate, as his successor. Stone returned to his old job as professor of physics at Caltech and began a career as advocate and fundraiser for a new generation of large aperture ground-based telescopes. Elachi continued the reforms Stone had set in motion in 2000, elevating Gavin's new Flight Projects Office to a full directorship reporting directly to him. Elachi also replaced much of the Lab's executive council and restored weekly senior management reviews, both efforts designed to reinforce technical oversight of the Lab's activities.

The initiation of the Mars Exploration Rover mission in August 2000 set JPL on a high-risk enterprise. The project team represented a conscious effort to try to blend the Lab's traditional technical conservatism—both Theisinger and his deputy, Richard Brace, were seen as traditionalists, not as faster-better-cheaper advocates—with the best parts of the more risk-taking subculture that had developed around Pathfinder. Most, though not quite all, of the Pathfinder gang wound up working for Richard Cook on the spacecraft development. But Mars Exploration Rover proved a far larger undertaking than Pathfinder had been, and more and more of JPL was pulled into it as the Lab struggled to meet its commitment to NASA.

Finally, it had not been the intent of either Dan Goldin or Ed Stone to put an end to the faster-better-cheaper era, but the "road map" missions that emerged from Hubbard and Naderi's planning did so. Their new program did not merely repair the underfunding and understaffing of the defunct Mars Surveyor program while maintaining missions of the same scale. The progression of Mars Exploration Rover, Mars Reconnaissance Orbiter, and the Mars Smart Lander reflected an era of expanding technological and scientific ambitions—and of increasing resources. That era would last only a few years, but some great successes could be achieved before shrinkage began again.

Margins on the Final Frontier

When Mark Adler and his team had formulated their Mars Mobile Pathfinder, they had only signed up for a 30-day surface mission for the rover. The short expected lifetime was a product of the bitter martian cold. The rover electronics had to be kept above about –55°C, and the batteries above about –20°C, through a combination of electric heaters, 1-watt radioisotope heater units, the amount of heat produced by the vehicle's electronics, and the insulation provided by its warm electronics box (WEB). Because the rover was solar powered, less and less power would be available as the mission progressed and the martian dust built up on the arrays. The changing Sun angle would have the same effect. At some point, there would not be enough power to keep the electronics from freezing.

The idea of a 30-day mission, though, wasn't attractive to either NASA or Steve Squyres, principal investigator for the rover's scientific payload. It might take a week to get the autonomous rover disentangled from the lander carcass and off doing science, leaving it with a bare three weeks of useful life. NASA would be spending a lot of money to get to one or two rocks during that time.

But what could Adler and his group really sign up to do? In a normal project, they would have a team analyze this question quantitatively, accounting for the capabilities of available technologies and the vagaries of the Sun's position relative to likely landing spots during various mission durations and the effect changing Sun positions would have on the rover's power and thermal needs. And this analysis would be done before the project even received NASA approval, during what's known as a Phase A study, lasting about a year. The Mars Exploration Rover (MER) mission never had a formal Phase A study. It just had the few weeks of study that Adler and company had put into it in May and June.

When the mission was going head-to-head with the Mars Surveyor Orbiter proposal at the "shootout" meeting in the Pasadena Hilton in July 2000, Mars program chief scientist Jim Garvin told Squyres that the 30-sol mission life was

Mars Mobile Pathfinder concept, from Mark Adler's May 3, 2000, presentation.
Mark Adler et al., Mars Mobile Pathfinder '03, May 3, 2000, MOAR-050300.ppt, Manning materials, Historian's Mars Exploration Collection, JPL. Courtesy Robert M. Manning.

putting them at a competitive disadvantage.[1] They had to at least do better than the Sojourner rover's 86-day life. Pete Theisinger and Rob Manning then had a conversation with Adler in the hallway about what they could sign up to. The short study they'd had performed showed a positive power margin at sol 91, so they decided they could accept a 90-day mission.[2]

This proved a fateful decision. The 90-day mission requirement drove redesign of the rover, increasing its mass. And the mass increase forced a redesign of the Pathfinder entry, descent, and landing (EDL) system. By the time the MER team got to their preliminary design review in October, at which their major designs were supposed to be finished, they already knew that, instead, their design was badly broken. It would take them more than a year to fix it.

Rover Growing Pains

Between NASA's approval of the MER mission in July 2000 and the middle of the following year, the rover grew substantially. In turn, the rover's growth in both volume and mass "broke" the original idea of using the Pathfinder lander, aeroshell, EDL subsystems, and cruise stage essentially unchanged. There were several causes of the rover's expansion, but a big one was thermal management. Keeping the vehicle electronics alive long enough to do their various jobs proved a difficult design challenge.

The first design pass at what the team had dubbed the Mars Geological Rover had segregated the vehicle electronics into three groups: electronics only needed during cruise, those needed for EDL and the first few days after landing, and those the rover needed to operate after it left the lander. The cruise-only electronics went on the cruise stage, which, like Pathfinder's, was discarded just prior to entry. Systems engineer Joel Krajewski remembers that in order to minimize the rover's volume so that it would fit into the Pathfinder tetrahedron, their intent was to put all electronics needed for only the first few days on the surface into the lander.[3] This meant a bunch of small triangular boxes that would have to fit in between the lander's structural members. And all the electronics for the rover itself were concentrated into a single, integrated rover electronics module (REM), whose thermal design principle they called an "engine block." Simply by being densely packed, the REM would warm and cool slowly, minimizing the rate of expansion and contraction of its component circuit boards.

But the electronics boxes on the lander turned out to be a big problem. Because they were actually pretty small and relatively thin—the lander structure was only a few inches deep—they lost heat quickly. They had to be kept above –55°C when the nighttime temperature might be –100°C, so they needed heaters. And the heaters needed power. The thermal analysis performed on the design suggested that they'd need seven batteries to keep the lander electronics alive overnight. The batteries themselves had to be kept between 0°C and 20°C, so they could only go into the rover's own WEB.

They'd only intended to have two batteries in the rover, not seven. Krajewski rather ruefully remembered, "[We] couldn't find a solution that didn't have us increasing our battery size inside the rover belly by a factor of three or some horrible number like that, which would defeat the whole purpose of pulling the electronics out of the belly."[4]

The analysis that revealed this problem was finished in early August 2000, at the same time the project was reviewing the two-rover option.[5] The team tried to salvage the architecture by studying various ways of insulating the lander electronics to reduce the power needs. By the project's preliminary design review during the second week of October, barely three months after they had started, they already knew that this approach wasn't feasible either. There just wasn't enough volume available for insulation. They had to move the electronics that controlled the pyrotechnics, the airbag retraction motors, and the lander petal motors into the rover's WEB. So the WEB grew. In late November, deputy spacecraft manager Barry Goldstein sent an e-mail out to the design team establishing the larger rover as the new baseline design.[6]

In part because the team knew they didn't have a workable design in October, the preliminary design review didn't go very well. But only in part. Theisinger's rapidly expanding team had a related, but much bigger, challenge. Because of their expanding rover, they had "negative margin" on their landed mass. Tom Rivellini had tested Pathfinder's airbags to a landed mass of 410 kilograms, and as of their preliminary design review, without the rover redesign and resulting mass increase having been completed and understood, MER faced a landed mass of 401 kg. The JPL design principles that were just being finalized required them to show a landed mass margin of about 50 kg at this point in the project, and they didn't have it. Nor did they have the required margins on their launch or entry mass.

Years later, Manning would recollect that "tortuous arguments were put forth as to why the system would NOT grow much above 450 kg and why the airbags could handle it if the touchdown velocity were somehow reduced."[7] The review board didn't believe these arguments. Theisinger had decided to write his concluding slides after the main body of the review was over; on the morning of the last day he got up and told the board that the team would hold a second preliminary design review (a "delta PDR") in February. They would do a thorough "mass scrub" and present a design with adequate margins. He remembered later that the board had been very nervous about the mass trouble and that it "helped our credibility a great deal by simply saying, 'We see it too.' "[8]

The second reason that the review didn't go well for Theisinger's crew was that their "Level 1" requirements were still in flux and needed to be settled quickly. These requirements specify, at a high level, what a mission is supposed to accomplish. JPL's systems engineers used these as the basis for specifying the detailed technical capabilities the vehicles had to have to achieve them. There were a couple of Level 1 issues still unresolved. David Lavery, their program executive at NASA headquarters and a big supporter of robotics, wanted one of the rovers to demonstrate the ability to drive substantial distances on Mars, and he sought a requirement that one of the rovers be able to drive at least a kilometer. Neither Theisinger nor Squyres wanted to have to do this. The rovers were going to be slow, and stopping to make scientific measurements would make them even slower. They wanted the achievement of a long driving distance to be contingent upon doing a certain number of measurements. NASA eventually agreed to their position, and the requirement was weakened to 600 meters with a "goal" of 1 km. The Level 1 requirements also specified visiting "at least eight" different spots for detailed investigations.[9]

The other issue was more difficult. Theisinger had worked out a set of "minimum mission return" and "expected mission return" criteria with Steve Squyres, criteria that were being used to help define the rover's capabilities while the Level 1 requirements were still in flux. But as of the October review, the project was "unable to construct a 90-day mission scenario that meets the current definition of Expected Mission Return for either of the Mars Exploration Rover missions given the current flight system capabilities."[10] The problem they faced was in getting the rover's data back to Earth.

The rover was being designed to use both X-band "direct-to-Earth" transmission and the UHF relay on 2001 Mars Odyssey to return its data, but with a strong preference on the engineering team's part to emphasize UHF. The direct-to-Earth transmitter produced so much heat that it would overheat the WEB after about an hour and a half, which wasn't enough time to return all the data. The UHF system generated much less heat. But Squyres was unwilling to depend on the as-yet unlaunched Odyssey.

Discussion between Squyres, Theisinger, and NASA officials led to an understanding that the direct-to-Earth would be the primary means of communications, although it was not explicitly spelled out in the approved Level 1 requirements. Instead, the flight system (Level 2) requirements were revised to specify that the rover had to support four hours of continuous direct-to-Earth use, which was the approximate length of time needed to return each sol's data.[11] That made the mission independent of Odyssey, but at the cost of still more mass and design complexity.

The delta PDR was scheduled for the first week of February 2001, less than three months away. Theisinger created two special engineering teams ("tiger teams") to redesign the cruise stage and lander to recover mass margin and to redesign the EDL subsystem to accommodate the bigger, heavier rover and increase mass margin.

The rover design hadn't stabilized as of February; in fact, it kept growing until July 2001. Again, the thermal needs of the rover's electronics were a big culprit. The key to the rovers' surviving the required 90 sols on Mars was the amount of heat dissipated into the warm electronics box. It didn't matter whether the heat was generated by battery charging, running the electronics in the electronics module, the power amplifiers, or by the 1-watt radioisotope heater units they were planning to put on the rover batteries. At the same time, energy consumed outside the WEB by the wheel motors, panorama camera (PanCam) and mini-TES (thermal emission spectrometer), or the robotic arm, was a dead loss as far as the rover's thermal state was concerned. (The mast supporting the PanCam and mini-TES also acted as a big heat "chimney," pulling warmth out of the WEB.)

The project's thermal engineer, Keith Novak, had built an analytic model of the WEB to use in evaluating its likely thermal performance under various Mars surface conditions. Using Jake Matijevic's surface operations scenarios as the basis for likely energy consumption by the rover, by December 2000 he'd found that the WEB wouldn't be able to keep the electronics warm enough beyond about sol 60. The WEB needed a heat input of around 530 watt-hours a day to keep the batteries and electronics from getting too cold. But the solar arrays had been sized for much less than this, around 470 watt-hours/day, toward the end of the 90-day mission.[12] There was no physical space left inside the WEB in which to put more insulation, so that wasn't an option. Neither was making the rover bigger. It already didn't fit the Pathfinder lander design. The only possible solution was more power.

The increased need for power forced the project to increase the solar array area by 25 percent, which just added to the team's design problems. The larger panels had to be folded into a lander that was already full. And, of course, they added to the mass problem. The new array sizing need was acknowledged during the project's thermal subsystem preliminary design review in January and accepted—with great reluctance—at the delta PDR. But the project didn't have a design for the new array, and wouldn't until July. Instead, the Lab's mechanical designers were consumed with the effort to redesign the entire spacecraft to accommodate the rapidly growing rover.

Will It Fit?

The rover's growth posed three big threats to the project. The first was that the mechanical engineers wouldn't be able to squeeze the rover into Pathfinder's lander and aeroshell. The second was that if they could make it fit, the resulting flight system might be too heavy for their Delta 7925 to send to Mars. And the third problem was that if they could fix those two issues, they still might not be able to land it safely on Mars due to the higher landed mass. This last fear was reinforced after their first airbag test in December 2000, just as the tiger teams were finishing their pieces of redesign. The flight spare set of Pathfinder airbags failed catastrophically at slightly *less* than the Pathfinder lander's mass, undermining confidence in the basic soundness of the Pathfinder design.[13]

The rover's size growth was one of two threats to the team's ability to reuse the Pathfinder lander design unchanged. The second was the need to increase the size of the parachute by some 40 percent. NASA had mandated a daytime (on Mars) landing to ensure the return of all the EDL data, whereas Pathfinder had landed before dawn. The daytime atmosphere was less dense than the nighttime atmosphere, reducing the parachute's performance. And the landing sites that Steve Squyres had in mind were at higher altitudes than Pathfinder's had been, again resulting in reduced parachute performance. The growing rover's larger mass also required a bigger parachute; for all these reasons, the parachute had to enlarge quite a bit.[14] Therefore its canister had to grow; of course, in the volume-constrained lander, it had to fit into space that was already taken by the rover.

The lander-rover integration tiger team addressed the growth by redesigning the lander tetrahedron so that it was no longer symmetrical. They "clipped" off the top of the tetrahedron to widen the opening enough to accept a wider parachute canister, and they widened the base petal to accept the longer rover. The EDL tiger team then modified the backshell and heat shield to accept the redesigned lander. The Pathfinder backshell had been a 70° blunt cone with the pointed end cut off, and engineer Dara Sabahi's tiger team didn't change its conic nature. Instead, they widened the narrower end of the cone by the simple expedient of cutting off the pointed end closer to the wide end. Then they put the length they'd removed from the narrow end onto the wider end.[15] The result was an increase in total volume within the aeroshell. This change also forced them to make the heat shield slightly wider, to fit the wider backshell.

The new, larger lander and revised backshell design cost the project a mass increase. Since the main purpose of the tiger teams had been to reduce mass, the two groups had to pay for the mass increase by finding or creating even larger mass reductions. They proposed to Barry Goldstein and flight system manager Richard Cook changing the materials of the lander and cruise stage. The Pathfinder lander's primary structure had been aluminum, which was simple to manufacture and cheap. And its cruise stage had been a combination of aluminum and titanium. These weren't necessarily the most mass efficient materials, though. Composite materials, such as those used on Mars Global Surveyor, were often lighter but more difficult to work, and more expensive. They thought replacing the lander primary structure with a composite would save them about 24 kg, while imposing increased cost and schedule risk—they'd lose, they figured, four to six weeks.[16] While they'd been hoping for a 30 kg reduction, Cook, Goldstein, and Pete Theisinger agreed to this change. There seemed no other way of reducing the lander mass further.

The cruise stage tiger team also adopted a composite structure and proposed two more changes in their search for lower mass. Mars Pathfinder's cruise stage had used four identical titanium fuel tanks containing a total of 100 kg of fuel. The MER mission design didn't need that much fuel; reworking the trajectory design left a need for only about 62 kg. Two somewhat larger tanks could contain that amount, simplifying the fuel system overall. The Mars program office had started funding the development of composite fuel tanks to further reduce mass in support of the 2003 Mars micromission effort, and Sabahi's EDL tiger team recommended adopting them for MER, too.[17] Changing the structure and tanks to composites promised to shave about 25 kg off the launch mass, while the fuel reduction saved another 38 kg. That was a significant savings, although again it came with higher costs and schedule delays.

The redesign effort between October and the end of January succeeded in raising the project's mass margins above the required 15 percent, satisfying the review board.[18] To track mass growth more rigorously, Theisinger empowered a "mass czar," an engineer named Grace Tan-Wang, who would chair what was known informally as the Mass Tribal Council. The council was to meet weekly to review all proposed mass increases and to recommend allocations from the mass contingency reserves held by Cook for the flight system and by Theisinger for the project overall.

The rover solar array story was finally completed when a mechanical engineer suggested a design for the additional array area that met with quick approval.[19] Secondary arrays attached to the rear of two lateral panels with hinges; stored for the cruise to Mars, they would be folded flat against the top surfaces of the primary lateral arrays. Once on the surface, they would be unfolded after the rover had lowered the primary panels into place. They were an elegant solution to a difficult problem.

But the deployable arrays added mass and complexity, which the project couldn't afford. And shortly before their critical design review, the project management team tried to remove them to stay within their landed mass limits, despite vigorous lobbying by Steve Squyres.[20] The effort fizzled a few days later when the solar panel's vendor told them that the arrays would be able to host fewer individual solar cells than they'd originally thought, leaving the rover without enough power to meet the Level 1 requirements. So the secondary arrays were put back on. The project sacrificed one small instrument, a "SunCam" that was to locate the Sun as a check on the rover's location and orientation in the name of mass recovery.[21] But otherwise they had to absorb the solar array mass increase from their margins.

Yet the rover design still wasn't done. The larger solar arrays actually created a new problem: too much power early in the mission, before dust built up. The larger arrays would overcharge the batteries, which would damage them. So they had to add what are called "shunts," a set of resistors on the outside of the arrays that would consume the excess power (and generate heat). The control circuits for the shunts also generated heat, but inside the WEB. And they would cause the WEB to overheat if the vehicle was also using its direct to Earth transmission for more than about two hours (the power amplifiers generated a lot of heat).[22] Theisinger got the requirement changed to four hours/sol total, and the rover operators had to break the communications up into two-hour windows.

With all the changes to the rover design, its allocated mass ballooned from 130 kg just before the two-rover decision in August 2000 to 175 kg at preliminary design review in October. It topped out at 185 kg at the critical design review in August 2001. They'd lost months of schedule in the redesign effort, and their state of being perpetually behind became a long-running joke. Grace Tan-Wang put a gag article in the project's not-entirely-monthly newsletter, the *Mars Explorer Times,* about new instructions to JPL's security force to keep project members on-site. Next to a photo of a compact car with JPL staff members stuffed in its trunk, she wrote: "The recently upgraded barbed wire fences have proven quite effective at retaining workforce at JPL. Occasionally we need to pepper spray someone attempting to leave without authorization, but overall the employees have been compliant."[23]

More seriously, getting all the detailed redesign work done caused Sabahi to remove the mechanical design team from all their other institutional responsibilities at JPL for a month. "In MER I realized that actually most of their time goes to deal with processes that are mandated. And the amount of time that they can spend to actually do design and delivery is a fraction of their time. So I gave a challenge to my division and the flight system. I said that the engineers that are working on the rover structure, for thirty days, they are to be excused from all meetings, all requirements, everything. Thirty days they do nothing but design. And I basically say, 'If that fails, I will throw in my badge.' And I meant it."[24]

Reengineering the Pathfinder Terminal Descent

In parallel with the effort to design the rover and manage its mass troubles, the MER team also found that it had to reengineer the Pathfinder entry, descent, and landing hardware and process. This realization came early in the project, but the redesign effort took them into 2003. Adapting the system for the growing mass of the rover was only one of their challenges. Another challenge was self-imposed. The project's engineering team wanted a quantitative standard against which to assess their probability of achieving a successful landing; Rob Manning argued for, and got, a Level 2 requirement specifying a standard of 95 percent. ("Or, a 5% chance that the conditions that Mars would throw at us would exceed EDL's capability," he wrote later.)[25] As Manning's EDL team worked through their studies, adherence to the 95 percent criteria forced them to redesign almost everything.

The first set of changes had been made to the parachute and its support equipment very early in the project, to enable landing at somewhat higher altitudes and during daylight hours. The less dense atmosphere at higher altitudes, in turn, had created a "timeline threat." The timing of events in the EDL design was critical, and early simulation studies showed that Pathfinder's parachute would probably leave the vehicle with too high a terminal velocity. There might not be enough time to carry out all the later events that had to happen. So they'd accepted the need for a much larger parachute, which would slow the vehicle more.

The MER team also knew very early that they were going to have to scale up the solid rockets in the backshell (the rocket-assisted deceleration, or RAD, motors). The RADs had to zero out the vehicle's vertical velocity, 70 or 80 meters/second, right before the computer cut the bridle, dropping the lander on the surface. The amount of momentum they had to remove depended on the terminal velocity on the parachute and the mass of the lander. They couldn't predict either of those things very accurately until the lander design was more mature. So they decided not to decide how much more powerful the RADs had to be until the last possible moment.

They had hoped, though, to preserve the airbag design that Tom Rivellini had agonized over so much during 1995. But the team also needed to prove that the design would accept the higher lander mass without much alteration, so they decided to hold a series of tests using the Pathfinder flight spare bags out at NASA Lewis Research Center's Plum Brook station. This first test series was designed to validate an assumption Manning had made from basic theory. What the airbags did was absorb the kinetic energy of the impact, which is a product of the lander mass and the square of the impact velocity. That should mean that the airbags were much more sensitive to impact velocity than to lander mass—so if they could lower the impact velocity a little, they could raise the lander mass quite a bit and still have the same amount of kinetic energy to absorb.[26]

The engineer assigned to the job of EDL systems engineer was Wayne Lee. Lee had come to JPL in 1993 the week before the Mars Observer loss. He'd later been assigned to shadow Manning on Mars Pathfinder during its EDL development and served as the systems engineer for the defunct sample return lander. Lee's first task had been to find the old Pathfinder airbags, which were supposed to be in an off-site warehouse full of stuff. After traipsing around the place and not being able to find them, he discovered that JPL's public services office had given them to the National Air and Space Museum in Washington, DC, where they were on display.

Lee and the EDL lead mechanical engineer, Adam Steltzner, a young-Elvis look-alike, repatriated the airbags and took them out to Plum Brook in December 2000, right before Christmas. The first drop test they did was at Mars Pathfinder's mass and a velocity slightly below the maximum for which Tom Rivellini had qualified the airbags. It didn't go well. According to Lee, "We do the first airbag test and, poof, the thing blows up on impact. I just remember staring at that, thinking, 'Well, there went a million dollars,' because it cost us a million dollars to get the facility and all the test platforms and rocks ready for that test." They found three separate holes in the air bladder. On Mars, they would have killed the mission.

But Lee, Steltzner, and the airbag contractor's engineers convinced each other that the bags had failed due to their age and exposure to ultraviolet light. The material they were made of was known to weaken substantially from prolonged UV exposure, and the bags hadn't been protected from it at the museum. Tests made later showed the fabric was more than 50 percent weaker than when it had been new. In the interim, they repaired the old Pathfinder bags and carried out three more tests at the higher mass they expected MER to have, about 510 kg. Rocks punctured the air bladder in each test. But the bags suffered *less* damage at the higher mass, and they took that as evidence that their basic idea was sound.[27]

In parallel, Manning and Dara Sabahi were also thinking about ways to reduce the horizontal component of motion the airbags had to withstand. They knew from Pathfinder's test program that the airbags' Achilles heel was being dragged across rock faces. Dragging tore the abrasion layers away from the bladder, leaving it vulnerable. And they believed from the thousands of simulations carried out during Mars Pathfinder that most of the horizontal velocity that the MER landers might experience would be induced by the RADs firing just before impact.

The RADs induced horizontal motion if they fired when the backshell was at an angle to the martian surface (the "local vertical"). This could happen because the lander configuration during descent was, in essence, three bodies connected by cables. If the spacecraft entered during strong "shear" conditions, with winds at different altitudes having different velocities, its three parts (lander, backshell, and parachute) would oscillate like a string of pearls. If that happened, when the RADs fired, they would be at a large angle to the local vertical, inducing a horizontal velocity that would tear the airbags. Simulations showed that being 20° off the local vertical induced a horizontal velocity well beyond the airbags' capability.[28] The backshell was that far out of alignment in only a few percent of the simulated entry cases, but even that was too much risk. The requirement that they have at least a 95 percent probability of success forced them to think of ways to eliminate that few percent.

The wind model they were using for the first year of the project, which was based on the winds at Kennedy Space Center, often had strong shear. Manning's crew had used this "Cape model" on Pathfinder in lieu of anything else, even though Mars isn't much like coastal Florida. As it became clear that this shear problem was a big vulnerability, the EDL team approached a couple of JPL's meteorologists about the model. They were David Kass, a recent hire, and John T. (Tim) Schofield, who'd been working with Dan McCleese on the repeated efforts to get his pressure modulated infrared radiometer instrument to Mars since the 1980s and who had been the investigation scientist for Pathfinder's meteorology package. Kass remembers that the EDL engineers were hoping that the Cape model was too conservative and that they could safely reduce its shear component. At Caltech and Colorado State University, Kass identified a couple of other "mesoscale," or regional-scale, models designed for Earth and started working with their designers to convert them for Mars. Mesoscale models needed decent topography, so adding the topography for the various proposed landing sites was part of the effort.[29]

The effort to get a better wind model promised to take about a year, so the EDL engineering team pursued a second track to solving the wind sensitivity problem, too. Since the basic problem was that the winds might push the backshell out of alignment with the local vertical, could they find a way to shove it back toward the vertical just before the RADs fired? In November 2000, Sabahi suggested what he thought would be a simple solution: in the event the backshell was far out of alignment with the local vertical, have one of the three rockets fire slightly early. One rocket firing early would force the backshell to start moving back toward the local vertical. This would reduce, though not completely eliminate, the RAD-induced horizontal motion. In order for the scheme to work, though, the rover had to know what the backshell's orientation was in order to command the appropriate rocket to fire early. So they had to add an inertial measurement unit to the backshell. By the delta PDR in January 2001, they had added this "differential RAD" firing plan to the baseline design.[30]

Sabahi's simple solution turned out not to be simple, though. There was immediate concern over the possibility that the early firing would cause part of the "triple bridle," the pairs of cables that ran between the edges of the backshell, and the lower, "single bridle" section connected to the lander to go slack. Because the RADs were mounted at an angle to the center of mass, firing only one would tend to cause the backshell to rotate around the center of mass toward the side opposite the rocket that was firing. So the bridle line opposite the firing RAD would go slack. Then, when the RAD on that side fired a half second later, the slack bridle line would suddenly go taught again, and it might break.

How serious this problem actually was proved very difficult for the EDL team to know. They had to meet the 95 percent probability of success requirement, which meant they had to show Pete Theisinger and the review board that they could produce a credible analysis of the problem. But they couldn't figure out a simple series of tests or analyses that would be convincing even to themselves, let alone everyone else. The off-axis thrust induced some nonlinear dynamics that couldn't be modeled easily, leaving them with a lot of uncertainty.

Miguel San Martin, a tall, balding Argentinian who was their guidance and control guru, disliked Sabahi's differential RAD idea because of this torque problem. He proposed an alternate solution that eliminated this uncertainty, but at the cost of precious mass. A separate set of three small solid rocket motors that were aligned with the backshell's center of mass would impose the horizontal velocity the team wanted without causing the backshell rotation they didn't want. The motors would cost them money and mass, and they'd have to cut holes in the backshell for the motors to fire through. But the approach would solve the horizontal velocity problem imposed by backshell oscillation without the uncertainties that plagued the differential RAD idea.

In April 2001, San Martin's concept formally replaced differential RAD.[31] The project named it transverse impulse rocket subsystem (TIRS). The addition of TIRS promised to take some of the pressure off the airbag team, but at least in the near-term, it had no effect. In May 2001, Lee and Steltzner took the refurbished Pathfinder bags out to Ohio again. This time, they replaced the abrasion layers with new ones made of a thinner material. The experiment was intended to see if layers of a thinner material might provide the air bladders with adequate protection with less mass. These May tests had somewhat ambiguous results. The thinner material did seem to provide results similar to Pathfinder's heavier abrasion layers, but the airbags still failed. Two of the three test drops suffered fatal bladder damage.[32]

In October, finally equipped with new airbags, the team returned to Ohio a third time. They had decided to try different abrasion layer variations, including both the heavier Pathfinder fabric and the lighter experimental fabric. On the very first drop, the new bags failed catastrophically. And on the second, too. In fact, a single rock became the demon of the test series. Each of the rocks on the test platform had been marked with colored chalk so the team could tell which rock did damage; the rock that did all the damage in this series had been colored black. This black rock blew out the bags on five out of eight drops.[33]

This test series threw the project into a crisis. The EDL team had convinced themselves that the Pathfinder flight spare bags had failed the previous two test series because they were old. But these bags were new. Their failure raised questions about MER and Pathfinder. Maybe Pathfinder had just gotten lucky. If the Pathfinder design had been robust, these new bags should have survived these drops. They hadn't been carried out at extreme conditions. In fact, the team had lowered the impact velocity to 21 m/sec from Pathfinder's tested maximum of 26 m/sec. And yet, in Manning's words, "Rocks cut right through."[34]

And if the airbag failure wasn't bad enough, by early November 2001 it was becoming clear that TIRS was not enough to solve the horizontal velocity problem. David Kass's wind modeling effort during the year had shown the opposite of what the project's engineers had wanted to hear. The Pathfinder wind model wasn't too conservative—it wasn't conservative *enough*. The two wind models Kass had fostered generated winds that were worse for every proposed landing site except Meridiani. But the vertical shear problem wasn't the primary culprit any more; TIRS was adequate to resolve that issue even with the new simulated winds. The problem now was that Pathfinder's Cape model had very little steady-state wind; in other words, few conditions in which the winds blew continuously. It simply had gusts. But these new Mars mesoscale models showed that heating and cooling of the surface at the landing sites produced steady-state winds of sufficient energy to impose horizontal motion on the descending spacecraft, too.[35] (Meteorologists call these winds "katabatic" winds.)

The MER lander had no way of sensing these steady-state winds and responding to them, so the project was suddenly back where it had been a year before. In principle, there was a way, though, that TIRS could be used to fix this problem, too. What the EDL team needed was a sensor that could measure the lander's horizontal velocity with respect to the ground.[36] The Viking project and Mars Polar Lander had used a Doppler radar for this purpose, and at the preliminary design review a pair of very senior JPL reviewers had advocated adding one to MER, but neither Rob Manning nor anyone else could figure out how to squeeze one into the MER lander, let alone test it and its effects on the other electronics in the 19 months they had left before launch. But if they could attach such a sensor, they could use it to fire one or two of the TIRS motors to tilt the backshell away from the direction of motion. Then, the big RAD motors' firing would cancel out some or all of the wind-induced motion.

Manning left a review of the ugly October airbag tests thinking about how to resolve this wind problem and ran across Miguel San Martin out on the sidewalk. San Martin was about to leave for Buenos Aires for a vacation. Manning asked him if he could spend some time looking for a velocity sensor that would work for MER before he left. San Martin said no. He didn't need to. "Just give me two pictures," he said.[37]

Two pictures would tell San Martin what the horizontal velocity was just as well as a Doppler radar would. Objects on the surface that appeared to move from one image to the next would actually represent the spacecraft's own horizontal motion, if the lander's swinging on its cables could be subtracted. Onboard image processing would be able to extract the bits of image that showed motion and calculate its magnitude and direction. But it meant they needed another camera.

Manning immediately realized that, in terms of hardware, another camera was easy. They'd kicked the SunCam off the mission, but they'd been too busy to remove all of its supporting electronics from the avionics design. He also knew that Joel Krajewski had routed a camera cable out of the rover to the lander, hoping to get permission from Theisinger to add a rover stand-up camera. So putting the SunCam back on with a different lens, this time stuck on the bottom of the lander, wouldn't be technically difficult. Getting approval, however, would be politically difficult. It was very late in the development process to be adding yet another new piece to the vehicle's EDL suite that would have to be developed, integrated, and tested. Manning had to have a convincing story to take to Theisinger.

So he decided to start studying the problem. San Martin was going on vacation, but he told Manning to contact a recent JPL hire named Andrew Johnson, who'd come from Carnegie-Mellon University's machine vision group. Johnson had been working on a technology task to develop an imaging system that would enable landing spacecraft on comets. This imaging technology was already pretty mature. By 2001, Johnson had a small, automated helicopter containing the imaging system, an inertial measurement unit, and GPS flying around the Lab.

Manning sent Johnson a long e-mail explaining MER's need and the constraints a landing imager would be under. The putative descent imager had under a minute to do everything, and had to do it while consuming no more than about 20 percent of the spacecraft's CPU time. Johnson thought it was possible.

Manning told Richard Cook that he was going to assemble a small, clandestine study group around this "descent image motion estimation subsystem" (DIMES), as he decided to call it; the group became the "DIMES underground."

The DIMES underground worked fast. By November 6, they'd identified the basic technical issues, including the processing overhead. The rover's computer had a lot to do in the last 2km of the descent, and the biggest concern was that they'd overload it.[38] During November and early December, the group also ran experiments in their machine vision lab to determine whether their processing algorithms could handle the image smearing that would result from the lander's swinging on its bridle. They used a digital camera swung above a sandbox on a small gantry; they designed the motion to replicate the motion seen in the EDL Monte Carlo simulations.

On December 13, they formally pitched the DIMES idea to Cook, estimating a cost of $1.5 to $2 million. At this point, Theisinger found out about it and, as Manning had guessed, he wasn't happy. But he also let the effort continue, while continuing to keep it from JPL senior management. Theisinger recognized that it might be a valuable addition to the project, but that the review board wouldn't likely see it that way until it was much more mature. To them, it would look like more work for a project that was already overworked and behind schedule.

But the wind sensitivity problem made its way up the JPL, and NASA, chain of command, even if DIMES didn't. Jim Garvin participated in Matt Golombek's site selection workshops during the year, and the project didn't hide the wind problem—it was an important facet of site selection. As concerns over the wind issue grew, Garvin and Dave Lavery, the program executive, had Golombek essentially restart the site selection effort, this time looking for a second "wind-safe" site. The "hematite site" that Odyssey had found in Meridiani Planum was already acceptable from the standpoint of winds, altitude, and power availability, but the second-most-favored site was not. The landing site Squyres thought was most likely to get him the evidence of water was a place called Gusev Crater, which from orbiter imagery looked like it had once been host to a surface-water lake. It even had what looked like an old river channel running into it. But Gusev was windy, according to Kass's wind models. Given the vulnerability of the landing system to winds, Garvin and Lavery insisted Golombek's site selection team find a replacement for Gusev with winds no worse than Meridiani's. They eventually found one, a site called Elysium Planitia. Like Meridiani, it was a huge, flat region. Unlike Meridiani, it showed not a hint that liquid water had ever been on its surface. So Steve Squyres *hated* it. So did Golombek, and even Ed Weiler at headquarters. MER might be able to land safely there, but it wouldn't deliver the scientific returns they were hoping for. Squyres promised to buy the entire EDL team beer if they could make it possible to land at Gusev Crater instead.[39]

The terrible airbag tests and increasing awareness of their wind sensitivity led the project team to reopen the airbag design, too. The previous experiments in October had been minor variations on the Pathfinder design, with differing number and thicknesses of abrasion layers but without fundamental changes to the bladder design. A fundamental redesign had been a taboo subject within the project; the airbags' flight heritage was rooted in Pathfinder, and doing something really different meant a much more extensive (and expensive) test program. But Cook and Goldstein and Sabahi and Manning no longer thought minor changes would suffice. In a December workshop, they widened the range of possible redesigns considerably.

"Wigs, double bladders, bulging lobes, and gossamer bladders were all on the table" at the workshop.[40] It was stressful for everyone involved because the team knew they had to leave the meeting with a design that would absolutely, positively solve the problem—a "kill-all" solution, Sabahi called it.[41] The EDL team agreed to four new designs, including Sabahi's "kill-all" solution, a double-bladder set of bags. In fact, they agreed to two sets of double-bladder bags, using two different fabrics for the bladders. The contractor's staff in Delaware worked through Christmas to get the bags built; Sabahi trooped out to Ohio with Wayne Lee and Adam Steltzner in mid-January to set up the next set of tests.

They made a few key changes to the test methodology. They made several more copies of the notorious black rock from the October tests and attached them to the platform to make the tests still more challenging. They'd also hired the JPL photo lab to install high-speed cameras under the platform so they could watch the instant of impact more closely. They even cast a copy of the black rock in clear plastic and mounted a camera, with lights, inside it, so they could have a "rock's eye" view of their villain in action.

They weren't disappointed. Their rock camera caught several instances of airbag failure in gory detail during the 18 tests.[42] But the series also showed that the double-bladder design really would solve their problem. After the test series ended in late February 2002, the EDL team had a second airbag workshop to settle on the final design. On March 20 they adopted a double-bladder design made of Vectran, with six of the lighter-weight abrasion layers.[43] And they approved another test series for that August to further examine the chosen design, including the performance of design changes made to strengthen the locations where the bags attached to the lander.

March 2002 also saw DIMES "come out of the closet" and get exposed to JPL management and the project's review board.[44] In April, the DIMES underground faced an implementation review at which there was considerable hostility. Two members of the board wanted it dropped immediately, believing it would distract the project management from the larger MER effort. But Theisinger and Cook made clear that DIMES hadn't been baselined for flight yet; it was in development and could still be dropped or flown depending on its success. And they pointed out that the majority of the work being done on it was being performed by engineers who hadn't already been attached to the project, so they weren't being diverted from some other MER task. The board report concluded that "because DIMES offers a way to mitigate the clear vulnerability of the MER EDL system to sustained horizontal winds, continuing to pursue DIMES could result in a significant overall improvement in the probability of MER landing and mission success."[45] So DIMES stayed alive.

In May, Johnson selected a set of test sites in the California desert that seemed good analogs to the half dozen landing site finalists. Squyres particularly liked a site in the California desert, telling Johnson, "Well, if you show that this works at Ivanpah dry lake bed, it'll work at Mars."[46] In June, they flew an engineering model of the camera around the sites via helicopter to capture real images of terrain-in-motion for validation. They also built a set of simulated images, using Mars orbiter camera images of the potential landing sites and mathematically "smearing" them the way they would appear to the moving DIMES camera. During the summer, Monte Carlo simulations using the Mars orbiter camera images suggested that the system would probably have errors well under the 5 m/sec allowed by the project's requirements; in October, a final set of field tests confirmed that it did. In fact, only a small fraction of the imager's calculated velocities had errors over 2 m/sec.[47] The team passed their delivery review in mid-December, although clearing that hurdle didn't yet guarantee inclusion in the mission. They still had to perform well in their assembly, test, and launch operations (ATLO) systems tests.

The last bit of EDL redesign was in the parachute. The Pathfinder chute had to be scaled up for MER, although not to the size the Viking parachute had been, but Lee, Steltzner, and the parachute cognizant engineer, Robin Bruno, didn't think they'd have to do a significant redesign. They were disabused of that notion in May 2002. They needed to verify that the larger parachute would be strong enough to withstand the MER lander's larger mass, so they had built a set of six chutes to take out to Idaho for drop tests. To simulate the Mars atmosphere's effects on Earth, they had to use a large test mass—about 1,360 kg (3,000 pounds).[48] Lifting a mass that large by helicopter is rather dangerous, and they'd had to wait weeks for safe flying weather.

When they finally got good weather, the first test drop was a disaster. Their test body had a high-speed camera pointed up at the parachute, so they got a close-up view of their parachute inflating and then disintegrating. Worse, it wasn't the result of a freak accident or a manufacturing defect. Four more parachutes disintegrated as the team watched with growing horror.[49]

Worse still, Lee laughed later, was that the test methodology wasn't even stringent enough! They had needed such a heavy test body because on Mars, the atmosphere is so thin that the parachute opening doesn't slow the vehicle down significantly. But the much denser Earth atmosphere will slow the vehicle during inflation unless the vehicle is super heavy—that's why there's always a sudden jerk when a skydiver's parachute starts to inflate. It turned out that their big test body hadn't been big enough. The parachutes had started to slow their giant lawn dart as they'd inflated (and started disintegrating), so even if the parachutes had survived, the tests would have been invalid. They were not adequate enough reproductions of Mars conditions.

So MER had two new problems. The parachute design had failed, and the parachute test methodology had also failed. The team had to fix both in a hurry. ATLO, the point at which flight hardware was supposed to be finished and delivered for testing, had already started back in February; they had 12 months before launch.

The solution to the methodology problem came when Lee visited Ames Research Center to check up on the heat shield testing. He asked the heat shield engineer what one of the giant, but strange-looking, buildings at the center was; it was a 40-by-80-foot wind tunnel that had been built for rotorcraft development. Lee and Steltzner had talked about using a wind tunnel for parachute testing earlier but hadn't thought anyone would let them fire the mortar that deployed the parachute inside a tunnel; if they couldn't fire the mortar, the test would violate the "test as you fly" commandment and be invalid.

But Lee was desperate enough to call the tunnel's manager and ask. "I thought they were going to tell me, 'You're crazy. There's no way we're going to let you deploy a parachute in the wind tunnel.' They started thinking, and they said, 'You know, we could probably do something like that.' "[50] The tunnel could easily provide the wind velocity they needed. And the parachute wouldn't affect the giant fans that produced the wind, so they wouldn't have to worry about the deceleration problem. The wind speed would remain constant no matter what the parachute did. The tunnel had been built to test helicopters with their engines running; because experimental rotor blades disintegrated on occasion, the tunnel was also very strong. Their little parachute mortar couldn't hurt it.

The possibility of using a wind tunnel engendered quite an argument. One problem was that qualification testing hadn't been done this way before. Another was that it required a considerable investment of time designing the necessary test fixtures, building them, and testing them. And if the test methodology turned out not to be adequate, there would no longer be an alternative. Robin Bruno remembered that they simply couldn't afford the manpower to keep a parallel testing track going. They were going to lose six months of their schedule just doing all the preparatory work.[51] But Lee gradually won the parachute team over to the wind tunnel strategy and arranged for test time in September.

The contractor's lead engineer, Allen Witkowsky of Pioneer Aerospace, explained later that the parachutes were failing because the location of the area of peak stress had changed when they'd altered the Pathfinder design, and they hadn't realized it:

> We were chasing failure modes. We'd see lines break, so we would strengthen the line joint, and then it would move to just up from the lines there would be a break, and this was all in the band region. We'd fix that, and then it would break a little bit further up. We were just chasing the failures, trying to fix them incrementally, saying, "Well, this must be wrong," and, of course, we're all being rushed to hurry up and get the design finished so they can start qualification so we can get it on the spacecraft and so forth and so on. At some point we realized, "Let's take a step back. What's going on here? Why is this happening?" And we realized that the location of peak stress had changed.[52]

For Pathfinder, it had been in the disk. For MER, it was in the band.

In late June, the EDL team held another stressful workshop to design a new parachute. Since they hadn't yet figured out what the problem actually was, they decided to have three different designs built. One was what Steltzner called a "brick," heavily reinforced everywhere and made of a thicker, stronger fabric. But because the parachute canister was fixed in volume, the reinforcement forced him to make the parachute smaller in drag area. The "brick" would also be less stable, and would allow more oscillation to occur during the descent. So if that were the one that had to go to Mars, it would make some of the potential landing sites unachievable, including, probably, the one Squyres most wanted, Gusev Crater.[53]

They also chose a lightweight design that was only reinforced where Witkowsky's engineers thought the failures had started, based on the videos of the four failures. It would have the best performance but might not be strong enough. And they designed one that was a compromise between the "brick" and the "high performance" parachutes.

June was also the month in which the project had to commit the TIRS motors to manufacturing, which meant having to decide how powerful they had to be. That was a problem, since if the less-stable "brick" parachute was chosen as the flight design, the TIRS motors would have to be more powerful than if the high-performance chute went to Mars instead. They couldn't wait for the September tests to answer this question, so they decided to build the bigger TIRS motors that the brick would need. It seemed the lowest-risk choice.

Steltzner took the three new parachute designs up to the Ames wind tunnel in September. These tests gave him a new problem: squidding. It occurs when a parachute deploys but doesn't open fully. Instead, the lower section of the parachute, the "band" on the MER parachutes, just oscillates around. Squidding happens when the amount of air going in through the parachute base and out through its vent is the same; the parachute can only inflate if more air enters the base than escapes through the vent. On the first test he did, using the intermediate strength chute, the parachute opened partly, squidded, and only fully opened after they'd started shutting the test down. Steltzner was horrified. The parachute had been strong enough, at least.[54] But no one had ever seen a Mars parachute squid before. Steltzner thought it might be because this parachute had used an impermeable fabric in its band for higher strength. That could have kept the parachute from inflating properly. The brick had the same impermeable fabric in its band, so for the next day's test he chose the weakest of the three, which used a more permeable fabric in the band.

The next day, he had to watch as the "high performance" chute also didn't open properly. Instead, it squidded more than a minute before it finally opened. But again, the chute stayed intact. So at least he knew that the weakest chute was strong enough—if he could figure out how to make it open properly. Steltzner decided the vent looked too big. He measured it and discovered it was. There had been a miscommunication somewhere. He had one of Pioneer Aerospace's parachute technicians hand-stitch in a patch to shrink the vent and retested it. This time, it opened properly, and it still didn't disintegrate. The test wasn't entirely valid, as Steltzner hadn't been able to repack the chute into its container and refire it from the mortar. Instead, he'd had the tunnel staff do a "dis-reefing" test in which the chute was held shut by a lanyard while the tunnel was started up and then released.[55] But it convinced Steltzner that if he could get the parachute built correctly, he had a sound design.

The parachute team returned twice to Ames in October for more tests, this time with few problems. In the first week of November, the team chose the "high performance" parachute for the flight to Mars, hoping to preserve access to the Gusev Crater landing site. The parachute was finally flight qualified in January 2003, five months before launch, two months after the contractor had started building the flight parachutes, the same week the project shipped its first batch of flight hardware down to Kennedy Space Center.

The project's airbag challenges also ended satisfactorily in late 2002. Tests of the new double-bladder design between August and October went well, with the bags remaining intact at impact velocities up to 26 m/sec.[56] Combined with the successful parachute tests in January, these ended the hardware redesign effort for the EDL team. They didn't end the team's work, though. They still faced a number of significant hurdles, starting with the master EDL simulation, which they needed for the systems test program and for site selection, which was supposed to conclude in March 2003. The master EDL simulation wasn't ready in time, though, so the workshop focused instead on generating a science priority ranking, resulting in confirmation of Meridiani Planum and Gusev Crater as the highest priority sites.

By February, Lee's team had the master simulation running well enough to begin analyzing the EDL system's performance at each of the four remaining sites. These showed roughly what the engineers had expected: the Meridiani and Elysium sites were statistically indistinguishable at 95–96 percent probability of successful landing, while the Gusev and Isidis sites had 89 percent and 91 percent probabilities, respectively.[57] Higher winds at Gusev were the culprit, again as expected, while the Isidis site had lower winds than Gusev but more rocks. Their work gained Pete Theisinger's agreement to recommend Gusev Crater as the primary landing site for MER-B (which they expected to launch first) and Meridiani as the primary site for MER-A.

The next step in site selection was a peer review scheduled for the end of March. NASA had formed an independent review committee to rake through all the work done by the project's engineers and by Matt Golombek's two years of science workshops. The chairman was Tom Young, and he threw a few more monkey wrenches into the process. He thought that the Meridiani site was too high-altitude for the qualification limits on the parachute. So to save Meridiani, the team had to promise to requalify the parachute for larger opening loads. That, given the three months left before launch, meant doing the tests after launch.

Young liked Gusev Crater even less. David Kass had produced a set of "worst case" wind profiles for the EDL simulation effort, and when these profiles were fed into the master simulation, they had generated terrible results. The landing failed nearly half the time. Worse, the results were so recent that Rob Manning and Wayne Lee couldn't explain why the system was failing so badly. So the end result of the peer review was that Meridiani was conditionally approved for MER-A, depending on the results of the post-launch parachute requalification, but Gusev wasn't. At launch, MER-B would be targeted so that it could go either to Gusev or to the "wind safe" site that Squyres hated, Elysium Planitia, with the choice to be made before August 1.[58] But if only one of the two vehicles launched successfully, it would be sent to Meridiani to maximize the chance of success.

After the review, Manning and Lee dove into the master simulation's reams of data to try to understand the system's behavior under the "worst-case" winds. They concluded that TIRS was the problem: it was overpowered, an artifact of the team's decision to adopt the "high performance" parachute design after having chosen the larger TIRS motors to support the less stable, but stronger, brick parachute. The worst-case winds weren't actually all that bad, and the greater stability of the "high performance" parachute they had actually chosen meant that the backshell misalignment generally wasn't that large when TIRS fired. But because the TIRS motors had a fixed size, their firing shoved the backshell too far in the opposite direction. Then, when the RADs fired, they accelerated the spacecraft too much in the direction opposite the winds. The net result was airbag failure on impact. Manning ruefully characterized TIRS as a hammer that turned out to be too big.[59]

It was too late for the project to order a smaller TIRS rocket. So in May 2003, the EDL team started working on what they decided to call "dual TIRS." A new set of algorithms would decide whether to fire the TIRS rockets 0.2 seconds after firing the RADs (in the event of very high winds) or 1.1 seconds after firing the RADs (in the event of only moderate winds). The delayed firing reduced the effective velocity correction produced by TIRS from about 23 m/sec to about 11 m/sec.[60] It was yet another complexity added on to a design that, years before, had been chosen for its simplicity.

By the time MER reached the launch pad, the old Pathfinder heritage was long gone. Its EDL system was entirely new (they even redesigned the mechanism that lowered the lander down the single bridle), and much more complex. They'd been driven away from Pathfinder by higher masses, by their desire to make the Gusev landing site, and, most importantly, by their increasing realization of how

Mars Exploration Rover costs (millions of U.S.$)

Development	Operations	Launch	Total
625	75	100	800

Source: NASA Mars Exploration Rover Launches press kit, June 2003, www.jpl.nasa.gov/news/presskits.cfm (accessed 7 February 2012).

Note: Operations cost is through the end of the primary mission and includes science costs.

sensitive terminal descent was to the martian wind. Their only real alternative to redesign had been to accept higher risk of failure. Both the 95 percent success criteria and the political environment made accepting greater risk impossible. In addition to the painful Climate Orbiter and Polar Lander losses burned into the minds of everyone at JPL, in January 2003, space shuttle *Columbia* had disintegrated over Texas during reentry, killing its crew and further focusing NASA on risk. MER *couldn't* fail.

All the changes cost money. Pete Theisinger had to ask for additional funds twice. In October 2001, shortly before the project's critical design review, he'd asked for $50 million. JPL Mars Exploration director Firouz Naderi and NASA's Mars program manager Scott Hubbard reprogrammed the money from the Mars budget, cutting funding for some technology developments and from the Mars Smart Lander, effectively postponing it from 2005 to 2007. After the ATLO readiness review at the end of January 2002, Theisinger had to ask for another $50 million. Glenn Cunningham, who had chaired NASA's independent review board, caused a lot of heartburn on the project by recommending to NASA that the second rover be dropped so that the overworked project could focus its attention on doing one rover well (and so NASA would save the $65 million cost of the second Delta II, which could be sold).[61] But after a tense meeting in Washington on March 26, 2002, Ed Weiler, the associate administrator for space science, approved the second funding increase.[62] That second increment came with a warning that there would be no more—next time, the project would lose one of the rovers.

ATLO

The redesign of nearly everything during 2001 meant that the project's assembly, test, and launch operations phase didn't start on time. Scheduled for late January, it slipped into February and finally started March 6, 2002. The delivery needed to start ATLO was the spacecraft's "brain," the rover electronics module, because everything else had to be built around it. The ATLO manager, Matt Wallace, a former nuclear engineer and member of Pathfinder's ATLO team, knew by the end of January that he wouldn't be able to get one of the two flight REMs even by early March. He thought that getting the ATLO team working on realistic hardware mattered more than having the actual flight hardware. The ATLO team needed to get familiar with the complex procedures necessary to assemble the MER spacecraft, and they needed to check out the operation of the ground support equipment and cabling required for the testing procedures. So he decided to start ATLO with an engineering model of the module that would eventually become part of one of the project's testbeds. It showed up at JPL's high-bay clean room on a Friday afternoon. In the first of many unpopular decisions forced by the schedule, he made the team work that weekend.

Wallace, Richard Cook, and Pete Theisinger had also understood months before that they couldn't hope to get both their vehicles to the launch pad on time without far more work hours than were possible on a normal weekday schedule. They'd baselined dual-shifting for the entire duration of ATLO. Each shift worked ten-hour days, five days a week. But Wallace didn't schedule weekend work in advance—weekends were part of his schedule margin.

But even dual-shifting through the entire yearlong ATLO wasn't enough time, as delivery dates kept slipping. Sometime in December or early January, Wallace had concluded that there still wasn't enough time to complete the entire testing sequence on both vehicles. On Mars Pathfinder, the ATLO progression had been to do functional testing on each piece of the spacecraft as it arrived, integrate the piece, and make sure it connected properly. For electronics, the ATLO team also had to make sure each component communicated properly with all the other electronics it was supposed to. Then, when the entire vehicle had been assembled—"stacked" is the proper term—it went into a cruise thermal vacuum testing phase in JPL's 25-foot space simulator. The spacecraft was commanded to do all the things it would have to do during its flight to Mars, so the cruise thermal vacuum test served as both a thermal design test and a system-level test. Anything seriously wrong with the design from either standpoint should show up.

Because Pathfinder had also been a lander, though, it had to go through a Mars surface environmental test. So after the cruise thermal vacuum test, it had been removed from the chamber, torn back down to the lander with its rover, and then while the cruise stage and aeroshell were reworked, the lander and rover went through the thermal vacuum test again, this time being commanded to do their surface mission activities. Finally, the vehicle had been shipped in several pieces to Cape Canaveral, where the radioisotope heater units and batteries had been inserted into Sojourner, Sojourner had been sealed up in the lander, the airbags were attached, and the whole thing was put back inside the aeroshell before being attached to the cruise stage.

And that's what Wallace had planned to do with both MER spacecraft, initially. But shortly before the ATLO readiness review in January, he realized it was no longer possible. After losing the first two months of 2002 due to late deliveries, he didn't have enough calendar time, let alone work hours, to complete two sets of environmental tests. He remembered that JPL used to build two spacecraft identical spacecraft at a time routinely, although it hadn't done so in a generation, and he started to wonder if his predecessors had bothered doing the full set of tests on each. It really didn't make sense to do the whole suite on both, as many of the tests were of the design, not of the workmanship of the individual pieces. He couldn't see any reason to do the design-oriented testing on both vehicles, as they were the same design.

And he found that those predecessors had thought the same way. JPL's engineering directorate maintains a library of old project documents that he consulted, and he went and talked to the old-timers who were still around to understand their reasoning. What they had done was to build a third spacecraft, which they called a "proof test model" and was never intended to fly, and subjected it to the full set of tests. The two flight spacecraft got a much smaller subset of tests to assure that they would behave just as the proof test model had. They had worked this way so that all the mistakes got made on the proof test model, not the flight spacecraft, and because some of the mechanical tests were potentially damaging. As the planetary budget shrank beginning in the late 1970s, both the model and the duplicate spacecraft—seen as luxuries in Washington, DC—went away. So the idea had fallen out of use.

Wallace couldn't have a third spacecraft built up as a proof test model—although since Theisinger, like Rob Manning, believed that being "hardware rich" would actually make the test program go faster overall, he did have the project build three complete rovers, a fourth mobility test rover, a spare lander chassis, three sets of flight avionics, and two sets of nonflight avionics for the test beds—but he did see that his predecessors had put some thought into how to test multiple spacecraft efficiently. Their process had been, in essence, to put the first spacecraft through every test, and to figure out what minimum set of tests was necessary to prove that the second one was enough like the first that it would behave the same way in space. So what was that minimum set of tests?

Wallace asked Manning. Manning arranged a meeting out at his house on December 18, 2001, where they, Richard Brace, Richard Cook, Dara Sabahi, and a few others went through the whole test program to figure out what they had to do to convince themselves that the second spacecraft would be basically sound.[63] They couldn't go quite so far as not performing thermal vacuum testing on MER 2 at all, but they realized that they didn't need to subject the rover and lander to both the cruise thermal vacuum test and the Mars surface testing. They could put the lander and rover through the surface thermal vacuum test to prove that its thermal performance was as predicted by the thermal models (and also just like MER 1); having verified the model was right, they could put the cruise stage and aeroshell through cruise thermal vacuum testing *without integrating the lander and rover.* They could simply put one of the engineering model REMs into the aeroshell on an easily made test fixture to provide the "brain" needed for the testing. They called this the "cruise stage/aeroshell subassembly" approach, or CSAS. This CSAS test would verify the thermal performance of the cruise assembly. And they'd ship the two big subassemblies to the Cape separately, first fully assembling MER 2 in Florida.

This produced big manpower savings: about 3,000 work hours at JPL.[64] The key to the savings lay in realizing that both vehicles had to be shipped to Kennedy Space Center in pieces anyway. The rovers had radioisotope heater units buried under their batteries, and the Department of Energy would only deliver them to Kennedy, not to JPL, so building both spacecraft up at JPL meant having to tear them almost all the way back down again before shipping so the heaters could be inserted in Florida. Not doing a complete buildup of MER 2 at JPL produced the huge savings. In fact, it meant that MER 2 would ship before MER 1 and be launched first, at least if nothing went horribly wrong in testing.

Wallace didn't have a hard time convincing Theisinger, who knew they were in deep schedule trouble. He had a little more trouble with JPL's director of flight projects, Tom Gavin, who made him have the plan vetted by some more JPL old timers before agreeing to it. The independent review board accepted its necessity but recommended canceling one rover instead; as noted above, NASA's Ed Weiler approved it anyway.

Wallace got the first flight rover electronics module on August 9, three weeks after his "drop dead" date; he got the second one in mid-September. The second REM also caused him the worst moment he remembers during ATLO. He was driving home one evening when the vehicle's test conductor called him on his cell phone. "Man, Matt, we just blew up the REM." He turned around and went back to the lab, thinking they'd finally lost one of the rovers. There was no time to build another REM. It turned out not to be quite that bad, however. The test team had connected power to a place it shouldn't have gone, causing some sparking and smoke. But the actual damage turned out to be repairable.[65]

The REM "explosion" wasn't the project's only near-disaster that got resolved after much heartache. The solar array vendor for the rovers went bankrupt, and the team had to find a second source that could deliver on very short notice. The composite fuel tanks that were supposed to save them mass in the cruise stage never appeared; the vendor had lost the personnel who knew how to build them. The cognizant engineer, suspecting this might have happened, had identified titanium tanks that would do and could be obtained quickly, at the cost, of course, of mass (about 7kg). In June 2002, they became the new baseline.[66] The lightweight composite landers arrived extremely late and turned out not be lighter after all. Theisinger ruefully explained later that while the composite structure was lighter, the metal attachment points needed to protect the composite from other metal parts—the rover's wheels, the electronics boxes, and so forth—wound up raising the mass back to what the aluminum landers would have weighed anyway.[67] As a result, the project's mass margins wavered between zero and negative numbers for most of ATLO.

MER 1 was completely assembled in early October 2002. It wasn't in perfect shape; its REM had developed problems in testing, but the team needed to keep the system-level environmental tests on schedule. The ATLO team put the vehicle "stack" through a set of vibration and electromagnetic tests prior to sending it up to JPL's space simulator for two weeks of cruise thermal vacuum testing. MER 1 passed the thermal testing with only some minor problems (several heaters and thermostats were miswired, and some of the thermal control software was buggy), a considerable relief to the team.[68]

The MER 2 rover was fully assembled by October, too, and was intact enough by mid-September to start taking pictures of its surroundings in the assembly bay—even one of MER 1's cruise stage hanging nearby. It performed the first surface thermal environment test starting December 11. The testing again showed that the thermal design was sound—within 3°C of its predicted performance, in fact—even though it revealed a host of other problems. These were gradually, and somewhat painfully, fixed over the couple of months left before MER's pieces had to start making their way to Florida.

The first increment, MER 2's cruise stage and aeroshell assemblies, had gone at the end of January 2003 with about 20 people from JPL; MER 2's rover and lander, and MER 1's cruise stage and aeroshell, had gone the last week of February with about 35 more engineers and technicians. MER 1's rover went with about 15 more people.[69] The first week of March marked the last shipment of MER hardware to Kennedy Space Center. The launch period opened May 30; to get all this done in time, Wallace had scheduled staff for two shifts, seven

Mars Exploration Rover (MER) management posing with MER-1 in JPL's 25-foot Space Simulator, the Lab's largest solar thermal vacuum chamber. *Left to right*: Rob Manning, Richard Cook, Matthew Wallace, Richard Brace, Peter Theisinger, Barry Goldstein. JPL image D2002_1106_B54, Courtesy NASA/JPL-Caltech.

days a week. His "margin" by this time was the potential to go to three shifts, but that was it.

By this time, the ATLO team knew that all the efforts to keep the vehicle mass down had paid off. The final launch mass for MER 1 was 1,064 kg, including 52 kg of fuel; the team had projected 1,078 kg for the launch vehicle manufacturer, so they wound up having to add ballast to raise the mass for launch. The entry mass turned out to be 832 kg, and the touchdown mass 539 kg. The launch mass had grown 20 percent above Pathfinder's, while the landed mass had increased

47 percent—the team packed a lot more stuff into the same volume and taken a lot of mass out of the cruise stage, mostly fuel and "excess" aluminum from the structure, in partial compensation.[70]

Two days of thunderstorms delayed MER 2's launch to June 10. But the launch itself was uneventful. MER 1, however, suffered a lengthy delay due to a bizarre launch vehicle problem. Cork insulation on the outside of the first stage hadn't been applied correctly. The launch vehicle manufacturer's engineers scraped it off and applied new cork. But that wouldn't stay on. The second launch window opened June 25, and they struggled with the cork until July 5, when a new problem emerged with a battery. They finally gave up on the cork, accepted the risk it might fall off, replaced the bad battery, and got MER 1 on its way to Mars on July 7.[71] The rovers were renamed *Spirit* (MER 2) and *Opportunity* (MER 1), the result of a naming competition that drew nearly 10,000 entries.[72]

Conclusion

The Mars Exploration Rover project had started from a false assumption, that the fairly simple Pathfinder EDL system could be reused without much change for a much different mission. That assumption might have been true if NASA, and Steve Squyres, had been willing to accept a shorter "required" mission length, more risk of a failed landing, or less windy landing sites (or some combination thereof). But they hadn't. Instead, mass growth in the rovers needed to provide the proper thermal conditions for a longer mission, combined with the team's desire to get Squyres's instruments to the attractive but windy Gusev Crater landing site, produced rapid evolution of the EDL system into something rather more complex.

The redesigns and delays led to cost growth. MER overran its development budget by 17 percent. The money mostly came from delaying the Mars Smart Lander mission, whose engineers were pulled in to help MER out of its schedule crises anyway. NASA policy toward cost had changed substantially since the loss of Surveyor 1998; Pete Theisinger had put it this way during CDR: "The Project's approach is to spend money to maintain schedule and product quality—and the Program Office and JPL management support that."[73]

MER's development had consequences for later projects beyond funding. The airbag test horror show triggered a major change in the Mars Smart Lander effort. When Tom Rivellini went back to Smart Lander after helping MER out with its airbags, he started advocating landing on the wheels, the idea he and Dara Sabahi, Rob Manning, and Steve Jolly had back in 2000.

Launch hadn't ended the MER team's work. They still had testing to do on the airbags, backshell, and parachute. Their cruise software was complete, but the EDL and surface operations software were not. Many members of the development team had to be retrained for operations, although that retraining had to be delayed until the testing and software was finished. And they still had a huge amount of systems testing to do on their testbeds. They had a busy seven months before their first landing day.

Sending a Spy Satellite to Mars

The Mars Exploration Rover development occurred in parallel with a second project, the large Mars surveillance orbiter that MER principal investigator Steve Squyres had dubbed "GavSat." While the concept had lost the direct competition with MER, the idea of a big photographic reconnaissance satellite for Mars had been made one of the "road map" missions that NASA's Scott Hubbard and JPL's Firouz Naderi had laid out for the decade, and later in 2000 it had been established as a formal pre-project. Richard Zurek remained the project scientist, but JPL engineer James Graf replaced David Lehman as project manager. Graf had been manager of several smaller missions, including the QuickScat ocean winds measurement project, which had been done as a very short turnaround systems contract with Ball Aerospace, and he had been assigned briefly the "large lander" task that had tried to conceive better ways to land on Mars.

The project, eventually named Mars Reconnaissance Orbiter (MRO), was to be established as a systems contract placed with an aerospace company, as Mars Global Surveyor, the two Surveyor 1998 missions, and 2001 Mars Odyssey had been. NASA also wanted the existing contract with Lockheed Martin rebid, so it was not foreordained that Lockheed would be the contractor again. A few things were certain, though, as Zurek and Graf started to work on the proposal information package that had to be released to prospective bidders in April 2001. The formulation authorization NASA sent to JPL specified that Mars Climate Orbiter's scientific objectives had to be recovered by the new mission; it would conduct site characterization for future potential landers and would provide UHF relay services to future missions.

The project also could anticipate having much greater resources to draw on. The Mars program received an increase in its budget from $248.4 million in FY 2000 to $428 million in FY 2001. Some of this went to the MER project, of course, but a significant amount was specifically for Mars Reconnaissance Orbiter. Scott

Hubbard and James Garvin had gone to the NASA budget examiner, Steve Isakowitz, at the Office of Management and Budget in January 2001 and argued for an expanded mission scope.[1] If the Mars program were to do the big Smart Lander later in the decade and sample return in the 2010s, the program had to figure out where to send those missions in the late 2000s. The mission should be able to identify water-bearing minerals from orbit and map near-surface ice in pursuit of the martian habitability theme as well as identifying the locations most likely to be both safe and interesting. Garvin explained later, "We convinced Isakowitz that just like a big telescope can find the stars to look at, MRO is the eye in the sky over Mars to refine where we need to go. Instead of 30,000 sites that [Mars Global Surveyor] tells us are pretty cool, let's downsize it to the 100 top places to go on Mars for the next twenty-five years."[2] They got the MRO budget authority increased substantially, from an initial $455 million to $633 million.[3]

The availability of greater resources meant that the project had the potential to advance the technical art for Mars-bound spacecraft. Graf asked Zurek what the biggest technical problem scientists had with doing science at Mars. Zurek told him "data rate."[4] The data return limitation from Mars was a product of a variety of different restrictions in the "data pipeline" back to Earth: the ability of the spacecraft to provide power to its transmitter, the Deep Space Network's ability to receive and retransmit data to JPL, the ground data system at JPL itself. Innovation in each area would be necessary to expand the orbiter's data return.

Finally, Mars Reconnaissance Orbiter was to be the first systems contract project to be performed under the new design principles and flight project practices that JPL implemented in the aftermath of the 1999 failures. These standardized testing and review requirements, while also specifying technical standards and management practices. The design principles specified required margins in many areas that typically had not been addressed in the past; in addition to the traditional mass margins, they set margins for CPU cycles, memory capacity, telecommunications bandwidth, even power switches. They also weighed in heavily against single-string, nonredundant spacecraft systems without preventing them—certain things, like high-gain antennae, cannot be made redundant. But the new focus on rigor required justification and documentation of each accepted case of nonredundancy.

These changes, in company with the preceding decade of experience that Lockheed's Denver-based engineers had acquired in planetary spacecraft design and operations, meant that MRO would be a far different spacecraft than the Mars orbiters that had preceded it.

Designing Mars Reconnaissance Orbiter

In late 2000, Richard Zurek had started the process of determining the potential scientific mission for the orbiter. When Zurek had previously defined the mission in May 2000 during the "shootout" with the MER proposal, the potential launch date of 2003 had caused him to focus on existing instruments—spares of already flown instruments, in essence. When the MRO project was approved for a 2005 launch instead, Zurek gained enough time to use the slower, more formalized NASA announcement of opportunity process and acquire new instrument concepts that way. The first step was to form a science definition team (SDT) for the mission, which was co-chaired by Zurek and Ronald Greely, a geologist at Arizona State University. This team met several times between December 2000 and early February 2001 to hash out scientific objectives and a conceptual payload.

As had happened with Mars Observer's payload, the first order of business for the SDT was in ensuring that the Mars Climate Orbiter's investigations were reflown. Dan McCleese's deputy on his pressure-modulated infrared radiometer, Tim Schofield, had had the instrument completely redesigned to reduce its power and mass requirements after MCO's loss. That new instrument, Mars climate sounder, took the place of the pressure modulated infrared radiometer (PMIRR) in the payload. Mike Malin was also owed a reflight of his Mars color imager (MARCI). But this engendered some controversy in the group because the European Space Agency's Mars Express had a very similar camera going to Mars before MRO would. In its original form, MARCI would not provide new science. Instead, in the course of the SDT's discussions, it would be redefined.

One of the primary points of contention on the SDT was whether MRO should focus resources on a single primary investigation (i.e., high-resolution imaging) or pursue "exploration and discovery in a few carefully chosen [research] areas."[5] A mission dedicated purely to imaging could cover much more of the martian surface than could one in which spacecraft resources had to be shared by several different investigations. The majority of the committee favored a multidisciplinary mission that would simultaneously address several high-priority questions: recovering the atmospheric remote sensing experiments lost with Climate Orbiter, high-resolution imaging, flying a spectrometer like the visual and infrared mapping spectrometer deleted from Mars Observer a decade before, and perhaps even carrying a subsurface sounding radar to search for near-surface ice and possible liquid water. This comprehensive instrument suite would be deployed to "explore in detail hundreds of targeted, globally distributed sites."[6]

The team also had to resolve arguments over the desirable resolution of an onboard camera. Atmospheric and climate scientists on the committee favored a camera that could better study the polar laminated terrain, which had no Earth analogue. But the majority of the geologists wanted lower resolution in order to achieve a wider field of view, so that geologic units could be mapped over larger areas. They ultimately concluded that the high-resolution imager should be paired with a medium-resolution camera that would serve as a "context imager," allowing more of the surface to be seen, and the two imagers should be designed so that they and the spectrometer would view the same scene at the same time.[7] They hoped to bring more of Mars into view this way, not less. This is where Malin's MARCI came into the discussion. In its original form aboard Mars Climate Orbiter, MARCI had been two cameras: a wide-angle camera for imaging the atmosphere and a "medium-angle camera" for photographing large areas of the surface. A redesigned medium-angle camera could serve as the context imager that the SDT sought. The redesigned MARCI medium-angle camera gained a new name—the context camera—while the wide-angle camera stayed in the payload under the original name.

The SDT also supported including a subsurface sounding radar but only as a secondary priority, if it were affordable and wouldn't interfere with the other instruments. The European Space Agency's Mars Express mission was already slated to bring a low-frequency, deep-sounding radar called MARSIS to Mars in 2004, but the science team members thought it had some weaknesses. The Mars Express orbit was highly elliptical and would permit use of the radar only occasionally. The team members also thought it wouldn't operate well on the day side of Mars due to atmospheric interference. The MRO orbit was closer to circular and lower-altitude, and the spacecraft's increased data rate would enable much greater use of a radar. But the team was concerned that a radar aboard MRO might create unacceptable levels of electromagnetic interference with the other instruments, so they recommended that a radar should be flown to Mars only if would not impede the higher-priority MRO investigations.[8]

In June, the announcement of opportunity went out. It faithfully reflected the team's recommendations, soliciting proposals for a high-resolution imager with a surface resolution of about 60 centimeters/pixel from the orbit's highest altitude and a target mass of about 40 kilograms, and an imaging infrared spectrometer with a surface resolution of about 40 meters/pixel from the highest point of the science orbit. In the two months since the SDT report, NASA and JPL had found a foreign partner for the sounding radar, the Italian space agency ASI. ASI had also been involved with the development of the MARSIS radar for Mars Express.[9]

The announcement of opportunity also reflected an increasing commitment to education in the science directorate at NASA. It required the investigation proposals to address public outreach and education, specifying that 1–2 percent of the mission's budget, and about 5 percent of participating scientists' time, be devoted to it.[10]

Selection of the spacecraft contractor proceeded in parallel with the selection of the instruments solicited by the announcement. The large payload mass expected for the vehicle, about 85 kg total, combined with the availability of more money and the desire for higher data rates had led Jim Graf's engineers to assume they would be using either an Atlas III or an Atlas V launch vehicle. An Atlas III could send 2,000 kg to Mars in 2005, and the Atlas V slightly more, so Graf established the mass limit for proposed spacecraft as 1,800 kg (a little more than half of which would be fuel).[11] This was well over twice Mars Odyssey's 725 kg launch mass.

At Lockheed Martin Astronautics, which had to compete again for the Mars contract, the proposal team decided to develop a new orbiter that incorporated their lessons learned from the preceding four Mars missions, as well as one that could achieve the very high pointing accuracy, stability, and telecommunications requirements Graf's engineers had imposed. Steve Jolly remembers that the proposal team had a sense that the previous orbiters had not been very science friendly, and they set out to optimize the design for the scientific mission at the outset.[12] One feature they offered was derived from a Lockheed-built Earth observation satellite, IKONOS. IKONOS was programmed to calculate its own imaging maneuvers once given the latitude and longitude of the desired target, reducing workload for the ground operators. MRO would be able to do the same for Mars. They also included an even larger antenna than the announcement required, enlarging the notional 2-meter-diameter antenna to 3 meters was a cheap way to increase data return dramatically.

The lost Climate Orbiter mission figured strongly in Lockheed's preliminary design efforts, too. That vehicle had required frequent configuration changes during the cruise to Mars and during aerobraking, requiring more work on the part of the operations staff. It had also experienced never-resolved problems with the solar array gimbals (or their position sensors) during flight. So another design ground rule they adopted for MRO was to eliminate configuration changes during cruise or aerobraking. All deployments were to be completed right after separation from the launch vehicle, and the spacecraft would assume a flight configuration that would not change until it was in its final science orbit. One consequence was the need to design gimbal motors powerful enough to hold the solar arrays

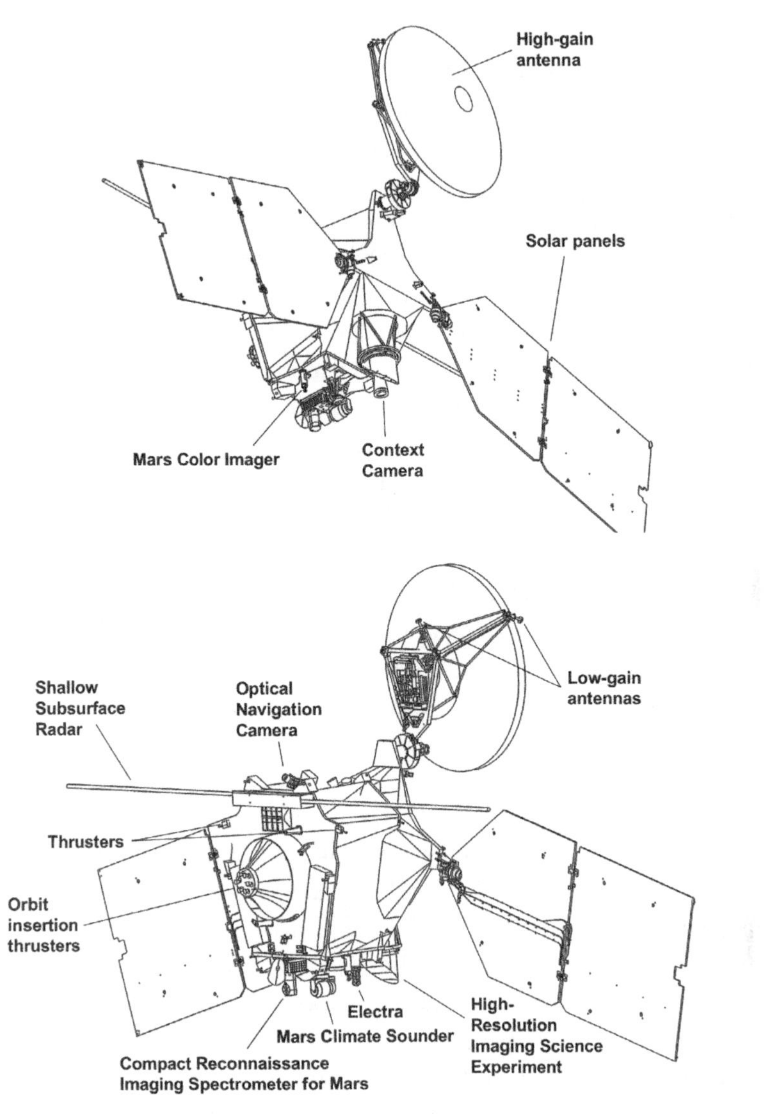

Mars Reconnaissance Orbiter spacecraft. The nadir deck points down in both images. Electra replaced the Mars relay that had flown on Mars Observer, Mars Global Surveyor, Mars Climate Orbiter, and 2001 Mars Odyssey. The optical navigation camera was an experiment intended to improve near-Mars navigation and orbit insertion accuracy. Note the fundamental differences between this design and that of Mars Climate Orbiter (see chapter 6) and the 2001 Mars Odyssey spacecraft (see chapter 8).

and high-gain antenna in position during aerobraking, since they would not have the aid of a mechanical restraint.[13]

The proposal team also decided to simplify the propulsion system by dropping the bi-propellant design they had been using for orbiters since Global Surveyor. The bi-propellant design supplied both hydrazine and an oxidizer to the combustion chamber during Mars orbit insertion to gain greater efficiency but at the cost of significant complexity (additional piping, valves, tanks, and pressure regulators). The job could be done with hydrazine only, the same way the attitude control thrusters worked, using a catalyst bed to create the reaction. The tanks simply had to carry more hydrazine. The team went further than that, though, designing a thruster cluster with one more engine than necessary so that MRO would be able to complete orbit insertion even after loss of a thruster. The design eliminated an important single-point failure possibility.[14]

MRO was designed with redundant CPUs and memory, with the ability to swap between them automatically in the event of a fault, eliminating another possible single-point failure. The Lockheed team also aimed to reduce the reboot time from about 90 seconds to 10, so that the vehicle would be able to complete its orbit insertion even if it experienced a CPU fault in mid-burn. After swapping computers, the fault protection system would be able to regain attitude knowledge automatically and restart the insertion burn without ground assistance.[15]

Another decision Jolly's designers made was to try to simplify access to the instruments and improve the vehicle's testability by using an open bay architecture for the vehicle bus. Jolly remarked, that way "you didn't have to remove ten things to get to one."[16] This design was possible because the vehicle could be much larger due to the larger, medium-class launch vehicle.

The contractor selection was based on "best value." Here the Lockheed team's decision to propose a larger antenna than that specified in the request for proposals worked strongly in their favor. It provided a large increase in data return for not very much more money. "We wanted the ability to evaluate how much they were asking for versus how much science we could return," Graf recalled. "There were some other very nice things that Lockheed did that enhanced the science that we were looking for without really driving the price up."[17] Lockheed won the contract over two other bidders with this approach.

The winning proposal for the high-resolution camera came from Alfred McEwen of Arizona State University. McEwen had forged an alliance with W. Alan Delamere of Ball Aerospace on an earlier proposal, a camera for a Europa Orbiter mission that was never funded. When the opportunity to propose the MRO camera came up, Delamere had suggested proposing a lightweight variant of Ball's QuickBird satellite camera.[18] Their proposed camera, the high-resolution imaging science experiment (HiRISE), would have a 0.5 meter diameter primary mirror and a resolution at the surface of about 0.3 m/pixel. It was also a "pushbroom" imager, building images up line by line. But in order to accumulate enough light to make a clear image given the relative darkness of the surface and the spacecraft's high speed over the ground (3.2 km/second), the camera in essence would image each spot on the ground 128 times. The optics placed the line image onto a set of 14 charge-coupled detector (CCD) arrays that were 128 lines deep. The camera's electronics had to shift the electrons generated in that line of detectors "down" one detector row at the same rate of speed as the spacecraft's motion across the surface (a shift every 74 to 100 microseconds depending on the spacecraft's altitude) to build up a sufficient charge. After 127 shifts, the resulting line of data went into its place in the camera's memory. (This process is called time delay integration). A complete HiRISE exposure would take between 6.2 and 12.7 milliseconds to complete. Optimized for the red end of the visible light spectrum, the camera also included some CCD arrays capable of imaging in the blue and near infrared so that the central portion of the scene would be in color.[19]

McEwen had responded to the announcement of opportunity's emphasis on public education and outreach by proposing to allow members of the public to use a downloadable client to suggest targets and present rationales for their choices. As security needs became better understood, the downloadable client, called HiWeb, was restricted to the science team while a public Web site, HiWish, permitted public participation and also allowed access to all of HiRISE's imagery.[20]

The infrared spectrometer NASA selected had been proposed by Scott Murchie of the Johns Hopkins University Applied Physics Laboratory. Murchie had been involved with analysis of data from the Soviet Phobos 2 mission, and he was also an experimenter on the MESSENGER mission to be launched to Mercury in 2004. For MRO, he had proposed a compact reconnaissance imaging spectrometer for Mars (CRISM).[21] CRISM took a very different approach to solving the problem of image smearing during the relatively long exposure times necessary. Instead of using the time delay integration method, the instrument's sensor head was mounted on gimbals so that it could track the surface beneath the spacecraft, compensating for spacecraft motion during the exposure. The instrument's projected best spatial resolution was about 40 m/pixel and could resolve more than 500 wavelengths to aid mineral identification.[22] The instrument could also operate in a lower-resolution, larger-swath mapping mode so that the science team could eventually produce a comprehensive map of Mars.

Murchie intended CRISM to look for outcrops of carbonate minerals on Mars. Finding them was very important to understanding the climatic history of Mars because carbonates form in liquid water. But the spacecraft sent to Mars thus far had not found carbonates, despite finding the necessary constituents. It was difficult to believe that Mars had never formed carbonates. Instead, Murchie thought that the carbonate outcrops were simply smaller than the effective spatial resolution of CRISM's predecessors. A consequence of his desire to identify carbonates was that CRISM was designed with a slightly wider spectral capability than that specified in the announcement of opportunity and had to be kept colder as a result.[23] It made for a technical challenge because MRO was expected to roll in order to make HiRISE images, which meant a radiative cooler for the instrument would sometimes look at Mars instead of deep space, and looking at Mars would cause the instrument to warm. An active cooling system could counter that tendency but would add complexity and mass.

By the time the instrument selection had been completed, a few more decisions had been made about the payload. The Italian space agency had tentatively agreed to provide a "shallow radar" (SHARAD), for the mission although the project team had not yet figured out how to accommodate its antenna on the spacecraft. The Mars Exploration Program office at JPL had added two technology experiments to demonstrate technologies that might be useful to future missions. One was an optical navigation camera designed to improve navigation accuracy close to Mars. The other was a telecommunications demonstration of higher-frequency communications. JPL's Deep Space Network had operated in a frequency region called X-band for decades, but shifting to the higher "Ka-band" offered potentially higher data rates and would resolve a problem of spectrum crowding.

Graf tried to reject the Ka-band experiment, because initially it was not funded. The Deep Space Network directorate at JPL had approached Graf about putting its hardware on MRO but had no money to pay for it. Graf said no. He didn't know that the Deep Space Network was also promoting the idea through JPL's administrative chain of command, and at a director's review he was told to find a way to accommodate the experiment. Eventually, the Space Operations Directorate at NASA headquarters, which financed the Deep Space Network, agreed to provide $10 million to cover the cost.

The selection of Lockheed and of the instruments was completed in November and December 2001, and the MRO project team started their accommoda-

Mars Reconnaissance Orbiter payload

Instrument	Investigators (institution)
Compact reconnaissance imaging spectrometer for Mars (CRISM)	Scott Murchie (Applied Physics Laboratory of Johns Hopkins University)
Context imager (CTX)	Michael Malin (Malin Space Science Systems)*
High-resolution imaging science experiment (HiRISE)	Alfred McEwen (University of Arizona)
Shallow subsurface radar (SHARAD)	Roberto Seu (University of Rome)
Mars color imager (MARCI)	Michael Malin (Malin Space Science Systems)
Mars climate sounder (MCS)	Daniel J. McCleese (JPL)

Source: Adapted from Zurek and Smrekar, "Overview of the Mars Reconnaissance Orbiter Science Mission."

*Rather than a principal investigator (the leader for an instrument chosen through a competitive process), Malin served as CTX "team leader" (the designation of the lead scientist for an instrument designated as a "facility instrument" by NASA and generally not competitive).

tion phase. There were a few outstanding issues that had to be dealt with in the project's first few months. One was that McEwen's camera required far greater pointing accuracy and stability from the spacecraft than JPL had required in its request for proposals. "We were absolutely horrified because we had specified the spacecraft at a whole different level. Headquarters essentially did not take into account what we told them the capability of the bus was," Graf remembered, in selecting the instruments. The spacecraft had to be substantially redesigned to suit HiRISE.[24] This was a product of its use of time delay integration as its imaging method. Each of the 128 line arrays had to see exactly the same piece of martian real estate as the line before it in sequence; any vibration in the spacecraft structure could displace the image enough to produce smearing.

Solving the problem ultimately required making several changes to the vehicle and to how it would be operated. MRO's intended orbit was elliptical, a 200 km by 400 km orbit with the orbit's periapsis (lowest point) gradually drifting around the planet. The low periapsis meant higher effective resolution for HiRISE, but it also meant more aerodynamic force on the spacecraft, more vibration, and more fuel used to maintain the orbit. A more circular, but somewhat higher altitude, orbit would provide greater stability and a longer potential mission life, so Jolly's mission designers proposed to the science working group raising the orbit to a 255 km by 320 km ellipse with the periapsis locked near the martian south pole. McEwen thought altering the focal length of his camera would recover the lost resolution, so he and the science team agreed to the change.

The second big change to MRO aimed at fixing the stability problem but increased its mass considerably. The structure had to be stiffened, and more importantly, the orbiter needed bigger reaction wheels. The larger reaction wheels were a mass cost of 28 kg.[25] This increase meant carrying about the same amount of additional fuel, so it was not trivial.

The third change was to the vehicle's science operations, and it required negotiation between McEwen and Dan McCleese, the principal investigator for the Mars climate sounder instrument. MCS was designed to scan across the spacecraft's ground track continuously, producing wide strips of data that would almost completely cover the Mars atmosphere every day. It had also been designed for a spacecraft that was "nadir oriented," one whose instrument mountings always pointed straight down toward Mars. But MRO would not do that; instead, it was going to roll frequently to point HiRISE and CRISM toward targets that were off-axis. That would produce data gaps for MCS. The instrument also used motors that would induce vibration in the spacecraft structure and smear HiRISE's images. Ultimately, McCleese and McEwen agreed to a set of flight rules that required MCS to stop scanning a minute before HiRISE took an image. Because MCS lost data when it stopped, they and the rest of the science team also agreed to limit the number of HiRISE rolls that could be commanded each orbit and each day so that MCS's scientific output was not degraded too much. This was not as great a sacrifice on McEwen's part as it might sound, since the spacecraft's onboard data storage capacity, though very large compared to previous missions, could still only permit a few HiRISE images per day.[26]

At the project's preliminary design review in August 2002, Glenn Cunningham, chairman of the standing review board that JPL had assembled to oversee the MRO project, declared that "development of the HiRISE, CRISM, and SHARAD instruments are the project's highest risk items."[27] Graf recalls one of the board members commenting that he hoped Graf had lots of money, as the instruments were certain to overrun. (The project's payload development budget at the time was $88.3 million.) There was less concern about the spacecraft development, except for the ability to return, and manage, the huge stream of data that the project planned to send back from Mars. Data processing and management drew the board's second-highest risk.

During 2002, MRO's mass grew relentlessly, just as it had to Pathfinder and MER. The spacecraft was a new development, and new spacecraft nearly always wind up overweight; NASA's decision to select HiRISE with its far-out-of-specification need for pointing accuracy and stability further exacerbated that tendency. But as it turned out, both HiRISE and CRISM had mass growth of their own to contribute the MRO's troubles. The growth during left Jim Graf with the possibility that his spacecraft would be too heavy to send to Mars unless he either spent a lot of money that he did not have on mass reduction efforts or dropped one of the heavier instruments.

He was saved from having to do either by Lockheed's launch vehicle division, which made him an offer he couldn't refuse.[28] In July 2002, the Kennedy Space Center, which handled launch vehicle procurement for NASA, had chosen Lockheed's Atlas III over the Delta III. The Atlas III could throw slightly more mass toward Mars than the Delta III, raising the allowable injected mass for MRO from 1,975 kg to 2,000 kg. That was enough to restore his mass margin almost to the requirements of JPL's design principles, which required 20 percent margin at the preliminary design review. But between the July review and the end of the year, mass growth had reduced the margin to about 7 percent, an unacceptable level.[29] It had to be at or above 10 percent by the project's critical design review in May 2003. Lockheed's launch vehicle division, which was organizationally separate from its spacecraft division (but also located in Denver) had approached Graf about switching to the company's new, slightly more powerful, Atlas V. Switching to the Atlas V would add another 125 kg to the spacecraft's allowable mass. But even better, Lockheed did not ask for more money. The Atlas V would not cost the project any more than the Atlas III would.

What "kept him up at night," Graf explained, was the rocket's newness. The variant Lockheed offered, the Atlas V 401, had just made its maiden flight in August and would make its second flight in May 2003, right at his project critical design review. MRO would only be the third Atlas V 401 flight, and the fifth Atlas V of any version. Going to the Atlas V put the project at some additional risk, due to both the vehicle's short track record and the launch vehicle certification process, which would be handled by Kennedy Space Center and out of his project's hands. Failure of the second launch might well cause NASA to refuse to certify the vehicle until more launches took place, wrecking his schedule. But it got him out of a big mass bind. In March 2003, he told the project science group that they were switching to the Atlas V.[30]

Phoenix: The First Mars Scout

As Graf was settling on his final launch vehicle, NASA was selecting the first of the competed Mars Scout missions envisioned under the strategic plan adopted in 2001. Like NASA's Discovery program, its Scout program solicited proposals for entire missions via an announcement of opportunity. The rules required that a single principal investigator submit the proposal; in practice, potential principal investigators assembled proposal teams to help put it together. The Scout guidelines limited mission cost to $350 million, more than twice the funds available to the lost 1998 missions.

The Scout announcement of opportunity was released in May 2002, with the first round of proposals due August 1. The selection was to be made in two parts. In the first round, NASA would select four proposals. The proposers would receive $500,000 to fund more detailed, six-month long studies due in to headquarters in July 2003. In August 2003, Orlando Figueroa, who had succeeded Scott Hubbard as Mars program director, and associate administrator for space science Ed Weiler would make the final selection.

The Mars Exploration Program Office at NASA headquarters received 25 proposals for missions in response to the Scout announcement. The four Figueroa chose for detailed studies were varied. One, an atmosphere sample return mission called SCIM, was to use an aerogel-based collector to obtain samples of Mars's atmosphere and return them to Earth. Another, an investigation called ARES, from NASA Langley Research Center, would deposit an airplane in the martian atmosphere to conduct magnetic and low-altitude atmospheric chemistry surveys. A third, proposed by a JPL scientist, was an orbiter aimed at better understanding of the photochemistry of Mars's atmosphere. These three missions had in common that they were primarily aimed at understanding the martian atmosphere. The outlier was Mars Phoenix, a reflight of Polar Lander's mission, proposed by Peter Smith of the University of Arizona, formerly principal investigator of the imager for Mars Pathfinder and an important member of Alfred McEwen's HiRISE team.

All four of the chosen finalists had proposed JPL as the project management center and Lockheed as the implementing contractor, requiring JPL management to do some internal reshuffling to isolate the four teams. Diana Blaney, who had been serving as project scientist for the first round proposals for both Phoenix and ARES, decided to stay with ARES. Smith then asked a former student of his named Leslie Tamppari to take on the role of project scientist. Barry Goldstein, who was serving as deputy spacecraft manager for Mars Exploration Rover and had worked with Smith before on the payload for Polar Lander, was asked to take on the Phoenix proposal management job as an additional duty. He took it because, as he put it, "I never have a second chance to say no."[31] Lockheed's Edward Sedivy, who had been chief engineer for the Surveyor 2001 lander, would reprise that role for Phoenix, providing technical continuity for the project.

The basic concept for the Phoenix mission was to take the 2001 lander hardware that was in storage in Denver, recondition and assemble it, test it rigorously, fix the problems already identified in the Polar Lander failure investigations and whatever new ones were discovered in the process, and send it off to Mars. Orbital mechanics dictated that Phoenix went to Mars's north pole instead of the south pole, where Polar Lander had been aimed, but the vehicle would carry copies of Polar Lander's instruments and two more built for the 2001 lander. It would have a short, busy 90-day primary mission of trench digging and soil analysis, with perhaps an additional 60–70 days possible before the polar winter set in and lack of sunlight killed the lander. The landing site would be in a region around 70°N 240°E, which was apparently flat and in which Boynton's gamma ray spectrometer showed a strong water-ice signature.

As a surface mission, Phoenix was aimed directly at the Mars program's overarching water strategy. Smith made that clear on his concept study cover, which baldly stated: "Phoenix: 'Follow the Water' "[32] That no doubt helped his chances with NASA headquarters. But at least equally helpful was the low cost risk. With the spacecraft hardware more than 70 percent complete, the flight software about 85 percent complete, and three of the six instruments already delivered, the proposal seemed to have little chance of experiencing a large cost overrun. The Scout selection was also the year in which the Mars Exploration Rover mission had experienced $100 million in overruns in its race to the launch pad, and Smith and Goldstein had emphasized the state of completeness in their proposal. They proposed a cost to NASA of $318 million.[33]

The mission's principal strength was also its main weakness: high heritage. It shared a design that had been called into question by the Polar Lander failure, rendering it suspect. Addressing the landing leg problem that had been fingered as the most likely cause of the loss only required a simple software fix, but the review boards had called into question the basic soundness of the pulse-jet terminal descent system, too. If those reviewers prevailed, there would not be a Phoenix mission. Because the lander was built around its propulsion system, if the pulse-jet system really was untenable, then the whole project would have to be scrapped. What Goldstein proposed to do to resolve that big question was build and test a "hot fire testbed" immediately, during the project's Phase B, before the critical design review and the project's entry into what is normally the major construction and testing phase.[34] Demonstrating that the terminal descent system was controllable, worked without experiencing damage from water hammer, and did not produce shockwaves substantial enough to damage the lander structure would retire a very big risk early in the project.

The Phoenix mission was selected in early August 2003, a month after the Mars Exploration rovers had been launched.[35] Destined for their own launch

Mars Phoenix payload

Instrument	Investigators (institution)
Mars descent imager	Michael Malin (Malin Space Science Systems)
Surface stereo imager	Mark Lemmon (Texas A&M University)
Thermal evolved gas analyzer	William Boynton (University of Arizona)
Microscopy, electrochemistry, conductivity analyzer (MECA)	Michael Hecht (JPL)
Robotic arm and camera	Ray Arvidson (Washington University in St. Louis) Uwe Keller (Max Planck Institut für Aeronomie)
Meteorological station	Jim Whiteway (Canadian Space Agency)

Source: Adapted from Smith et al., "Introduction to Special Section on the Phoenix Mission," and Phoenix Landing press kit, May 2008, www.jpl.nasa.gov/news/press_kits/phoenix-landing.pdf.

four years later, the project's engineers would find that many more things had been wrong with Polar Lander than had been discovered by the various failure review committees. They would also put a final end to the faster-better-cheaper era's martian saga.

Whither the Mars Smart Lander?

The year 2003 also saw a major realignment in the Mars program's next strategic mission after MRO, Mars Smart Lander (MSL). After the large lander study of 2000 had closed down, Tom Rivellini had run a technology task designed to investigate further two of the approaches conceived during it, the pallet lander and the 4pi airbag lander. His engineers had built scale models and tested them in a variety of simulated martian terrains to help them understand how each would perform. In part due to the problems MER had with its own airbags, the pallet lander had become the baseline design. As the mission concept stood in 2001, Smart Lander's goals were to demonstrate active hazard avoidance during terminal descent, precision landing, and increased mobility, as well as the capacity to land such a large vehicle safely. It was not yet conceived as a high-science return mission.

In parallel with the engineering studies of the lander, a science definition team had been established to develop a scientific rationale for the mission. Chaired by Raymond Arvidson of Washington University in St. Louis, it was chartered in April 2001 and completed its report in October. This team argued that MSL should achieve "major scientific breakthroughs" while preparing the way for sample return.[36] The Smart Lander mission, with a notional cost of $750 million, was too expensive an opportunity to waste on an engineering demonstration.

Because the technological purpose of MSL was to demonstrate the ability to safely land a large payload as a precursor to carrying out sample return, the project's large lander—about 800 kg in mass—was to have about 100 kg available for payload. For the sample return mission, that 100 kg would primarily be allocated to the ascent rocket that was supposed to place the collected samples in a low Mars orbit. But for the Smart Lander mission, it was available for scientific instruments. The science definition team took the availability of a large payload mass as an opportunity to propose two very different, but equally ambitious, missions: a mobile geobiology explorer and a multidisciplinary platform. The geobiology explorer would feature a large, autonomous rover aimed at looking for the chemical signatures of past life on Mars, while the platform would be a fixed lander designed to drill 5 to 10 meters beneath the surface and analyze the resulting samples. Either would be, in their words, "the capstone mission for this decade."[37]

A variety of factors biased the MSL decision toward the rover. The pallet lander was designed to facilitate rover egress. Scientists wanted regional-scale surface mobility for sample return anyway, to ensure they could obtain the most scientifically desirable samples. And JPL engineers, having originated Mars surface mobility, wanted to expand their capabilities. During early 2002, the pre-project team JPL assembled performed trade studies on four variants suggested by the science definition team report, lander versus rover and solar versus radioisotope power.

The expansive possibilities, unsurprisingly, produced a rapid cost escalation, and later in 2002 a "design to cost" study had to be convened to relimit the options. This project science integration group (PSIG) used the results of a recent Mars Exploration Program Analysis Group study to argue for a mission focused determining the habitability of *ancient* Mars. Finalized in June 2003, this mission concept was based around the large rover with a somewhat smaller payload than the geobiology explorer. Reflecting their desire for breakthrough science, the PSIG argued for significantly greater money for the mission: "The PSIG and the MSL Project doubt that the resources, as presented to PSIG, for MSL will be sufficient to fund the payloads needed to meet the science floors of scientifically supportable missions."[38]

At the same time, the pre-project's engineers were souring on the pallet lander concept. Like the airbag landers, the pallet promised an expensive developmental testing program because numerical analysis was not sufficient to fully characterize the pallet structure's ability to collapse predictably while still protecting the fuel tanks and assorted propulsion hardware. Rover egress from the pallet at a variety of slopes and rock abundances would also have to be tested. In contrast, the "landing on wheels" approach, which JPL's engineers had started to refer to as "Skycrane" after the Sikorsky S-64 Skycrane helicopter, had begun to look more and more promising. It eliminated the developmental test program the pallet lander required and simplified some other engineering problems—load paths through the structure, packaging and integration of a radioisotope thermoelectric generator power source if NASA chose that path, and so forth. The Skycrane was a much more tractable problem analytically, since at the moment of touchdown, only the rover made contact with the ground. The descent stage stayed hovering above it until the cable connecting them was cut, then it flew off and crashed. It did not have to survive the crash or protect a payload. The descent stage and rover could be tested separately, and there was no egress problem. Their rough costing exercise put the Skycrane's development cost at about one-third that of the pallet lander.[39] By their October 2003 mission concept review, the pre-project had settled on the Skycrane approach.

The Committee on Lunar and Planetary Exploration of the National Academy of Sciences had reviewed NASA's Mars strategy during 2002 as well, and it joined the MSL science definition team and PSIG in advocating for remaking Smart Lander into a high science mission. In particular, they urged "the use of nuclear power sources, if at all feasible, on advanced Mars lander missions" to achieve longer mission lives and increase the science return. Accordingly, the MSL rover gained two radioisotope thermoelectric generators for power.[40] The decision to use radioisotope power also contributed to a decision to delay the mission's launch to 2009.

Reflecting the changed nature of the mission from technology demonstration to high-science return, the pre-project was renamed Mars Science Laboratory.[41] In later years, NASA and JPL management would regret keeping the same acronym, MSL, for two rather different mission concepts. But in 2003, it seemed clever.

Building Mars Reconnaissance Orbiter

While Smart Lander was being transformed into Science Laboratory, James Graf's engineers were working to deliver their flagship-scale orbiter to Mars. Mars Reconnaissance Orbiter, fortunately for the Mars program's budget, had neither the developmental challenges that MER had nor the conceptual turmoil of MSL. For most of the development period, getting the instruments assembled and tested was the project's chief challenge.

The ambitious nature of the instruments chosen for MRO had caused Graf to establish a review panel focused solely on the payload. He picked Fred Vescelus, a jolly instrument specialist who'd been at JPL since the late 1960s, to chair that panel. He wound up going onto HiRISE almost full-time after the project's critical design review because Ball Aerospace was getting further and further behind schedule on the camera. The principal challenge they faced was with the electronics design and integration. The camera's electronics had to run much faster than even HiRISE's QuickBird predecessor while still keeping all fourteen CCD arrays perfectly synchronized. The camera also had an unusual thermal design: it would heat up rapidly while making an image and had to shut off when it got too warm. The heating problem was the principal limitation on the size of a HiRISE image, not the amount of memory the camera had. More challenging for the project was that the heating tended to change the circuit timing. The circuitry ran faster cold than warm; if the camera heated unevenly across the detector arrays during an image, timing drift made a mess of the result. When the camera was first assembled and tested in June 2004, there was a great deal of image corruption. Resolving it delayed HiRISE's final delivery to the project until after the spacecraft had already been moved to Kennedy Space Center for launch. The final programming for the focal plane array's electronics was loaded at Cape Canaveral as well.

CRISM shared some similar electronics troubles, as well as difficulties with its complex thermal design. CRISM had a set of active Stirling coolers, which removed heat from the detectors and transferred it to a radiator mounted on the anti-sunward side of the orbiter so that it would never see the Sun. Some of the orbiter's rolls to point HiRISE at various targets would cause the radiator to face Mars briefly, however, and Mars is a heat source, too. Seeing Mars would cause the radiator to be less efficient and result in the instrument warming. But for the first year or so of the project, there was controversy over how much the occasional view of Mars would affect the instrument's temperature. Thermal models from JPL, the Applied Physics Laboratory, and Lockheed produced widely different results, which took time to understand. As of the instrument critical design review in May 2003, the discrepancies still had not been resolved.[42] Some of the instrument's electronics drew more power than expected and generated more heat, while the Stirling coolers were less efficient than hoped.

Murchie's team also ran afoul of the management changes that followed the Surveyor 1998 failures. The MRO announcement of opportunity had permitted principal investigators to propose their own management methods, and Murchie had planned to use the Applied Physics Laboratory's less intrusive, lower-cost management methodology for CRISM. That turned out to be unacceptable to JPL in the new environment, and Vescelus, as chairman of the instrument review board, had criticized the lack of a full-time instrument manager and senior management oversight at the critical design review.[43] Reinforcing the instrument management, of course, cost more.

The SHARAD radar gave Graf's engineers some additional headaches. By its nature, SHARAD would broadcast a lot of electrical energy in close proximity to the vehicle, making shielding of the vehicle's other electronics from it more difficult. The initial antenna design placed it on an outrigger trailing behind the spacecraft to reduce the interference hazard to the spacecraft, but proved mechanically unviable. The antenna got moved back onto the spacecraft body, and Lockheed's engineers wound up wrapping copper tape over the spacecraft's wiring to shield it instead. SHARAD was not the only electronic interference problem, as it turned out, though. Murchie's cryocoolers interfered with Electra, the UHF relay package, and so did McCleese's Mars climate sounder. For that reason, ultimately, the MER and Phoenix projects would largely rely on Odyssey, not MRO, for their UHF relay needs so as not to impair MRO's science mission.

Toward the end of the project's development phase, it ran into one big technical snafu reminiscent of the diode problem the Surveyor 1998 project had run into: the field programmable gate arrays (FPGAs), special solid state devices that had been built into the spacecraft's electronics and its instruments, were found to have a strange problem. Once in awhile, an FPGA would suddenly lose its programming. About a year before MRO's launch, another NASA project had discovered the problem, but Graf's engineers initially thought their testing would catch what Graf later called the "weak sisters," parts that might be bad. But then CRISM had one array fail even after passing all the tests, invalidating the team's assumption. Replacing all the suspect FPGAs with a new, untested design was not appealing, since it meant that all the circuit boards that had already been tested would have to be retested. And that still would not really assure the team that the new FPGA design was any better than its predecessor. "We didn't have any clear indication or testing to guarantee that they wouldn't have the same problem," Graf recalled.

Ultimately, Graf had all the FPGAs replaced with the new design. "The logic was," he said, "we know that the old design was faulty. We couldn't say why. So let's try the new design." Only the SHARAD radar flew with the original design, because the International Traffic in Arms Regulations prevented JPL from telling the Italian contractor about the FPGA problem. The change-out took a lot of work but fortunately did not introduce any new problems.

The project's final year before launch presented Graf the only cost increases that he couldn't fund out of his project reserve. These were largely a product of unexpected demands from senior management at JPL. The Lab's chief engineer, Brian Muirhead, had set in place a certificate of flight readiness (CoFR) process modeled on that used by launch vehicle manufacturers. It was designed to ensure that JPL's flight projects met a specific level of rigor in testing, following up on anomalies reported in testing and in documentation. Like his Pathfinder colleague Tony Spear, Muirhead did not think the extensive design reviews that were JPL and NASA's traditional solution to maintaining quality were terribly effective (and they were very expensive). Instead, he thought testing rigor was the best way to ensure product quality, and he hoped his certification process would enable reduction of design reviews in the future as JPL and NASA leadership adjusted their own thinking.[44] Muirhead's CoFR process was built on a set of audits that reviewed a project's compliance with the new design principles and flight project practices, its own "incompressible test list," which he as chief engineer negotiated at the beginning of a project, and gauged that project's progress against what was left to do. By MRO's third audit, in June 2005, the spacecraft and instruments were built and well along in testing, with the exception of HiRISE, still struggling with timing troubles.

What was not going as well was closeout of the "paper," the documentation of all the anomalies that had appeared in testing and been addressed.[45] In the new management environment that prevailed after the Surveyor 1998 losses, informal responses to problems and undocumented modifications were no longer acceptable. The flagship scale of MRO made the formal documentation requirement challenging, and the project was falling behind during early 2005. Typically flight projects begin to shrink their workforce during the last nine months or so before launch; spacecraft are generally nearly finished by then, and the test workforce is smaller than the design and manufacturing workforce. But Graf actually had to bring more systems engineers in to get all the paperwork done.[46] And he had to ask NASA for more money to pay them.

Graf's worries about the newness of his chosen Atlas V proved unfounded, and MRO launched successfully on August 12, 2005, after two days of minor technical delays. The flight to Mars was also relatively uneventful, with HiRISE proving to have a detector anomaly that could be worked around. But the Ka-band telecommunications experiment that Graf had grudgingly accepted never took place. The switch that was supposed to connect the Ka-band transmitter to the antenna failed in a partly open position, presenting some risk of disabling the spacecraft's telecommunications system if the Ka-band system was powered up, so it was not. The now more than $700 million mission was too much to risk on a telecommunications experiment.

Conclusion

The MRO spacecraft and instruments cost about $450 million to develop, not counting the launch, science team, and operations costs; overall, it was a $720 million mission through the end of 2010.[47] Its size, complexity, capability, and cost reflect both the end of the faster-better-cheaper era's focus on low cost and the return of scientific ambition. MRO and its contemporary in the development cycle, Mars Science Laboratory, reflected "flagship" class missions of the type that NASA administrator Dan Goldin had tried to bring to an end. The scientific community NASA served had explicitly argued for the revival of these high-science missions; both MRO and MSL also served engineering needs and as technical precursors for the still far-in-the future sample return mission. In short, they represented a confluence of interests.

Mars Reconnaissance Orbiter costs (millions of U.S.$)

FY	2002	2003	2004	2005	2006	2007	2008	2009	Total
Development	70.0	146.9	182.4	104.2					503.5
Operations				4.2	24.1	23.7	17.8	11.6	81.1

Source: NASA FY 2005 Mars Exploration Budget proposal, www.nasa.gov/pdf/55390main_07%20MEP.pdf.
Note: Costs are in real year dollars.

From the outset, Mars Reconnaissance Orbiter was intended to be part of an infrastructure at Mars for conducting surface science. HiRISE's engineering purpose of landing site safety analysis would be quickly put into use for the Mars Phoenix mission and for surface operations by the Mars Exploration Rover mission when MRO finally reached Mars.

Robotic Geologists on the Red Planet

On February 1, 2003, NASA's space shuttle *Columbia* disintegrated over Texas, with the loss of all seven astronauts aboard. *Columbia* had been on one of the last scientific missions scheduled for the shuttle program, which by then was almost entirely devoted to completing the International Space Station. As occurred after the *Challenger* accident in 1986, shuttle flights were suspended indefinitely while the cause of the accident was investigated. That meant NASA's only visible activity during the more than two-year halt to the shuttle program was by its robotic science spacecraft. In particular, it ensured that the public eye would be on JPL's Spirit and Opportunity rovers, which would reach Mars in January 2004.

The *Columbia* tragedy set in motion a reconsideration of U.S. space policy, which for two decades had been centered on the space shuttle. While Dan Goldin had left NASA in 2001, his successor, a former secretary of the Navy, Sean O'Keefe, had continued Goldin's drive for a new direction for the human space program. But what emerged was not a Mars-directed policy. Rather, it was one aimed at establishing a permanent, sustainable human presence on the Moon, through the use of lunar resources—a mining colony, in essence. Announced on January 14, 2004, President George W. Bush's "Vision for Space Exploration" called on NASA to retire the space shuttle in 2010, end support for the International Space Station in 2017, and develop a new heavy lift launch vehicle, crewed Moon lander, and base infrastructure for missions in the 2020s.[1] As often happens in NASA, the new policy also swept into it some other planned missions that were not essential to the new destination: a large, multi-mirror space telescope to replace the Hubble Space Telescope, named for NASA administrator James Webb, and an enormous, nuclear-reactor powered vehicle called the Jupiter Icy Moons Orbiter.

What this ambitious program did not sweep in was a much larger NASA budget. The agency budget would rise about 17 percent over five years, and then be held to the level of inflation.[2] The George W. Bush administration had responded to the September 11, 2001, terrorist attacks by invading Afghanistan, and then had marketed an invasion of Iraq in 2003 during the same year the Vision for Space Exploration was being prepared in Washington. These two wars cost tens, and eventually hundreds, of billions of dollars, so the administration did not expect Congress to allow a large increase in funds for NASA. What NASA headquarters needed from JPL in January 2004 was a highly visible success to help sell even this small budget increase.

This pressure to succeed was compounded by new competition in Mars exploration. In 1998, Japan had launched its first Mars mission, referred to as both Nozomi and "Planet B." Primarily an atmospheric science mission, Nozomi experienced various difficulties on the flight to Mars, and did not reach the Red Planet until late in 2003. It failed to execute its orbit insertion sequence correctly, though, leaving it stranded in a solar orbit.

The European Space Agency had also gained approval for its first Mars mission in 1999, an orbiter based on the Rosetta comet mission named Mars Express, and launched it from Russia's Baikonur complex just a few days before the two Mars Exploration Rover missions were launched. Mars Express carried instruments rebuilt from the failed Russian Mars '96 mission, a new low-frequency radar called MARSIS that had some significant JPL participation, and, of great interest and concern to JPL, a low-cost lander called Beagle 2.[3]

Mars Express itself was a relatively inexpensive mission at around $185 million, but Beagle 2 was dirt cheap. Managed by Colin Pillinger of the Open University, Milton Keynes, UK, the Beagle 2 consortium spent about $60 million developing its lander. It was accordingly small, 33 kg, less than a tenth of the MER landers' mass. But it was ambitious, with a sampling arm and drill, mass spectrometer, gas chromatograph, X-ray spectrometer, and a Mossbauer spectrometer. Like the Viking landers, it was to look for the chemical signals of life on Mars.[4]

The launch of Mars Express and Beagle 2 on June 2, 2003, posed a difficult political dilemma for both NASA and JPL. What if Beagle 2 worked? MER had spent about $680 million on the two spacecraft; the Pathfinder lander had cost $170 million; even the failed Polar Lander had cost more than Beagle 2. Nobody at JPL thought they knew how to do a successful Mars lander for $60 million, while everyone understood that if Beagle 2 succeeded, there would be enormous pressure to try to replicate it. Faster, better, cheaper would return with a vengeance.

But while Mars Express promised competition, many participants also saw it as a cooperative venture. NASA managers had negotiated an agreement with the European Space Agency to include a UHF relay on the mission so that it could also serve as relay for MER and future surface missions in the event that 2001 Mars Odyssey failed. Mars Express and Mars Reconnaissance Orbiter had overlapping team membership among the scientists, which would facilitate instrument testing before launch, ground system testing during cruise, and targeting interesting bits of Mars once in orbit. Competition or not, Mars Express would also become part of the research infrastructure for Mars.

Landings

On December 19, 2003, ground controllers at the Mars Express mission control facility in Darmstadt, Germany, commanded Beagle 2 to separate from the orbiter and make its descent to Mars. The small, disk-shaped lander was supposed to have a four-day flight before reaching the Mars atmosphere; it would actually land on Christmas Day. But Beagle 2 had no way of communicating during its short flight. Like Polar Lander, it was designed to be silent until it was safely on the surface. If it didn't make it, the spacecraft's operators on Earth would never know what happened.

Mars Express itself went into orbit successfully on December 25, but Beagle 2 remained silent. It was never heard from again, despite efforts very much like those JPL undertook to contact Polar Lander.[5] On February 6, the Beagle 2 management team declared the mission lost and began an investigation. One possibility for Beagle's demise was a problem that troubled engineers at JPL, too. The martian atmosphere was acting up.

During the final week of December, another global dust storm occurred on Mars. Like the one Mars Global Surveyor had encountered during its aerobraking, the storm caused the atmosphere to expand and become less dense. The less-dense atmosphere reduced the parachute performance, creating the possibility of exceeding the time margins they'd built into the entry, descent, and landing sequence—a "timeline threat," in their jargon. The MER EDL team spent part of the week arguing among themselves over whether to change the timeline; they could, and ultimately did, send commands designed to raise the parachute deployment altitude.[6] That had not been an option for Beagle 2, which could not be reprogrammed before touchdown.

Another, rather more difficult to diagnose, problem emerged that last week of 2003, too. On Sunday afternoon, December 28, the MER team had two landing simulations fail in their flight software testbed. The failure wasn't discovered until the next day, and the testbed supervisor, Jason Willis, did not find out about it until December 30. The testbed operator mentioned it to spacecraft chief engineer Rob Manning that morning; Manning mentioned it to Willis just as the rest of the project management team was filing into a conference room for what was supposed to be the final EDL "robustness review" before the first rover landing on January 3. Willis found that many of the pyrotechnic firings that were supposed to do things like inflate the airbags and cut the bridle had not occurred. It wasn't at all clear why, though. It looked initially like the signals from the CPU that were supposed to arm the pyrotechnics weren't getting through the circuitry that actually controlled the firing events. This had also happened much earlier in the project but had been traced to faulty ground support equipment and fixed. At least, Willis thought it had been fixed. So he wasn't sure it was a problem.[7]

The tests in question had been initiated by one of the project's lead avionics designers. Back at the end of July, he had identified a potential problem with the pyrotechnics in the event of a CPU reset during EDL. Like Pathfinder, MER had a set of "backup timers" that would carry out the pyro firings if the CPU reset. On MER, these backup timers had been implemented first in software and then converted into a field programmable gate array. Once created, the FPGA couldn't be reprogrammed—it became, in essence, a hard-wired special purpose computer. What the designer had discovered back in July was that MER's pyro gate array contained an error. Some lines of computer code that the software testing engineers had needed during ground testing back on Earth had been left in and shouldn't have been. The code had the effect of making the array wait 32 seconds after a CPU reset before it would take over and start issuing the pyro firing orders, but it was supposed to take over after only a few milliseconds.[8]

This timing error only mattered to the success of landing during about two of the six minutes of the EDL sequence—for a minute around parachute deployment and during the minute before rocket-assisted deceleration motor firing. In all the systems testing they'd done, the EDL team had never experienced a CPU reset during these vulnerable windows, so it didn't seem to be a major problem. There had been some arguments within the project over whether they should even address it. But ultimately, Willis explained later, the team decided that they couldn't risk not trying to fix it. As he put it years later, "The consequences of not trying to do something, of having something go wrong and us not doing something, knowing that it was there . . . it was deemed that we needed to do something."[9]

What they did to solve the FPGA problem caused the problem they discovered at the end of December. One thing they'd done was to desynchronize the pyro timer gate array from the CPU. The array had its own internal clock, but it wasn't as stable as the CPU's internal clock, so the two would slowly drift apart in time. And what turned out to cause the pyro failures in December was that in cases where the CPU clock and the FPGA clock had drifted very far apart, the gate array wouldn't accept the first of the two commands necessary to fire the pyros, the "arm" command. And if the pyros weren't armed, they wouldn't fire.

But nobody understood that at the end of December. It took Willis's testbed crews running all-out until 1:00 a.m. on January 2 to re-create the problem. This was because the project's two testbeds, the flight software testbed and the cruise/EDL testbed, had different timing drift. The flight software testbed experienced "positive" drift, with the FPGA clock getting ahead of the CPU, while the cruise/EDL testbed had a negative drift rate—and the problem only occurred, it turned out, with a positive drift rate. So the trouble only ever appeared on the flight software testbed, which wasn't the primary testbed for the EDL simulation effort. But Spirit's electronics, according to the test logs, had a positive rate—it was more like the flight software testbed.[10] So they did, in fact, have a problem.

Finally, after much analysis, at an 8:00 p.m. meeting the same evening in one of the project's office trailers, Willis explained what the fault really was to the team. He and Manning both remember that Glenn Reeves, their software development manager, already knew how to fix it. The gate array only had to receive the "arm" signal once, so all they had to do was send the spacecraft a command sequence that would arm all the pyros during the last telecommunications window about four hours before entry. But this was a fairly drastic thing to get approval for in the risk averse climate of 2004, and they only had one day left to do it.

The morning of January 3, Manning and Reeves discussed the basic question of whether the command would work with Kim Gostelow, a longtime JPL software engineer who'd actually written the EDL code. At this meeting, there was some confusion over language. Manning remembers asking if each of them thought sending the arming command would be a good thing to do; Gostelow, he recalls, thought that it would not be, initially.[11] But what Gostelow actually meant was that he didn't think it would be adequate to solving the problem—it wouldn't be good enough. He didn't mean that it would actually be a bad thing to do. It took the three of them nearly two and a half hours to agree that arming the pyros manually wouldn't be harmful, at least, and it might solve the problem.

Next they had to present it to their review board, at 10:30 that morning. That meeting went badly. Manning, Reeves, Gostelow, and Willis had all been deeply involved in figuring out what the problem was and understanding that it was rooted in very subtle timing differences. The review board hadn't been. After a half hour, the meeting relocated to JPL director Charles Elachi's office, with still more people, including associate administrator Ed Weiler and the NASA chief engineer. They were even further removed from the problem. They accepted the decision, though not without concern.

Two hours later, the decision was re-reviewed. After trying to explain the problem again, Manning remembers ultimately having to just say "trust us." It was, everyone involved remembers, a very nervous day.

The person who actually had to approve the commands was Mark Adler, who'd become one of Spirit's mission managers after launch. His job wasn't to determine whether or not the commands would fix the problem; it was simply to ensure they wouldn't cause new problems. "I only had to be convinced that they wouldn't make things worse," he remembered later.[12] He was the mission manager for Spirit's EDL; his flight director was testbed supervisor Jason Willis. Adler went about getting convinced by having a long talk with Willis about what Willis had put the testbeds through. About four hours before the start of EDL, Adler signed off on the commands.

For the Spirit landing, MER had abandoned the Pathfinder faster-better-cheaper approach of monitoring the landing from a corner of their office block. Instead, they'd set up a much larger, more formal, mission-control-like space in JPL's Space Flight Operations Center. Wayne Lee and Rob Manning split the duties of narrating the events: Lee calling out the predictions and Manning the confirmations.

The EDL sequence started with an automatic turn to point the heat shield toward Mars. The turn completed successfully at 7:26 p.m. Pacific time on January 3; it was followed by venting of the Freon coolant and ejection of the cruise stage. These were pyro firing events, and since they'd made the last-minute change to the pyro programming, they were also the first real evidence that they hadn't made a catastrophic mistake. But both completed on time, and their telecommunications monitor received the semaphore signal that confirmed cruise stage separation. In fact, the team received the semaphores for every event down to radar lock with the surface. After that, they lost the signal for an agonizing moment before regaining it briefly; Manning thought it was the signal for bouncing, but they lost it again too soon to confirm. The team cheered anyway—any signal at all by then meant the lander had survived its first impact. There was also nothing they could do about it but wait.

It was another 16 nervous minutes before they got a signal again. As Manning said at the press conference later, with Pathfinder they'd had a signal the entire time it was bouncing, so the absolute lack of a signal made him nervous indeed. By the time they got a signal again, Spirit had come to a stop and started to deflate the airbags. They wouldn't have details, though, until Mars Odyssey flew over two hours later and Spirit sent it a data dump via its UHF transmitter. By then, it should also have deployed its solar panels and taken a series of low-resolution images with its navigation and hazard avoidance cameras. That data came back successfully, too. And when the surface operations team's imaging lead put the first images up on the surface operations area's wall-sized projection screens, the team erupted in hugs, cheers, and not a few tears.[13]

At the press conference following the landing, NASA administrator Sean O'Keefe dramatically announced, "This is a big night for NASA. We're back! I'm very, very proud of this team," and he poured cups of champagne for the officials assembled on the Von Karman Auditorium stage. He presented JPL as a learning organization, contending that its success in 2004 was built around having learned from its mistakes in 1999. Richard Cook, who of course had been on that same stage in 1999, taking the heat for two successive failures, put it a little differently: "I really like doing it when it works like this."[14]

Joel Krajewski's timeline for getting Spirit detached from its lander carcass and onto the surface of Mars was 9 sols, with two "decision cycles" by the team on Earth per sol. Many events had to occur in a strict order to meet that schedule, and the team had to verify that each event had happened before authorizing the next. Their landing site turned out to be significantly warmer than expected, so they ran up against the rover's thermal limitations a few times, very nearly overheating the rover electronics while trying to get precious data back. They had also planned to rely primarily on the direct-to-Earth transmitter for most of the effort, but in fact used the UHF links to Odyssey and Global Surveyor a great deal for their higher data return capability. Using the UHF gave them three decision cycles a day, as it turned out, but that didn't get them off the lander sooner.

Instead, they had a few troubles along the way. The largest challenge was airbag "bubbles" blocking the forward egress ramp they'd wanted to use. The team reactivated the retraction motors to eliminate them. That didn't quite work over two successive attempts, so they had Spirit make a 120° turn on the lander deck to use one of the side ramps. And, of course, they first had to test the maneuver on their mobile surface testbed setup at JPL. These small troubles, and their own conservatism, meant the team didn't get Spirit onto dirt until sol 12.

At the same time, the EDL team had been sent off to a conference room for a week to reconstruct and analyze Spirit's EDL performance. It wasn't enough to know that the vehicle had landed safely: they could simply have been lucky. And, it turned out, Spirit had been rather lucky. Their analysis quickly revealed that its descent rate limiter, the device that controlled the lowering of the lander on its bridle after heat shield ejection, had let the lander down far too slowly. Deployment had taken twice as long as it should have. It had been survivable because the team had designed such large margins into the EDL timeline (they hadn't just been lucky), but it also wasn't acceptable performance. They launched a fast series of additional tests in the 25-foot space simulator with their engineering model, but they couldn't get it to reproduce whatever happened on Spirit. They also found that Spirit had deployed its parachute at a lower altitude than expected, which they chalked up to the atmosphere's being even less dense than they had planned for a couple of days before.

Their data also showed that all the effort expended to mitigate the wind hazard at Gusev Crater hadn't been wasted. The DIMES camera, which measured horizontal wind forces, had, in fact, triggered the TIRS motors, to cancel out wind-induced motion. From its image pairs, DIMES had calculated a horizontal velocity of 11 meters/second, and a sudden gust of wind near the RAD firing altitude had produced a backshell tilt that would have imposed an additional 12 m/sec when the RADs fired. The combination would have put them at the bare edge of their tested airbag performance. The resulting TIRS firing had cut the horizontal velocity to about 11.5 m/sec, well within their capability. And as luck would have it, the wind had been pushing Spirit toward a deep crater that it might have bounced into; the TIRS firing shoved it in the opposite direction, so that it bounced away instead.[15]

As a result of the reconstruction effort, the project leaders decided to raise the parachute's deployment dynamic pressure trigger for Opportunity's landing. In effect, this would raise the deployment altitude slightly, regaining some of their timeline margin against the descent rate limiter problem that they couldn't figure out. The additional testing they had done in 2003 had strength-qualified the parachute to the higher recommended pressure.[16] At a meeting of the Governing Program Management Council on January 16, JPL's senior management agreed to this change.

Things had gone spectacularly well for the team, despite all these little troubles. In addition to getting Spirit off the lander carcass, the operations team had been able to get most, though not quite all, of the required minimum mission success criteria measurements in their first two weeks on Mars. They had gotten the required high-resolution panorama down while still strapped to the lander.

They'd also acquired three "stares," as they called them, from the mini-TES (thermal emission spectrometer) to examine the mineralogy of the site around Spirit. Once off the lander, they'd rolled a few meters, deployed the robotic arm, and taken soil measurements with the arm's instruments. Then they'd rolled a few meters over to a rock they'd named Adirondack, hitting it on the first drive.[17]

Their pattern of success changed dramatically on Spirit's sol 18, their second day of measurements on Adirondack. Jennifer Trosper was in the mission manager's chair, when the morning data download didn't happen, seemingly due to rain at the Deep Space Network site in Canberra. But then Spirit hadn't responded to a Mars Odyssey overflight later in the day. That was worrying. A Mars Global Surveyor overpass at 2:00 a.m. at Spirit's landing site got a signal, but no data—and it cut off 11 minutes early. A series of increasingly distraught meetings took place during which the engineers tried to figure out, with essentially no data, why Spirit continued not to respond to repeated communications attempts. They got a single beep the next day. At that point, they only had three days before Opportunity's descent. The beep told them that Spirit was in some sort of fault mode. But it had several, and they couldn't tell which one.

Glenn Reeves, their software manager and architect, had a thought on sol 19 that maybe something was causing Spirit's computer to reboot continuously. That would explain the symptoms, but there were many things that could cause the problem. The team got a little more information out of an unexpected 11-minute transmission sent via Spirit's low-gain antenna on sol 20. It suggested that Reeves was probably right. Spirit's CPU had detected a fault on booting up and, as programmed, had waited 15 minutes before shutting down again. Unfortunately, the data didn't tell them what the fault was. But it carried a time stamp that indicated that the spacecraft clock had reset, which would only happen if the vehicle had completely depleted its batteries sometime during the night then partially recharged them after sunrise. Mark Adler's crew was on duty this time, and they decided to command another communications period. This one succeeded, and they finally got solid telemetry data. It showed that the battery state was very low, as if the spacecraft had stayed "awake" most of the last two nights. They then tried to force a shutdown for the rest of the day and upcoming night, to recharge and preserve the batteries. It didn't work.

Reeves had another idea on sol 21. The data sent down the day before was all "real-time," data not stored in the flash memory from the previous several days' activities (even though it was supposed to be sending that data) but only what was in its volatile memory. The same thing had happened before in ATLO, when the flash memory's file system had gotten corrupted. He'd written a bit of software that would force a reboot while also preventing the CPU from trying to access the flash memory. If they could get this command, called "INIT-CRIPPLED," into the spacecraft, Spirit would boot up reading only from its read-only memory, which held uncorrupted software (unless it had been physically damaged, of course, but they had no indication that it was). Adler's shift on sol 21 spammed Spirit with the command, followed by shut-down commands, trying to force the vehicle to quit trying to reboot and stay shut down for battery charging. That finally worked—fortunately, as Spirit's sol 21 was also Opportunity's landing day.

The Spirit fault was so distracting to the team that MER project manager Pete Theisinger had been forced to order Manning and Willis to go prepare to land their other machine the previous day. And they had to do it under a lot of scrutiny. NASA's bigwigs had flown out early for the Spirit crisis; Ed Weiler said later he'd come out to Pasadena prepared for a funeral. He didn't get one. Spirit didn't die, and Opportunity landed successfully, too.

It was a late-night (in Pasadena) landing, but the hour didn't keep anyone away from the celebration. Manning's EDL team, led by Adam Steltzner and Wayne Lee, turned the post-landing press conference in Von Karman Auditorium, already packed well beyond its 200-seat capacity, into an impromptu victory parade that went out on national television.[18]

Once his whole team had danced by the stage in Von Karman and shaken his hand, Rob Manning explained that Mars Global Surveyor had received Opportunity's UHF data stream all the way to the ground, and the Deep Space Network had received the direct-to-Earth signal all the way down, too, so there was already a lot of information about the vehicle's performance. Opportunity had been behind its ideal timeline—far behind, as it turned out, cutting its bridle barely an airbag diameter above the surface. It had also not come to rest on its base pedal and would have to self-right. DIMES hadn't triggered the TIRS rockets this time; as forecast by the wind models, Meridiani didn't have strong winds. The lander's horizontal velocity had been only about 2 m/sec.[19]

Luck cast even more favor on Opportunity than it had on Spirit. When the first imagery reached the surface mission operations area two hours after landing, a navigation camera image shot over the rover's nose showed that it had come to rest facing a rock outcrop (a layered outcrop, a geologist's dream) about 10 meters away. They'd scored what principal investigator Steve Squyres called "a three hundred million mile interplanetary hole in one," landing in a small, shallow crater.[20] The outcrop was the crater wall, the closest any Mars lander had ever come to bedrock.

The team's practice with Spirit helped them get Opportunity off the lander on sol 7, aided by a good airbag retraction and an almost perfectly smooth area surrounding the lander. There were no obstacles, and, in fact, not even small rocks until they rolled up to the outcrop. The only thing marring the operation to this point had been discovery of a heater stuck on one of the joints of the instrument deployment arm.[21] That was a problem; it consumed precious energy. Early in the mission, it wasn't a big problem, as they had plenty of power. But after the first month or so, as the solar arrays got dirtier and the Sun angle got lower, it would cause trouble unless they could find a workaround.

Opportunity spent its first 57 sols investigating the small crater it had bounced into during landing, and in particular the low outcrop that its first images had revealed only a few meters away. The team promptly named the crater Eagle. Opportunity almost immediately found the evidence of water that Squyres and his fellow scientists had thought they'd be most likely to find on the other side of Mars, at Gusev. Their first evidence came from tiny, spherical concretions that were scattered all around the rock outcrop—the team called them "blueberries," although they were actually gray. The Mossbauer spectrometer showed that these contained the hematite that Phil Christensen's TES instrument on Mars Global Surveyor had seen from orbit. The outcrop was composed of very finely layered, sulfate-rich deposits. Sulfates on Earth are commonly evaporites, formed by the evaporation of water that leached the sulfur out of volcanic rock somewhere else. So the science team interpreted the outcrop's layers as having formed through repeated events of flooding and evaporation.[22] Meridiani, which some team members had thought was the most boring place on Mars to drop a rover, had actually turned out to be interesting.

Spirit, despite having landed in what Squyres had been positive was a former lake bed, found no evidence of water at its landing site. Instead, it was in a plain of endless basalt. That finding was terrible for the Spirit team's morale; the engineers had worked very hard to make the Gusev site feasible and it didn't seem to be panning out. In late March 2004, just before the end of the primary mission, Squyres and his science team decided to abandon Spirit's landing site and embark on a 2.4km trek across the plains to a large outcrop the team had named Columbia Hills after the destroyed shuttle.[23] They needed a new goal, and the outcrop might reveal the signs of water that the scientists were certain existed somewhere.

Outside JPL's fence, MER proved popular. JPL had allowed public television's *Nova* series to film a documentary on the mission's development phase that was broadcast right after Spirit's landing. Called "Mars: Dead or Alive," it was a risky move since JPL couldn't control the outcome of the mission or the content of the documentary. But the Lab had once been more open to documentary efforts, and it turned out that the documentary gained wide attention. An IMAX movie, *Roving Mars*, based on principal investigator Steve Squyres's book, followed in 2006. In print media, MER garnered much less attention than Mars Pathfinder had, though in part the decline was likely a product of the shrinking of the newspaper industry (and of science coverage within newspapers), itself a result of the enormous growth of the World Wide Web in the preceding decade.[24]

Evolving Operations

During the Opportunity post-landing press conference, Craig Covault from *Aviation Week and Space Technology* magazine had revisited a question he'd asked Pete Theisinger right after launch. Referring to the enormous amount of overtime necessary to make it to the launch pad, Covault had asked Theisinger if he'd do it again. Theisinger had said "never again." Even in the euphoria of the dual landings months later, Theisinger still said no: "It took a lot out of our people, and out of our sponsor," an answer that drew applause from the audience in Von Karman, and a "what he said" out of Sean O'Keefe.[25] Theisinger's "no" also reflected the problem the MER project faced as the rovers failed to die at the end of their primary mission. The operations team had to find a way to keep going without overworking the team eating into their family lives, and breaking the Mars budget.

At the end of the 90-day primary mission, both rovers were still fully functional, excepting the bad heater on Opportunity. NASA funded an extended mission through the end of fiscal year 2004. September 2004 coincided with the martian winter solstice, the shortest day of the year and the least available power for both machines. It wasn't clear yet that either rover would survive it, although both landing sites turned out to be warmer than Keith Novak's thermal design had anticipated. That translated into less power needed for bare survival and with the first 90 days of surface data in hand, the operations teams were already pretty sure they could last through the solstice.[26]

The "wildcard" was that stuck heater on Opportunity. At 140 watt-hours wasted per day, it would leave the rover powerless even before the solstice arrived. They had an idea to "fix" it by ordering the rover to disconnect its batteries just before shutting down for the night, and reconnecting them when the

Sun lit up the arrays in the morning. That meant no survival heaters at all during the long night. But the thermal data showed the rover wasn't using its survival heaters overnight anyway, except on mini-TES, which was pretty thermally isolated from the rest of the rover's avionics (and their heat). So this "deep sleep" mode, they decided to call it, might kill mini-TES but save the rest of the rover. In March, they had started adding the deep sleep mode to the new software they intended to upload after the end of the primary mission.[27]

Planning for the extended mission also forced the MER management to revisit the Mars time issue. Three months on the crazily rotating schedule had been a much larger drain on the teams than anyone had thought when they'd decided to do it. It was particularly damaging to families, as not everyone on the two operations teams were young bachelors (and bachelorettes). The project management faced attrition of their now highly experienced operators who wanted their family time back. Working on Mars time also wasn't clearly necessary any more; as the teams had gained experience on their operations tools and with the daily planning process, the time needed to complete a plan had come down substantially. What had taken them 16 hours to do on their first few sols was taking them only ten by sol 90. The shrinkage in operations time presented the opportunity to switch to an Earth-time-based schedule without losing much science. They calculated that an Earth-time schedule of ten hours per day, seven days per week, would permit commanding on 30 out of every 37 days. In addition to salvaging family life and reducing attrition, an Earth-time schedule would also be less expensive to staff, with about 40 percent fewer people—not a minor consideration.[28] They also developed a less-expensive option, with only five-day work weeks, but that meant commanding on only 21 of the 37 days of the martian cycle.

While the switch back to Earth time was predicated on the wear and tear on JPL's engineers and their families, a similar calculation applied to the scientists of the operations working group, too. Initially, scientists had loved the idea of being all located at JPL in a grand collaboration, and they had made the science office manager, a dark-haired physicist named John Callas, into one of Pasadena's biggest landlords. He'd saved the project a lot of money by leasing blocks of apartments and condominiums—a hundred of them—because it was cheaper than hotel rooms for lodging the visiting scientists. But they also wanted to go home as the rovers failed to die sol after sol (and practically speaking, many had to leave, as some universities still make science faculty teach). The project still needed the scientists to perform their role in deciding what the rovers should do each day, so they began to work on providing the infrastructure to enable scientists to continue to participate in the science operations working group from their home institutions.[29]

NASA approved the more expensive, seven-days-per-week operations model for the first extended mission, hoping to maximize the return on the rovers before they started to fail. During this first extension, Spirit started to show the signs of aging. On its trek to the Columbia Hills, its right front wheel motor started to draw excessive power, indicating it was starting to fail. The team decided to start driving it backward, dragging the wheel without power to preserve the motor until they needed it to start climbing up the Columbia Hills. Their surface testbed in the Mars yard climbed very well, even on 30° slopes, with all its wheels, but not so well when it was dragging one. But Spirit's bad wheel turned out to be the only "failure" they had during the first mission extension; even when they implemented deep sleep on Opportunity's sol 101, mini-TES came through fine.[30]

Opportunity's science team had decided to send the vehicle off to another crater after 57 sols exploring Eagle crater. The new destination, Endurance crater, appeared to be much deeper, which meant potentially older, with different strata to investigate. The trek across the plains was a distance of about 800 meters. The rovers' height of eye of only about 1.4 meters meant that they relied on Mars orbiter camera (MOC) imagery to plan the path, but MOC images lacked the resolution to help avoid obstacles that were of the same scale as the rover. So the rover planners had to pick their way through the landscape using surface imagery, while relying on a track placed on the orbital images by a couple of JPL geologists to keep going in the desired direction.[31]

Opportunity reached the crater rim on April 30, 2004. The team drove it along the rim for a couple of weeks, imaging outcrops and looking for ways down into the crater that were safe for the rover to traverse. One criterion for the drive down was power; they wanted a route that faced north because they would get more power that way. Another, of course, was the terrain itself. But the drive along the rim had its own rewards. The team obtained one spectacular panorama of what they called Burns Cliff, which seemed to be a large layered eolian deposit (produced by the wind) that had then been altered into stone. They started climb testing with the surface testbed to verify that they'd be able to drive down into the crater and back out again on the crater's 20°-30° slopes, which took until the end of May. By that time, they had a Sun-washed north-facing route picked out and a plan that involved repeated stops on the way down to drill into the crater wall with the rock abrasion tool.[32]

It took the Opportunity team the rest of the (Earth) summer to drive down their route into the crater. The science working group couldn't resist stopping to drill each new strata as they encountered it; ultimately, they left a trail of 11 holes in the crater. The entire section turned out to be various sulfates, meaning that all the rock probably had been laid down in water. If there was an underlying basaltic floor, they didn't find it, and the rover came out of the crater on sol 315 as healthy as it had gone in.[33] It also encountered a phenomenon that proved a great boon to the project over time. While it was down on the crater floor, it experienced a couple of "cleaning events," when something cleaned the solar panels off overnight and power output jumped. The team speculated that "dust devils," tiny, rover-sized windstorms, might be the culprits. In April 2005, Spirit's cameras caught one blowing past.[34]

Spirit reached the Columbia Hills in early June 2004, just as Opportunity was starting its descent into Endurance Crater. During the long drive, the rover's planners had developed a routine that maximized distance traveled per day by using mobility upgrades to extend driving range. From the panoramic camera's height of eye, Spirit's planners could only accurately resolve obstacles out to about 70 meters (less if they only had the navigation camera images). So they commanded "blind drives," meaning the human operators in Pasadena did all the drive planning and hazard avoidance calculations, up to that distance, then had Spirit use its on-board autonomous navigation capabilities to go farther. Spirit had enough power to drive 1.5–2 hours per sol, and it could cover the first 70 meters in a little more than a half hour. Under autonomous navigation, Spirit could drive about 20 meters/hour, so a two-hour drive period could reach a total of 100 meters.[35] The 2.4km voyage took as long as it did because Squyres and the rest of the scientists understandably wanted to stop and take measurements along the way.

Spirit spent most of its first month in Columbia Hills "thrashing around" (in Squyres's words) a weird-looking rock called Pot of Gold.[36] The terrain caused a lot of wheel slip as the team tried to get close enough to apply the rock abrasion tool to it. When they finally hit it, Pot of Gold turned out to be the first rock they'd found in the Gusev region that seemed to have been altered by water. Then they set out, again along a north-facing route, to look for more potentially sedimentary strata, still chasing the history of water on Mars.

When FY 2004, and the first extended mission came to an end, the MER team and NASA started to face the question of how to keep the rovers funded and staffed for a very long period of time. If the winter solstice hadn't killed them, what would? JPL's engineers were increasingly convinced that the rovers could last a very long time. Opportunity's mini-TES was routinely surviving temperatures well below its qualification limit—the temperature JPL's test program had certified it down to—and the rest of its electronics weren't reaching dangerous temperatures, even in deep sleep. It is thermal cycling that kills electronics, and the rovers' electronics were much more resilient than expected.

The next mission extension, approved in September 2004, ran through March 30, 2005. It would end just as the two rovers entered the dangerous period for martian global dust storms, which tended to occur every six Earth years. At the same time, though, faced with possibly operating for years instead of months, NASA and the MER project leaders decided to reduce their operations costs further by switching to five-day work weeks. This adjustment reduced the operations teams by a little less than half. The seven-day week had required four sets of operators (two per rover) to staff a four-day-on, three-day-off work schedule. The five-day week required only two sets of operators (one for each rover), with only a handful of people who had to work weekends (but would have some weekdays off). This reduction necessitated changes in the way the project worked. The team would no longer command the rovers every day. Instead, they would make and upload command sequences intended to run over several days. In part, this was possible because they'd continued to improve the tools they used to build and test the command sequences. But they also had to reduce the complexity of their demands on the rovers in order to stay within their time and budget constraints. They would, in short, get less science done, but given that their project and its science had been intended to end in 2004, everything they did in the second mission extension was extra. For FY 2005 and 2006, NASA funded the project at about $32 million per year.[37]

Opportunity left Endurance crater on sol 315 and drove over to its heat shield, which had come to rest about 200 meters from the crater. JPL's engineers wanted to see how well the heat shield had performed and how deeply it had been eroded during entry—no one really knew how well the Viking-type heat shields worked because no previous mission had ever recovered one. And next to it, the Opportunity team found a nickel-iron meteorite that absorbed their interest for another few weeks. They then set a goal of moving toward a feature called Vostok, apparently a highly eroded crater, about a kilometer away. Spirit remained in the Columbia Hills area during the mission extension, finding more and more evidence of ancient water in thick, layered deposits exposed in the hills.

As it turned out, Mars didn't offer up a global dust storm in 2005, and with both rovers still healthy, NASA approved a third mission extension. Instead of running a few months, as the first two had, this one ran from April 2005 to September 2006. It was easier for NASA and the project to plan for a year and a half of operations than to make several shorter plans for the same period, and the rovers showed no sign of impending failure.

The third mission extension took Opportunity into what the team called "etched terrain" during the remainder of 2005 and also introduced them to a new threat to the rovers: sand traps. On April 26, Opportunity dug itself into a small sand dune 40 meters into what the team had planned to be a 90 meter drive. Some extensive testing with the surface testbed and simulated Mars "sand" made from playground sand and diatomaceous earth (normally used for pool filters) ensued while the engineers figured out how to command it out again; on June 4, they succeeded in getting it freed.[38] They studied the sand dune (Purgatory, they called it) to understand why this dune had caught them when they'd driven over many similar dunes without trouble, and they convinced themselves that the combination of slope and the nature of the sand had been responsible. They decided to avoid sinusoidal ripples with apparent slopes of 15° or more. These ripples seemed to be too soft for the heavy rover. Then they returned to exploring the etched terrain. After about six months there, they moved to another small crater, Erebus, looking for new exposures of strata. Ultimately, though, despite the crater's shallowness, they didn't enter it. It was full of the kinds of ripples the team was now afraid of getting stuck in. Instead, they moved around the north edge of the crater, where MOC imagery seemed to show much smaller ripples and a good deal of exposed bedrock, and headed toward a much more spectacular crater about 2 km away, named Victoria.[39]

Throughout the campaign around Erebus and the drive toward Victoria, Opportunity's planners continued to be challenged by the inadequate resolution of the MOC imagery. Mars Global Surveyor's operations team had come up with a way of improving the MOC's apparent spatial resolution by maneuvering the spacecraft while the image was being taken—a form of motion compensation. Their technique allowed the Opportunity team to identify rover-sized features, and even Opportunity itself, on MOC images, but only barely. And the stereo technique that geologist Matt Golombek and a colleague used to estimate heights from MOC images still couldn't generate slopes and heights on the rover's scale. So the MOC images could help them keep going in the general direction of Victoria, but they were not much help in avoiding future Purgatories. That had to be done using surface imagery.

On the other side of Mars, Spirit's team moved the rover into a feature near the Columbia Hills that they called the Inner Basin during the beginning of the third extended mission. Because Spirit was farther from the equator than Opportunity was, solar angle was more important, and the Sun reached its zenith in May 2005. By early the following year, they knew they needed to find a suitable north-facing, unobstructed hillside to park on for about eight months of winter, and this put a sizeable constraint on their operations. The team explored the Inner Basin formations until the end of 2005, and then started Spirit toward a "winter haven" spotted in orbital imagery, a hill they named for astronaut William McCool, who died aboard the shuttle *Columbia*.[40] But Spirit got stuck in a sand trap of its own, and by the time the team managed to extricate it, the balky right front wheel had failed entirely and they knew they couldn't make the distance in time. They picked a new winter haven nearby, a low ridge near a feature called Home Plate, and reached it in early April 2006. After that, Spirit was stationary for the next half year, taking long-duration spectral measurements with the Mossbauer and recording meteorology data with mini-TES.[41]

The Columbia Hills campaign demonstrated both the advantages and perils of using the autonomous driving functions. One feature added after the end of the primary mission, called visual odometry, served to check the rover's actual progress along a path, allowing the vehicle to know whether it was slipping or not in order to compensate. But it was very slow, allowing drives of only about 10 m/hour, causing the operators to develop complex command sequences that intermixed short blind drives over apparently flat or low-slope surfaces with short autonomous navigation drives on higher slopes where slippage might occur, informed by periodic visual odometry checks.[42] The complex sequences took more time and manpower to prepare and, of course, were more likely to have faults in execution.

Opportunity reached Victoria crater right at the end of the third extended mission in September 2006, on sol 952.[43] NASA funded a fourth extension, covering FY 2007. Opportunity spent the extension slowly exploring the rim of the crater, which was complex and made dangerous by the presence of sharp cliffs. Spirit spent it exploring the Inner Basin, which had turned out to contain both volcanic and sedimentary formations representing a difficult to disentangle geologic history. Both rovers were immobilized by a global dust storm during July 2007. Their operators imposed deep sleep on both rovers to keep them alive for what turned out to be almost a two month period. Opportunity's mini-TES was disabled by dust during this storm, and Spirit's was severely reduced in performance, but this proved the only damage from the experience.[44]

In September 2006, to facilitate more efficient operations, the MER project had uploaded new flight software to the two rovers with some new autonomy capabilities. One, visual target tracking, was very similar to visual odometry except that it tracked only a single target instead of dozens, allowing the rover to operate much more quickly. It could be used for a variety of things—moving toward or away from the target or bringing a rover close enough to an object to deploy its robotic arm. A second upgrade enabled automatically placing an instrument on the robotic arm on a target at the end of a drive. Before this, operators had been required to check the rover's position before deploying the arm, meaning that it took two command cycles to complete any measurement that required a drive followed by a measurement. Because they expected cost constraints to force reductions in the number of command cycles in future extensions, reducing the number of cycles necessary to perform a particular measurement was important to maintaining a reasonable operating tempo. The team also installed a global path planner developed by Carnegie-Mellon University. The rovers' initial autonomous planning capability could only "think" one or two steps ahead, so a rover could drive itself into areas it then couldn't get out of without intervention by its operators. The global planner (which wasn't really global, but regional) enabled it to keep track of a much larger volume of space and choose more efficient paths through it.[45]

This fourth mission extension inaugurated further shrinkage of the operations team. Funding was reduced by about a third from the previous mission extension, from $31 million in fiscal year 2006 to $22.6 million in FY 2007.[46] This was done by reducing the number of command cycles that the team would conduct each week. During the third mission extension, the team had commanded both rovers every weekday, and on Fridays they had prepared multi-sol command uploads to keep the rovers working over the weekend. Effectively, they were commanding ten times per week. For the fourth extension, they absorbed the cost reduction by proposing to command eight times per week and shrinking the science teams. This reduced staffing on certain positions and eliminated most, though not all, of the technical support that the project had been able to draw on.

Mars Global Reconnaissance

At the same time Opportunity reached Victoria crater, the Mars Reconnaissance Orbiter completed aerobraking and began its own science mission. It had gone into orbit without troubles on March 10, 2006; shortly thereafter, it had briefly powered up its instruments for what project manager Jim Graf called his "George Pace Science Upper."[47] Pace had been chairman of NASA's independent review team for MRO, and at one meeting Alfred McEwen and Rich Zurek had argued for taking science data right after going into orbit, so the principal investigators would have real data to work on with their ground software during the months of aerobraking prior to reaching their science orbit. McEwen in particular thought this was important, because the Mars Phoenix mission wanted stereo image pairs for landing site selection. Phoenix was supposed to land in May 2007, and MRO would not complete aerobraking until September 2006. McEwen did not think his team would be able to provide the imagery the Phoenix team wanted in time without the earlier images to work with.[48] And Pace agreed, over Graf's point that it wasn't in the budget or the requirements, so he recommended it to NASA, and NASA officials ordered that it be included.

The George Pace Science Upper was fortunate. One of the most critical parts of the MRO payload was its high-resolution imaging science experiment. The first images showed smearing, and some had simply missed the target.[49] The spacecraft team traced the trouble to an error in the on-board software that calculated the rolls necessary for the spacecraft to make to image targets that were not directly under it during its orbit. Once they fixed the bug, the problem was solved.

One of the first images back from MRO after aerobraking was a HiRISE shot of Victoria crater; it captured not only a clear image of Opportunity but also the rover's tracks and even its shadow.[50] The imagery from HiRISE allowed the rover planners to see everything around their vehicles, and stereo image pairs enabled derivation of the slopes of terrain features that were the same size the rovers were. Path planning for the rovers became far easier.

Achievement of the primary science orbit enabled the various instrument science teams to get started on their investigations. Guided by some suggestive observations from the OMEGA infrared mapping instrument on Mars Express, Scott Murchie's team used their CRISM spectrometer to obtain spectra from the Nili Fossae region on Mars. OMEGA had seemed to see a class of clay minerals that generally form in water, as well as locating several olivine-rich rock units. On Earth, the carbonates Murchie sought form as a weathering product of olivine-rich rocks, so Nili Fossae seemed a good region to look for carbonates. One of his investigation team members at Brown University made the detection; there were magnesium-rich carbonates in the region, but not in large amounts. On Earth, one often finds massive carbonate beds that span tens and even hundreds of kilometers in extent, the remnants of ancient ocean floors. At least in Nili Fossae, the carbonate formations spanned only tens of meters despite the presence of vastly larger olivine deposits. Rather than remnants of a past ocean, they were

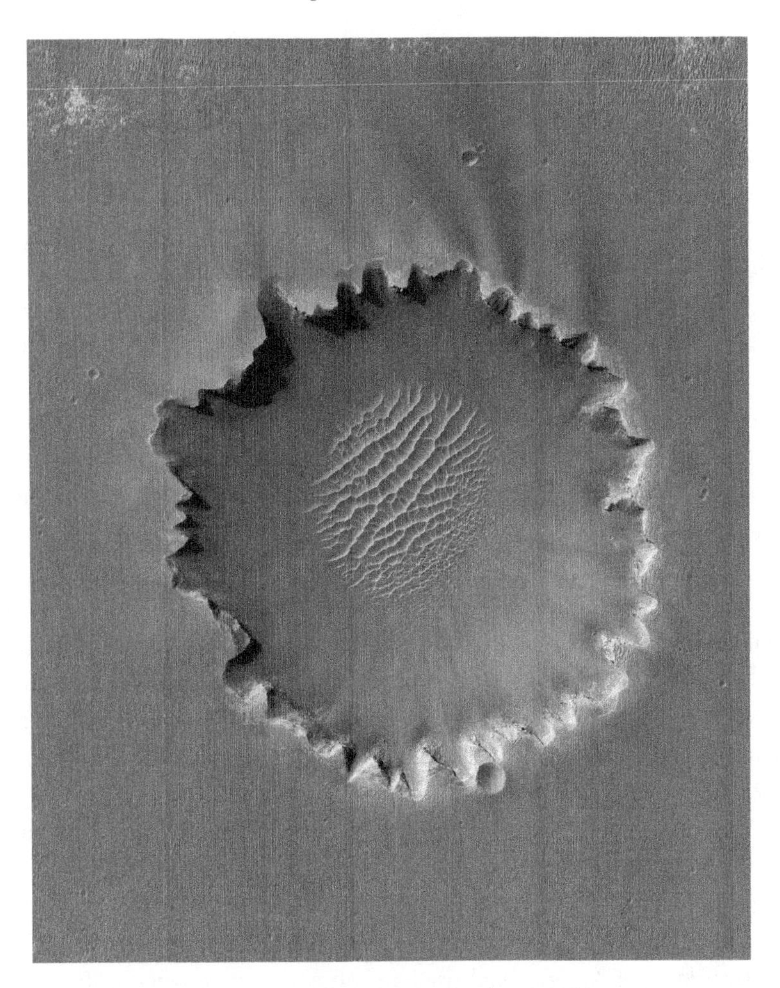

Victoria crater, imaged by the HiRISE camera on October 3, 2006.
JPL image PIA08813/UA image TRA_000873_1780, courtesy NASA/JPL-Caltech / University of Arizona / Cornell University / Ohio State University.

likely the product of groundwater percolation toward the surface, or possibly the remnants of short-lived lakes.[51]

The SHARAD radar found other indications of martian water, in the form of what seemed to be mountain glaciers in the southern mid-latitudes. Imagery from the various cameras sent to Mars in the previous decades showed many surface features that looked like dust-covered glaciers, but pictures alone could not make definitive identification. Radar, though, penetrates and reflects from water differently than stone; ice differently than water; and "dirty" ice differently than pure ice. So radar could distinguish between frozen groundwater (which would be "dirty") and glacial ice ("clean" because it originally evaporated from the surface and then precipitated as snow), in addition to simply distinguishing between ice and stone. SHARAD's data seemed to confirm that the suspected glaciers were, in fact, glaciers. In their publication, the science team argued this was possible due to the very high obliquity of the planet's spin axis—Mars's unusually large "wobble." In the geologic past, Mars's south pole would be pointed toward the Sun, causing its "seasonal" carbon dioxide ice cap to sublimate into the atmosphere and exposing its "permanent" water ice cap to the atmosphere and the Sun. Evaporating water would be circulated toward the mid-latitudes by winds and eventually precipitate out as snow in the Hellas mountains, producing mountain glaciers. These would eventually be buried by the frequent martian dust storms, insulating them from the Sun during Mars's more recent period of lower axial tilt.[52]

Rebalancing the Mars Program

While Mars scientists were carrying out their investigations, program officials had to wrestle with suddenly changed circumstances. In addition to the two rovers' successful landings, 2004 had witnessed publication of a spectacular claim about Mars that redirected scientific interest away from water and toward a very different molecule, methane. At the same time, the "Vision for Space Exploration" announcement meant suddenly shrinking budgets for Mars exploration, despite MER's great success, and many rounds of replanning within the Mars program.

The methane detection was highly controversial. An astronomer at the Catholic University of America had used an infrared spectrometer at the Canada-France-Hawaii Telescope on Mauna Kea to examine Mars's atmosphere. He and his colleagues found the spectral signature of methane in data collected on two days in January 1999, announced their results in January 2004, and published them in August 2004. That March, experimenters using data from Mars Express's planetary Fourier spectrometer also claimed to have found the spectral signature of methane in data taken from 16 orbits during January and February 2004. Both teams reported finding approximately the same concentration of methane, about 10 parts per billion on average, but the Mars Express team also believed their data showed large geographic variation in the concentration, from 0 to 30 ppbv.[53]

The findings were controversial for several reasons. The detections were marginal, in the sense that the signal apparently being measured was not much above expected noise levels. The ground-based measurement had the additional difficulty that Earth's atmosphere has three orders of magnitude more methane than the proposed level on Mars, and the terrestrial methane's effect has to be removed very carefully. A small error would leave an apparent signature of methane on Mars that was, in fact, spurious. The Mars Express data suffered from the problem that the instrument was not spectrally accurate enough to unambiguously separate a methane signal from that of similar molecules (such as ice)—it had not been intended for this application.

But even if one accepted that the Mars Express detection was sound, there was another problem. The experimenters' claim that they had found large geographic variation was contrary to well-established knowledge of the martian (and terrestrial) atmosphere. In the absence of an oxidizer, and Mars has no atmospheric oxygen, methane has a lifetime of centuries—340 years was the consensus opinion for Mars. On Earth, due to the oxygen that allows us humans to live, methane has an atmospheric residence time of about seven years. In both places, that long lifetime relative to the length of time it takes a piece of air to move all the way around the planet (no more than two weeks on both planets) means that methane should be found everywhere (it is on Earth but not, according to Mars Express, Mars), and it should vary in concentration from place to place by only a few parts per billion. In other words, a big methane "plume" should not be able to exist on either planet for very long. So atmospheric dynamicists immediately rejected the claims, setting up a conflict within the Mars community.

A third detection, by Michael Mumma at Goddard Space Flight Center using two different telescopes in 2003 and 2004 cemented the controversial status of the results within the Mars community. In a poster session at the American Astronomical Society's Division of Planetary Sciences meeting in September 2003 and again in September 2004, Mumma had presented results that seemed to show up to 250 ppbv methane, about ten times as much found by the other two investigation teams, also with large spatial variation.[54] But his results were not published beyond the conference abstract level until February 2009, drawing criticism.[55]

On Earth, the dominant source of atmospheric methane is biological, so the claims of methane on Mars automatically raised the specter of martian life. Earthly methane is primarily produced as a waste product by bacteria that live in the ground and in the digestive tracts of animals (and people); these methanogens largely, but not exclusively, consume plant matter. The connection between methane and life had led the Catholic University team to argue that Mars's methane might be the long-sought signature of life on Mars. But there is a nonbiological process at work making methane on Earth too, so the Mars Express team had been careful "to stress that the detection of methane does not imply the presence of life on Mars, now or in the past. It is one possibility, [but] other sources are at least as plausible."[56]

Back in 2002, the Mars Exploration Program Analysis Group had been tasked with creating a new strategy that would be somewhat flexible in its ability to respond to new discoveries. The committee had drafted a set of "pathways" in response; each pursued different themes depending on what was discovered during the 2000s. One pathway, "Search for Evidence of Past Life," extended forward from the discovery that Mars was once wet and warm (which the MER rovers had determined during 2004); another was the "Search for Present Life," extending forward from the discovery that Mars currently had active hydrothermal features (which Christensen's instruments had not yet found). Arguably, the methane discovery could trigger a course change to the Search for Present Life. But the Search for Present Life pathway involved a much different future mission set that was considerably more expensive than the Mars program could expect to afford. As the resulting report stated, this path "would be taken only following a revolution in policy and/or programmatic interest in Mars."[57]

But the opposite happened. High-level interest in Mars shrank over the next year, driven by the White House decision to send astronauts back to the Moon instead of to Mars. The policy change came without much new funding. The NASA administrator tasked the Science Directorate with creating a new lunar science program to support the return to the Moon program without any new funds; the money was carved out of the rest of the directorate's programs. The FY 2006 budget had projected the Mars budget increasing to nearly $1.3 billion per year by 2010; in early 2005, the Mars program office found that its share of the cuts would be almost $3 billion over the next five years, with the shrinking starting in FY 2007.[58] James Garvin, then still the Mars program scientist, remembered "the senior leaders of the agency said, 'Mars has to go.' It was ironic . . . but sometimes too much success breeds that kind of contempt."[59] Under the new budget plan, Mars would receive only $648 million in FY 2010.

Still worse was in store for the program. A large overrun in the Mars Science Laboratory project brought about a temporary increase in 2008, but shrinking then resumed. The actual FY 2010 budget was $438.2 million.[60] The combination of shrinking budgets and increasing costs led to a short-lived but dramatic controversy over the MER project in 2008. In his sixth extended mission proposal, John Callas presented options for further reductions in the MER budget. One of them was to shrink operations to six command cycles per week while still operating both rovers, drop the remaining technical and most management support, and cut the science budget by 14 percent. The other was to "hibernate" one rover, relying on a set of commands built into the vehicle originally to allow it to survive solar conjunction, a two-week period during which the Sun blocks communication between Earth and Mars, and only occasionally contacting it to refresh certain data tables that had to be kept up to date. The other rover would be operated, but with three command cycles per week, and the science team funding would be cut by half. The difference in funding levels between the two was significant; operating both rovers would cost $19.7 million in FY 2009, while the one-rover option would cost $9.8 million.[61]

In late March 2008, NASA's new associate administrator for space science, Alan Stern, formerly of the Southwest Research Institute, chose the less expensive one-rover option, and he made it retroactive, too, seeking a $4 million cut in FY 2008 to pay for a small portion of MSL's overrun. He apparently did this without consulting either the NASA administrator's office, key congressmen (the rovers had become very popular on Capitol Hill), or the MER project office. The consequences of a $4 million cut halfway through the fiscal year were severe. Callas remembers that it wouldn't have left enough funds to prepare the "hibernate" mode for the following fiscal year, so it was, in effect, a decision to terminate one vehicle immediately.[62]

NASA administrator Michael Griffin, the engineer who had tried to wrest Mars exploration away from Wes Huntress back in 1992 and who had succeeded Sean O'Keefe in 2005, found out about the decision from the news media on March 25. Steve Squyres was quoted as saying "if we had to take this 40 percent cut, there's no way we'd be able to operate both rovers."[63] Griffin responded via the press, too, telling *Nature* that the rovers wouldn't be cut. Stern promptly resigned; he e-mailed his decision around the Science Directorate leadership the morning of March 26.[64] Griffin asked Ed Weiler, who had moved to Goddard Space Flight Center to be its director, to resume his old job.[65] Stern left NASA on April 7, and MER survived for another year, albeit with a little less funding than Callas had requested. The agreed upon budget FY 2009 budget for MER came to $18.9 million.[66]

The longer-range consequence of the budget rebalancing act within NASA was to knock the long-desired sample return project off the table again. The agency had already committed to the ambitious Mars Science Laboratory mission, as well as to the first Mars Scout mission, Mars Phoenix. The budget out-years did not contain enough funding to sustain the Mars Scout program and embark on sample return prior to 2017, at the earliest.

Conclusion

As of May 2014, only Opportunity was still operating. In May 2009, Spirit became stuck in soft terrain near the winter haven of Home Plate. A six-month "Free Spirit" campaign, complete with T-shirts, by the small remaining MER team was unable to free it.[67] In January 2010, they stopped trying to extricate it, hoping to continue using it as a weather station, but they had not been able to maneuver it into a north-facing orientation to maximize its power as winter approached. On March 22, Spirit missed a communication window, and while the team kept listening for it, it was never heard from again.[68] Opportunity, though, kept moving toward a new goal, a 21 km wide crater called Endeavour, following a circuitous route to avoid some ugly sand dunes revealed by HiRISE. It reached the crater rim in August 2011.[69]

The two rovers' longevity was a direct result of the conservatism of the design assumptions made back in 2000 and 2001—particularly the thermal and power assumptions—versus the warmer environmental conditions the two machines encountered. The more moderate real-world conditions meant the vehicles were overdesigned; the tremendous effort the MER development team put toward figuring out how to fit larger solar arrays onto the vehicles and then to recover their mass margins and expand their mass capabilities, probably hadn't been necessary to meet the 90-day requirement. Instead, the bigger arrays provided the power margin that kept the rovers going through mission events that hadn't been part of the design requirements, like winter solstice and dust storms. And, of course, the team was also lucky. Mars kept cleaning their solar arrays off, granting the rovers occasional new leases on life.

The single-string nature of the two vehicles turned out not to be a significant liability, either. The nonredundant electronics didn't fail, partly because they'd been designed and built with substantial thermal margins and also because electronics in general had improved enormously since JPL had established its quality control processes in the early 1960s. (Consider Mars Observer, whose avionics had been initially rejected by JPL for poor thermal performance and workmanship and whose spares were reworked by Lockheed to create the Mars Global Surveyor mission.) Electronics failure had been common once, and JPL had put processes into place to minimize them, by parts screening and testing and the use of extensive redundancy. Mars Pathfinder had broken with that tradition of extensive redundancy while keeping the screening and testing, and MER continued in that new mold.

Still another contributor to the rovers' longevity was careful power management by the operations team, one aspect of which was the decision to rely almost exclusively on the UHF relay for data return and commanding during the mission. The UHF system used less power and offered a significantly higher data return than the rovers' direct-to-Earth capability, but it required the services of one or more orbiters. 2001 Mars Odyssey, in fact, returned the vast majority of the data from both rovers, due to MRO's electromagnetic interference troubles. The idea of a research infrastructure for Mars had paid off.

The MER project's success was rooted in testing, and the willingness to test to extreme conditions—pushing to find out what the engineering margins really were, not just what the team thought the margins were. When the Beagle 2 loss report was released by its commission of inquiry, it reinforced that point. Beagle 2's EDL designers had withdrawn from the program in 2001, the EDL system never received a deployment test, and the spacecraft was never subjected to system-level pyro testing. The original parachute and airbag designs both failed in testing (as, of course, MER's had too), but the replacement designs were not subjected to adequate retests. The inquiry concluded that the Beagle 2 coalition had neither enough time nor money to have done adequate testing; MER had been on essentially the same schedule but had received a good deal more money for the task.[70]

Mars Exploration Rover data return

Spirit (sols 001–2210, last downlink)		
System	Data (GB)	Total (%)
Mars Global Surveyor	0.35	1.6
2001 Mars Odyssey	20.72	96.1
Mars Reconnaissance Orbiter	0.12	0.6
Direct-to-Earth	0.37	1.7
Opportunity (sols 001–2680)		
System	Data (GB)	Total (%)
Mars Global Surveyor	0.39	1.4
2001 Mars Odyssey	25.77	94.8
Mars Reconnaissance Orbiter	0.58	2.1
Direct-to-Earth	0.43	1.6
Mars Express	0.01	0.0

Source: Data courtesy John Callas, JPL.

This conclusion, that thorough testing was paramount in robotic space projects, was reinforced by JPL's next landed mission to Mars, Phoenix. Expected, like MER, to be a rebuild based on "heritage," its engineers found that a sizeable amount of redesign was going to be necessary.

Reengineering a Spacecraft, and a Program

After Barry Goldstein left the Mars Exploration Rover project to take on the effort to refly the Polar Lander mission under the name Mars Phoenix, he and his engineering team first had to address the inadequacies the post-failure reviews had found in the Mars Surveyor 1998 project. There would be a formal review of those responses; in addition, he and the spacecraft manager he hired, Gary Parks, son of Robert J. Parks, who had run the Lab's Flight Projects Office for two decades, intended to apply the majority of their project resources to testing. They could afford to because the hardware was apparently complete and paid for by the Surveyor 2001 project. What they found in their own three-year run to the launch pad was an object lesson in what really had gone wrong toward the end of the faster-better-cheaper era. They would have to reengineer key systems to ensure mission success.

If the Phoenix project succeeded in reengineering its lander, salvaging Mars polar science in the process, the same could not be said of the Mars program. With the missions built into the 2000 "road map" nearly complete, Mars program officials began planning their next decade in 2008. Despite spectacular successes, though, political winds in Washington had shifted decisively against them. By 2012, their program was in turmoil, with no approved missions after Phoenix's successor in the Scout program, the Mars Atmosphere and Volatile Evolution mission, scheduled for launch in November 2013.

Redesigning a Built Spacecraft

The Mars Phoenix project team was faced with a task somewhat similar to that of the Mars Exploration Rover's entry, descent, and landing designers: to take a system designed for one mission and perform a very different mission. But they faced one big difference: their spacecraft had already been built. They had much less flexibility than the MER team had. The 2001 lander had been designed for a

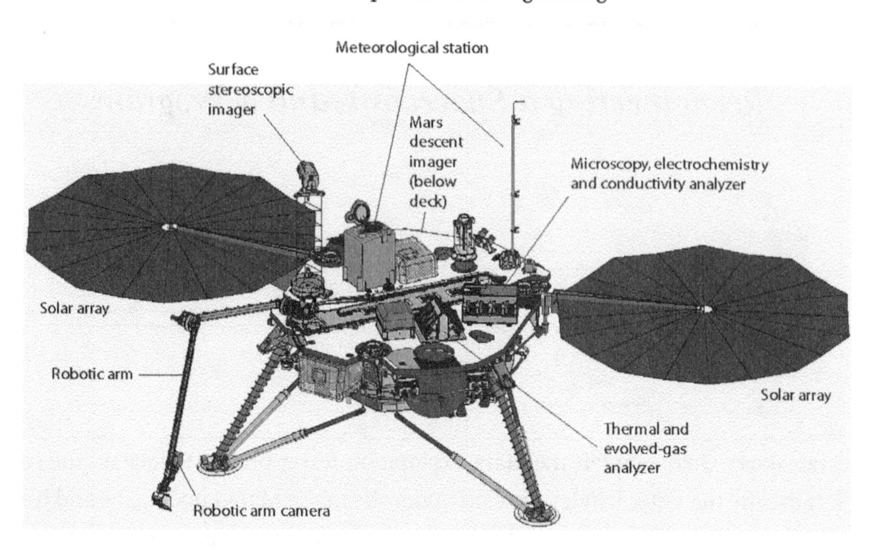

Mars Phoenix lander, in landed and deployed configuration.

mid-latitude landing site, not a polar one, so its thermal and power system designs had to be reevaluated and altered. One early decision was to replace the original lander solar arrays with more efficient triple-junction type cells to increase the lander's available power. But several other changes were made to the spacecraft and payload to accommodate the new mission and to respond to changing understanding of its environment.

One major change was to the cruise stage. The launch opportunity in August 2007 was less favorable than the 2001 opportunity the lander had been designed for, with Mars further from the Sun. This left the project's engineers with a power shortfall late in cruise. Their solution was to cover the sunward-facing part of the cruise stage with a third solar panel. This solved the power shortfall at the cost of requiring a complete reworking of the cruise stage. The solar panel generated heat while also preventing the other heat-producing components of the stage from radiating to space. Lockheed's lead thermal designer, Loren Zumwalt, had to redo the vehicle's thermal analysis, then redo the stage's layout. Some components had to be moved, heat pipes added, and thermal blanketing rearranged. The thermal rework cost Phoenix project manager Barry Goldstein about $2.5 million of his project reserve.[1] By contrast, Zumwalt recalled, not much needed done to alter the lander for the thermal conditions of its surface mission: "The lander came across pretty cleanly. The design that was intended for the equator went to the polar attitude with very few changes in terms of the spacecraft system."[2]

The Phoenix project also had to respond to the Mars program's demands for communications during all "critical mission events," a requirement imposed to ensure that in the event of a failure, JPL's engineers could reconstruct what had gone wrong from telemetry. This resulted in a major reworking of the telecommunications design of the vehicle. After the harsh criticism of Polar Lander's lack of EDL communications (an option not permitted to the earlier Mars Pathfinder), Surveyor 2001 project manager George Pace had planned to develop a "black box" that would survive a crash and send recorded telemetry back via an orbiter. But Goldstein's team did not like that approach. In addition to representing an expensive and risky new development, the black box consumed volume inside the aeroshell that the project did not really have.

Instead, the team tried to find a way to maintain continuous communications throughout EDL, despite the reality that their lander, unlike Polar Lander, had no direct-to-Earth capability. And while the team tried to figure out how to rearrange things inside the aeroshell to allow provision of an X-band direct-to-Earth link, they ultimately couldn't do it. The requisite antenna could not be stuffed into the aeroshell without dropping some other important equipment. This situation led them to depend on the three orbiters circling Mars for relaying EDL communications and to a search for an antenna design that would ensure that at least one of the three would be able to receive the lander's UHF signal. After several design iterations, they chose a conical antenna that was wrapped around the top of the backshell. It offered omnidirectional communications, relatively low impact on the existing vehicle systems, and a very high data rate. While the MER rovers had only been able to send semaphore tones during their landings, the Phoenix lander would be able to return digital telemetry continuously if everything worked as they intended.

While various aspects of the Polar Lander and 2001 lander hardware had to be reengineered to meet the needs of the Phoenix mission, the biggest challenge was ensuring a safe landing. Goldstein and Parks believed that merely responding to the specific findings of the various Polar Lander failure reports would not be sufficient. They began a set of inheritance reviews of the 2001 lander hardware and its supporting documentation, planned to build a hot-fire testbed to put to rest concerns over the lander's terminal descent system, and intended a thorough vetting of the rest of the EDL system. By the time they were done, fixing all the problems they found with the 2001 lander had driven the cost from $355.7 million (as proposed in May 2003) to $414.2 million.[3]

An early discovery by the project was that the 2001 lander structure was too weak to withstand the inflation of the parachute.[4] This derived from an error that the Mars Pathfinder team had made in calculating the parachute's inflation load but that was later caught and fixed. John McNamee's 1998 team had adopted the early Pathfinder calculations, not the reworked later ones, and the error had moved from there to the 2001 project. Simply put, if Polar Lander had lived long enough to deploy its parachute, the chute would have torn the tripod structure that held it to the lander deck off and floated away. It was a fatal design flaw.

Found in May 2004, the parachute problem forced the team to do a set of trade studies to determine the best solution. The options available to them were to have a new, smaller parachute manufactured, to strengthen the lander structure to accommodate the existing parachute, to deploy the parachute at a lower velocity (and therefore altitude), or to try a combination of these. They chose to shrink the parachute and reinforce the structure, leaving them with a terminal velocity that was higher than the 2001 lander had been designed for. That meant the lander would use more fuel in landing than had been allocated by the Surveyor 2001 project. Since Phoenix, unlike Pathfinder and MER, used the same fuel tanks for cruise to Mars and EDL, and the tanks could not get larger, fuel use during cruise had to be "squeezed." As Phoenix project engineer Robert Shotwell recalled, "There was a nice balancing act we had to do with propellant usage for cruise and propellant usage for EDL."[5] By the project's critical design review late the following year, they had adopted this suite of changes.[6]

A little later in 2004, the inheritance reviews held with Lockheed engineers out in Denver uncovered a more systemic problem. Lockheed, like JPL itself, required every piece of flight hardware that was delivered to the vehicle assembly floor to be accompanied by a package of documentation, analysis, and test data validating the hardware's conformance to its requirements. The 2001 lander project had already reviewed and accepted all this documentation and its associated hardware; Goldstein, Parks, and Lockheed project manager Ed Sedivy decided not to accept the 2001 project's certifications. Instead, they reran the hardware certification reviews. "What was amazing," Parks recalled, "was that it wasn't a few flaws that we found in the delivery packages; it was that the shortcomings were the rule, not the exception." Documentation was often missing, and testing was "wholly inadequate."[7] The review panel Parks had assembled for the telecommunications subsystem had made this point most strongly in their review outbriefing, stating, "There is [*sic*] voluminous missing background data on inherited components," and so the panel could not certify that the review objectives had been met.[8]

Parks, who had earlier been project manager for JPL's portion of the Herschel-Planck mission, reflected later that these inadequacies were a consequence of the faster-better-cheaper experiment. One principle of faster, better, cheaper had been reducing documentation and testing, enforced through demands for ever-lower costs. The elimination of institutional design and testing standards in pursuit of that principle had resulted in cut corners. But, he said, "the magnitude of the corners cut was beyond belief for me. That was the real eye-opener."[9] Goldstein and Parks challenged their Lockheed counterparts to locate the missing data and to re-create through retest that which could not be recovered. They held these delivery reviews again when the project began its assembly, test, and launch operations phase in the spring of 2006.

Goldstein and Parks's finding that much more work would have to be done to requalify the supposedly qualified hardware produced a major cost increase, estimated at about $20 million in early 2005. At the same time, Wayne Lee, who had become the manager of JPL's EDL group after completing his service on the MER project, was advocating increasing the Phoenix EDL task funding. Lee thought Goldstein's team had not planned for enough testing of the EDL system and its various components, and he had taken his argument up JPL's chain of command. In the risk-averse environment of the period, his arguments were accepted, and Goldstein made the pitch to NASA headquarters officials for more money. This gained Phoenix a further increase of $18.1 million. By May, the project's cost cap had been raised to $385 million.[10]

More troubling news reached JPL in the summer of 2005, as Loren Zumwalt's analysis of the vehicle's thermal state came together. The thermal model showed a large variety of violations, with some components, such as the star trackers, getting too warm and others, including the lander legs and their deployment hardware, far too cold.[11] The heat shield separation nuts were also too cold, and the thermal team had to add heaters to them. But by far the worst thermal exceedance, discovered sometime in the summer of 2005, was by the separation connectors that were supposed to release the aeroshell from the cruise stage prior to entry into the martian atmosphere. At the temperatures Zumwalt's model estimated for Phoenix's arrival at Mars, these were some 60°C below their limits. When Lockheed's mechanical test engineers tried to operate them at those predicted temperatures, they were, in essence, frozen solid.[12]

Like the lander's inadequate structural strength, this was a catastrophic design error. It meant to Goldstein and Parks that Polar Lander probably failed because these connectors never separated.[13] In that event, the cruise stage would have remained attached to the lander aeroshell as it entered the atmosphere, and the resulting atmospheric turbulence would have destroyed both vehicles. (This would also explain why the Deep Space 2 microprobes were never heard from. They were to be released by the physical separation of the cruise stage and aeroshell, which never happened, if Zumwalt's model were correct. They would also have been destroyed.)

Fortunately for Phoenix's engineers, this design flaw was relatively easy to fix. They added heaters to the separation connectors to warm them up to their qualified operating temperature range before entry. The heaters ate into their already thin power margins late in the flight to Mars, but they solved the problem relatively inexpensively.

By contrast, the hot fire testbed of the descent engines, the centerpiece of the project's response to the findings of the Polar Landing failure review board, was tested very successfully at the same time as the thermal problems were coming to light. Lockheed's engineers put it through two series of tests. One, a "cold firing" series used water instead of fuel to demonstrate the design's resistance to the perceived problem of water-hammer in the piping; after replacing copper fittings that should not have been in the testbed (hydrazine and copper react explosively in each other's presence), a second set of hot-fire tests was performed. These tests demonstrated that the system worked as intended and was controllable, even though it had a few minor problems.[14] Shockwaves produced by the thrusters caused some minor damage to nearby lander structure, and some of the fuel lines needed more secure mountings. After several complete "landing" tests, some of the thrusters also developed small leaks, causing some concern within the science team about possible hydrazine contamination of the landing site. The team strengthened the structure around the nozzles and remounted the fuel lines, but after analysis of the thruster leaks decided not to try to fix them. The testbed thrusters performed the equivalent of several Mars landings and in consequence suffered much more wear to their valves than those on the flight lander would, so the project's propulsion engineers did not think the flight lander would experience similar leakage.

Shotwell also had Lockheed investigate another potential failure mode that had, up to Phoenix, not been considered by any project: the possibility that pieces of the cruise stage could hit the aeroshell during entry. In the direct entry trajectories used since Mars Pathfinder, the cruise stage followed the aeroshell into the atmosphere on the same trajectory, and because the cruise stage was not designed to be aerodynamically efficient, engineers' expectations were that it would fall further and further behind while disintegrating. Shotwell suspected that expectation was probably not quite right, because the cruise stage contained some pretty heavy, solid pieces. Once released during the breakup, they might experience a reduction in drag forces, allowing them to catch up to the aeroshell and possibly hit it. So he funded Lockheed to do an analysis of the cruise stage breakup and the aerodynamics of its resulting pieces. Sure enough, the launch vehicle adapter, the telecommunications power amplifiers, and a few other items were heavy enough in mass and low enough in drag to catch, and even pass, the aeroshell. While this was still a fairly low-risk problem (the pieces would likely tumble and not follow the aeroshell trajectory exactly), in the risk-averse environment of the 2000s, it was unacceptable. The project solved the problem by adding a maneuver to the entry sequence that turned the vehicle off its entry trajectory just before releasing the aeroshell and then used the maneuvering thrusters in the backshell to put the aeroshell back on course after the cruise stage released it.[15] That way, the cruise stage would not be following the aeroshell. The maneuver was only possible because the 2001 lander had had attitude control thrusters in its backshell as part of its precision landing experiment. Neither Pathfinder nor the two MER entry vehicles were equipped with thrusters, and could not have performed any maneuver.

Those thrusters, though, gave the project team additional headaches. In simulating the aeroshell's aerodynamics, Langley Research Center engineers discovered that during a short portion of the vehicle's hypersonic flight, the thrusters could cause the vehicle to tumble. This happened because at certain velocities, the effect of the thrusters reversed. When a thruster fired to tip the aeroshell in one direction, the aeroshell would tilt the opposite way. (This had to do with the way the thruster flow fields interacted with the airflow around the backshell at different velocities.) The flight software would not understand that and would again fire that same thruster to compensate, eventually causing the vehicle to tumble out of control. This was actually a "known problem" within the aerodynamics community, first discovered during the X-15 flight program in the 1960s and long ago fixed within NASA's human flight programs. But the knowledge had not transferred over to the Mars program's robotic efforts in the 1990s. Discovered very late in the project, December 2006, only eight months before launch, this "hypersonic thruster efficacy" challenge was fortunately easy to fix. The team reprogrammed the spacecraft to turn off the attitude control thrusters during that short piece of the entry sequence and reactivate them after it had slowed below the dangerous velocity range.[16]

While each of these design flaws was serious, and in a couple of cases fatal, the system that caused the project the most difficulty, and ultimately cost so much to overcome that it forced NASA to hold a termination review late in January 2007 due to the excessive cost overrun, was the vehicle's radar. The Mars Polar Lander red team review board had fingered the radar as a potential problem because the off-the-shelf military radar that John McNamee's Surveyor 1998 team had chosen seemed to give misleading data over sloped terrain, so when the Phoenix project got started, Goldstein and Parks knew they would have to carry out a fairly intensive test program to retire this known risk.[17] One initial decision was to replace the two radar units bought by Surveyor 2001 because they had been built with parts that were not traceable back to screened, space-qualified lots. They ordered units built from screened parts and used the original flight units for the flight-testing necessary to help understand and resolve the slope problem.

The initial radar test took place in February 2006, at the manufacturer's plant in Minnesota. Dara Sabahi, the chief engineer, reflected later that they found five separate, unique problems with the radar that day.[18] He drafted several specialists from JPL's radar section to help understand and fix them; in March they found ways to either eliminate or mitigate four. But in retests done at Dryden Flight Research Center in April, the team found still more problems. The radar took far longer to lock on to the surface than it should, and worse, the beam that usually locked up first was not the vertical beam intended to sense altitude but one of the side-looking beams intended to measure velocity. This produced an incorrect altitude measure. The side-looking Doppler beams also had a positive velocity bias, sensing a meter/second or so of speed that was not actually present—an acceptable amount of error for the jet fighter the radar was built for but not for a Mars lander that could not tolerate more than 2 m/sec of velocity at impact. And there were other troubles, too.

Sabahi established a tiger team to begin digging deeply into the radar. Understanding why it had these problems was made somewhat difficult by the contractual arrangements involved. The radars had been bought "off the shelf" from Honeywell, which sold thousands of them to many customers; JPL was not a large client that carried much weight with the company. And the radar was proprietary, so JPL was in a weak position to be asking for access to the company's intellectual property. The legal challenges were made more difficult by the presence of Lockheed as a middleman, which, as the prime contractor, had actually purchased the radars, and, of course, the flight software that received and interpreted the radar signals was proprietary to Lockheed. So getting the JPL tiger team access to two sets of proprietary software was its own challenge. Sabahi credited his counterpart at Lockheed, Tim Gasparrini, for pushing the negotiations within his company forward so that the problems could be investigated and resolved.[19]

With only a little more than a year before launch, the tiger team's investigation pursued several tracks simultaneously. More flight and drop tests took place in April and May out at Dryden, and technical interchange meetings with Honeywell engineers began that month, too. From his pot of EDL robustness money, Wayne Lee had had the foresight to finance development of a simulator that would allow "dropping" the flight radar thousands of times in simulation relatively quickly, and assembly of that system also began in May.

The May tests at Dryden uncovered still more anomalies. Lockheed had built a test rig that enabled a tethered radar to be dropped from a helicopter at the right altitude and velocity for a Mars lander, permitting more realistic tests than those flown by Honeywell back in February. After 33 flight and drop tests, they had accumulated 13 different problems.[20] Due to the high risk posed by the radar's increasing troubles, JPL senior management formed a radar advisory group to oversee the tiger team's work.

Technical interchange meetings with Honeywell engineers determined that the company already knew about some of the faults and had fixed them in its current production units, but of course the units Phoenix had were built in the late 1990s for Surveyor 2001 and obsolete. And the project's two new radars were being built to replicate the 2001 model, not the current production model, because it was important to the project's engineers that all their radars behaved the same way. So fixes were already known for some of the flaws, narrowing what the team had to address.[21]

During the summer, the tiger team's engineers were able to fix most of the remaining problems by revising the radar's "firmware," software that was embedded in the radar's electronics, or by altering the lander's software to deactivate some of the radar's potentially problematic modes or to reject obviously wrong measurements ("data editing," they called this). They also chose to simply accept some of the radar's less-critical problems after analyzing them.[22] In September, they gained another new problem, however, when the side-looking antennae that enabled measurement of horizontal velocity failed in the spacecraft's thermal vacuum tests. They had been made of two different metals with different thermal expansion rates; in the extreme cold of space, the antennae became distorted and created unacceptable beam patterns. The project had to design, procure, and install new antennae by March 2007.[23]

In October, the radar team carried out a large series of new tests at Dryden designed to determine whether that had fixed most of the known problems.

Sixty test drops helped convince them that they'd fixed or mitigated most of the radar's performance troubles. The drops did not reveal any major new troubles.[24] On December 1, the tiger team hosted a review by the radar advisory group and an independent review team, who validated their conclusions.[25]

But the radar team was not quite out of trouble. Later that month, the radar simulation effort turned up a new problem: the radar could lock on to the heat shield and, thinking it had detected the surface close by, raise its pulse repetition frequency and trigger the final landing sequence. But the lander would be too far above the actual surface to land before running out of fuel. This was a problem that could not have been discovered by the drop test program, which had no way of simulating a heat shield dropping away from the radar. Sabahi, who had been keeping a "fever chart" summarizing the tiger team's activities and the associated risk, had initially painted December's risk level green. In January, he switched December to yellow, and he painted January red.[26]

At the same time, the project uncovered another, potentially very expensive, problem. Late in 2006, Lockheed's engineers had pulled the 2001 lander's backshell out of storage and found that its thermal protection coating did not meet the requirements. It was not the correct density. It had been put into storage in that condition; George Pace's engineers had documented that and, in the risk-taking posture of the era, accepted the resulting risk. Phoenix's engineers could not accept the risk without verification that it would not produce a failed mission, so they took samples of the material for arc jet testing at Ames Research Center. The team also attempted to replicate the material to see if newly made coating would pass the tests, but they found the material could not be replicated. Something was wrong with the manufacturing process. And when they put the vehicle into its first thermal vacuum test in December, the protective coating cracked—they could not know how deeply until they pulled it out again in January, disassembled it, and performed some tests. But by then, the project was out of money.

In October, with his financial reserves entirely eaten up due to all the known problems, and being fairly certain new problems would be found in the remaining eight months before launch, Barry Goldstein had presented to his principal investigator, Peter Smith, and NASA management a preliminary cost-to-complete estimate that required a budget augmentation of $26.6 million. Under the rules governing cost-capped missions, NASA was required to hold a termination review before approving additional funds. It took place at the end of January. By that time, a grassroots scrub of the entire project budget put the estimate at $31.1 million, or 8.1 percent above the $385.6 million cap.[27] At the review, Goldstein explained that the project's current completion plan required about 3,500 work-months of effort; remaining within the cap would require removing a little over 1,300 work-months from the schedule. "The associated removed work," he told his superiors, "would introduce unacceptable risk."[28]

By this time, NASA officials had become very uncomfortable about the Phoenix project. Dara Sabahi remembers that they increasingly did not want to fly the vehicle—too many severe problems had been found. But they had already spent a lot of money, and critics of NASA would doubtless use cancellation as a justification for attacking the agency.[29] Forcing Goldstein to finish the project within the cap was an even worse option. If the project did not finish the radar rework effort, or the backshell thermal protection replacement, or fix their latest problem (during the backshell separation test, the backshell hung up on the lander's thrusters and bent some of them) due to insufficient funds but launched anyway, failure of the mission seemed very likely. The critics would pounce. Science Directorate officials had no good options other than approving the increase, and they did.

Goldstein had gone into the termination review assuming he would be the review's casualty. Managers of JPL cost-capped missions who overran their budgets were generally removed; by 2007, this had happened to a few project managers in JPL's projects funded by the Discovery program. Because of the difficulty inherent in actually cancelling a project that was far along in its development phase, this was really the only way any kind of accountability could be imposed by either NASA or JPL management. "I survived, though," he recalled.[30]

During the remaining few months before launch, the project was able to clear up its known problems, and a few more. Radar tests in April 2007 indicated that the fix the tiger team had thought up for the heat shield problem, changing the pulse repetition frequency, would resolve the issue without introducing new ones. But soon after, they also found a new radar problem. The intended flight radar, newly built to meet the traceability requirements, took several tens of seconds to lock on to the surface again if it was suddenly shut off and then turned back on again ("power cycled") when it was at 250 feet altitude. While this was not supposed to happen, the radar's own fault-detection procedures could cause the radar to power cycle automatically. Investigation by Honeywell's engineers determined that the fault occurred in only a single temperature range, due to an error in the radar's firmware (somewhat ironically, the original pair of radars built for the 2001 lander did not have this error).[31] The team fixed this problem too, even though it was unlikely that the radar would reboot itself, at exactly that temperature and altitude, during the real landing. And they scheduled still more flight tests, this time at very low altitudes, for the following February to make certain there were no more low altitude anomalies.

The Phoenix team continued to find problems in the vehicle in the months remaining before launch, but none that were as severe as those found during 2005 and 2006. Another temperature-related radar flaw cropped up and was solved simply by heating the radar prior to entry. The lander's key instrument, the thermal and evolved gas analyzer (TEGA), built by Bill Boynton's team at the University of Arizona, developed an intermittent reset problem that the team thought they had resolved prior to launch, but which resulted in the instrument being reintegrated to the spacecraft very late down at Kennedy Space Center. And the circuit board that controlled data flow along the serial bus inside the spacecraft had a design flaw that caused it to reset when it received data from the science instruments at certain temperatures. If this happened during EDL, it could be severe—the flaw caused the spacecraft to lose data from its inertial measurement unit, which was an unacceptable risk. Mike Malin's descent camera, MARDI, was the only instrument in use during EDL, and the project's proposed resolution was to permit him only a single image, whose data would not be read out until after landing.[32] After a post-launch meeting in November, though, Goldstein and Parks convinced Peter Smith that even that one image involved too much risk. Parks recalled that while the risk was low, the availability of the high-resolution imaging science experiment (HiRISE) to provide the larger geologic context that MARDI's descent imaging was intended for tipped the decision toward simply leaving the camera off.[33]

Goldstein's engineers got Phoenix off the launch pad and on its way to Mars on August 4, 2007.[34] The launch itself was flawless, and so was the post-launch checkout, with one exception: TEGA had water contamination that made it unlikely to meet one of its requirements, to be able to distinguish between normal and "heavy" water. The project's mitigation plan was a series of long bake-outs during cruise, heating the instrument ovens to drive the undesired water out of the instrument.[35]

Landing Site Selection

Peter Smith, the Phoenix principal investigator, was responsible for the site selection, and he relied on the help of Ray Arvidson of Washington University in St. Louis to manage the process. Principal-investigator-led missions did not require the open community process that MER and Mars Pathfinder had used; instead, site selection was done through a pair of workshops that were largely internal to the project science team, with some assistance from the Mars program office.

Smith had initially proposed sites selected from a region around 67.5°N and 240°E, which was low-lying, smooth, and appeared in Bill Boynton's gamma ray spectrometer data to be rich in near-surface water ice. Because of the early suspicion that Phoenix would not be able to land successfully on high slopes, Smith and Goldstein had agreed that the site had to have slopes less than 20° at the lander's scale. And because they knew Phoenix, like any other legged lander, was vulnerable to rocks larger than the landing deck height, they also had limited the site's rock size to less than 0.4 meters.[36]

The science team's site survey had generated four areas around the region that seemed to have appropriate characteristics. Data from the Mars orbiter camera and thermal emission imaging system (THEMIS) suggested that individual sites existed within those areas that met the standards, and Smith and Arvidson decided to focus on what they called "Region B," a wedge between 120° and 140° east longitude in the interior of the Vastitas Borealis. Region B had the lowest elevation, providing the longest entry sequence (and valuable "timeline margin"). It was not, based on the MOC images, the region likely to have the lowest rock abundance, but the science team convinced themselves that the region would still be within the standard established by the Viking 2 landing site—preferred by the engineering team because it was a known, flat plain with many very small rocks.[37] Because of the deep cold of the northern polar plains, the THEMIS data was not definitive about the rock abundance, though, and Mike Malin suspected that the team's analysis was wrong. He thought dark areas in his images were likely to have lots of large rocks.

The landing site selection subteam chose a landing ellipse within Region B in the summer of 2006 and asked Alfred McEwen to image within it using his HiRISE camera while Mars Reconnaissance Orbiter was still in its "transition" orbit phase. This had to be done because by the time MRO reached its science orbit, there would no longer be enough sunlight for HiRISE to produce images at its highest resolution. McEwen's team got eight images within Region B during September and October. The images revealed that Region B, and especially the part of it they'd been focusing their attention on, was full of boulder fields. Rocks exceeding 8 meters in size seemed to be frequent; the mission design and science teams simply couldn't fit Phoenix's landing ellipse anywhere within Region B that didn't include at least one of these boulder fields. This produced what Matt Golombek termed a "freak-out period," for they had very little time to pick new spots in the martian high latitudes for HiRISE to target for new imaging before the Sun set and the north polar region began its long, cold, dark winter. So by the tenth science team meeting in mid-October, the team had abandoned

Region B and had reopened investigation of Regions A and D, far away from their original site and dominated by different geology.[38] These regions were higher altitude, requiring some rework of the EDL timeline by the engineers, but MOC and THEMIS data suggested were likely to be much lower in rockiness.

Arvidson put a graduate student of his to work automating the process of rock identification and counting, with the aid of programmers at JPL.[39] They relied on new HiRISE images of the Viking 2 landing site to calibrate the model, which was designed to automatically identify and count rocks within HiRISE imagery. This automation was necessary because HiRISE full-resolution images were enormous—whole walls of the MRO operations area at JPL were covered by printouts of single images—and the team lacked time and funds to do manual rock counting and then compare the results to the much larger supply of MOC and THEMIS images. While that model was being prepared and tested, the landing site effort turned to the THEMIS thermal imagery again, looking for areas of Regions A and D that appeared very cold in early morning, an indication of low rockiness.[40]

The area they finally chose was a broad valley in Region D, on the other side of Mars from their original site. One end of the landing ellipse just missed a large crater.[41] Tentatively chosen in February 2007, the landing area could not be imaged again by HiRISE until late in 2007—well after launch—because of the onset of winter. What confidence in the safety of the landing site they could develop came from the efforts to cross-compare THEMIS, MOC, and the handful of HiRISE images already available to validate the automatic rock-counting model that was in the works. New HiRISE images obtained in the martian spring of 2008 would confirm the site safety; the project's planned trajectory correction maneuvers would have to be used to retarget, if necessary.

Fortunately for the project, when the Sun finally rose again over the martian south pole, new HiRISE images confirmed that their valley (they named it "Green Valley") had low slopes and few rocks.[42] At the site certification review in March 2008, Lockheed's Tim Gasparrini presented a Monte Carlo analysis indicating a touchdown failure probability of under 5 percent for the site.[43] The rockiness of the Green Valley site was acceptable.

Operating Phoenix on Mars

Landing day for the Mars Phoenix was May 25, 2008, Memorial Day weekend in the United States. The operations plan for the mission was that landing would be monitored at JPL, but once safely on the ground, JPL engineers who'd relocated to the University of Arizona's new Science Operations Center would take over, with Lockheed engineers in Denver providing lander health, power status, and other engineering statuses remotely, just as they did for Mars Odyssey and Mars Reconnaissance Orbiter. Because Phoenix did not have direct-to-Earth communications, it would be operated solely via UHF relay. Barry Goldstein's deputy had negotiated agreements with the MRO, 2001 Mars Odyssey, and Mars Express flight teams to have them positioned to provide those relay services. They had also worked a deal with the National Radio Astronomy Observatory in Green Bank, West Virginia, to monitor the spacecraft's signals, just in case.

Joel Krajewski, who had been brought over from MER to reprise his role as lead for the transition from EDL to surface operations, had devised a timeline for the lander that began with signaling a successful touchdown, and then doing some internal housekeeping while waiting for dust kicked up by its thrusters to settle—venting pressure from its tanks, writing telemetry data to flash memory, and after one minute, shutting down its transmitters to save power. After 15 minutes, it would deploy the solar arrays, release the biobarrier that protected the robotic arm, and raise the meteorological and surface stereo imager masts. Then it would take a few images chosen to prove that each of these events had succeeded. Because MRO and Odyssey orbiters only overflew the landing site every two hours, the team in Arizona would have to wait to see these results. Touchdown would be about 4:53 p.m. at JPL (and about 4:00 p.m. local time on Mars); the teams at JPL, Arizona, and Denver would find out if the all-important solar panels had deployed, and they had a viable mission, a little after 6:00 p.m.[44]

Jim Erickson, who had become project manager for MRO, had been asked in February 2006 if there was a way to photograph Phoenix with HiRISE during its descent. The inquiry had provoked a good deal of examination within the MRO project. MRO's first priority was recording Phoenix's telemetry stream, and that function could not be put at risk. But getting an image would be very useful for failure reconstruction—if there were a failure—and it would be a great public relations coup if Phoenix landed safely. The telemetry requirement already put MRO in almost the right position to capture an image; the challenge was to get the spacecraft's slew rate right and HiRISE set up correctly. The camera's pushbroom nature made that task easier. It could scan a large part of the landing ellipse, so unless Phoenix was far off its trajectory, the HiRISE team was likely to get an image. Lockheed's engineers thought it could be done; Rob Manning, then in the Mars program office, really wanted to try to get one or two images; at the same time, in the more risk-averse climate in the 2000s, Erickson's team had to convince JPL reviewers that they had put sufficient rigor into their analyses and understood all the risks. He gained JPL's institutional approval after a meeting on May 1.

Phoenix's entry sequence started with the turn away from the entry trajectory to ensure the cruise stage did not follow directly behind the aeroshell, followed by cruise stage separation and the aeroshell's turn back to the proper entry alignment. Receipt of the signal that this had all been accomplished correctly was greeted by Barry Goldstein with a huge sigh of relief. By this time, he was convinced that this was the point at which Polar Lander had come to grief. After that, Richard Kornfeld read out the entry deceleration numbers as they came back down from Mars. Everything seemed perfect until parachute deployment, which was late—it felt painfully late but in fact was only seven seconds behind the expected timeline. This was the only glitch, though. The heat shield ejected

Celebrating the Phoenix landing. *Left to right*, Lockheed's Ed Sedivy, JPL's Barry Goldstein, and University of Arizona's Peter Smith.
JPL image PIA D2008_0525_T136A, courtesy NASA/JPL-Caltech.

properly, the troublesome radar found the surface correctly, the thrusters all worked, and the vehicle parked itself safely on the martian surface at 4:53 p.m. Pasadena time, to cheers at JPL.

A couple of hours after the landing, Erickson walked into the press support area that had been set up in JPL's Von Karman auditorium to begin telling people that the imaging experiment had actually worked out. HiRISE had caught Phoenix on its parachute, descending toward, it appeared, the very imposing-looking 700-meter-deep crater they named Heimdall.[45] This was somewhat of an illusion created by the angle between the two spacecraft; Phoenix was actually about 20 km in front of the crater.[46] It had landed 21 km "long," much closer to Heimdall than intended, nonetheless. Reconstruction of the entry sequence by a joint team from JPL, Langley, and Lockheed eventually suggested that the aeroshell had generated slightly more lift than expected, placing the lander farther down track.[47] Erickson commented afterward that this was the first time anyone had actually seen a Mars parachute work. "We could even see the risers on the parachute."[48]

Out in Arizona, attention turned to getting the lander into commission. Data back from Mars Odyssey on the first overpass after Phoenix's landing showed the solar arrays and instrument masts had all deployed correctly, and the vehicle was generating power. Krajewski's timeline had the first sol on the surface taken up by imaging of the spacecraft and its landing site, the second sol for unstowing the robot arm and the beginning of instrument checkouts, and sols 3–6 for continuing instrument and landing site checkout. By sol 7, he hoped to get the first surface sample into TEGA. The surface itself was flat, with polygonal structures like those typically found in Arctic tundra throughout the visible landscape. It was devoid of large, and even medium-sized, rocks. Instead, like Meridiani, it had scatterings of pebbles and small rocks, mostly smaller than the robot arm's scoop.

Krajewski's timeline did not evolve quite as expected after the nearly perfect landing day. The biobarrier did not completely retract but could be worked around. Commands sent up on sol 2 to release the robotic arm did not reach the spacecraft due to a glitch on MRO, so Phoenix kept running a backup command sequence that generated more images.[49] This put the operations team a day behind the timeline; meanwhile, another problem cropped up in the checkout of TEGA. The telemetry stream seemed to indicate that the instrument had an intermittent short circuit.[50] JPL's design requirements ensured that the short could not kill the spacecraft itself, but it could damage, or even disable, TEGA. As understanding of the nature of the problem percolated up through the project and then to NASA management, it became the single biggest threat to achieving a fully successful mission.

Discovering the magnitude of TEGA's electrical problem took time, though, and meanwhile the surface operations team pushed ahead. On sol 5, they had the robot arm camera take a picture of the surface underneath the lander. This produced a beautiful image, showing that the landing thrusters had swept away the dirt overlying what looked like a sheet of solid ice. While they had expected soil cemented by ice, the team had not expected an ice sheet. But the image showed a surface that looked like a smooth, frozen lake—or an ice skating rink. What else could it be but ice? But they refused that speculation, needing better evidence that the smooth, shiny surface was really water ice. The image validated the choice of landing site, though. The icy layer they wanted was only a few centimeters below loose dirt.[51]

One week after landing was the first time they were able to dig up a sample, photograph it, and dump it in a designated area. But some of the dirt stuck to the scoop, and they decided to make another attempt. They needed a precision delivery because TEGA had a second problem: not all of the ovens opened completely. A machining error had caused some of them to open only partially, including the first oven they needed to use. After the second dump experiment completed successfully, the team ordered Phoenix to dig its third sample and

Ice beneath. The Phoenix lander's robot arm camera took this image under the lander shortly after its descent. The silver objects at the top of the image are the lander's descent engine nozzles. The project scientists interpreted the white patches as ground ice blown clear of loose surface material by the descent thrusters.
JPL image PIA10741, courtesy NASA/JPL-Caltech / University of Arizona / Max Planck Institute.

dump it into a TEGA oven. The sample did not go in. The arm delivered it properly, as it turned out, but a screen placed over the oven intended to break up clumps prevented the soil from entering the oven. The martian soil's clumpiness proved too great. Vibrating the screen the next sol for 20 minutes did not work, either.[52] The robot arm team began working on a new delivery method in their testbed at the university, using the rasp on the scoop's bottom to vibrate the scoop over the oven in the hope of sifting dirt, instead of dumping it. That seemed to work, and was first used to sprinkle some soil for study by the lander's optical microscope.

Meanwhile, on sol 16, Mars threw Boynton's TEGA team a bit of a curve ball. The day after the sprinkle technique was tested, the team ordered another shaking of the TEGA screen, and this time it worked. The soil sample that had been sitting on top of the fourth oven's screen suddenly fell into the oven.[53] Nobody knew why. But the team ordered Phoenix to start testing that sample on sol 18.

In digging its first trench, the lander had turned up some bright chunks of material that some science team members thought had been ice, and these chunks—rather, their disappearance over the next four sols—became the subject of the project's first big post-landing press conference. On June 20, the science team announced that the chunks must have been water ice. Carbon dioxide ice would not have still been present at the landing site's temperature, and the other alternative, some kind of salt, would not have simply sublimated away. Only water ice would have gradually vanished over a few days. "It must be ice," said Peter Smith at the news conference. "These little clumps completely disappearing over the course of a few days, that is perfect evidence that it's ice. There had been some question whether the bright material was salt. Salt can't do that."[54]

Their next step was using the new sprinkle technique to feed Michael Hecht's Microscopy, Electrochemistry, and Conductivity Analyzer (MECA) instrument, which was designed to examine soil chemistry and electrical conductivity.[55] This worked well. But TEGA's short-circuit problem reared its head again on sol 26. It appeared that the shaking they had ordered on oven #4, hoping to get the sample to drop in, had exacerbated the short circuit. The science team delayed delivery of the next sample to oven #5 while the engineers tried to develop a better diagnosis of the problem's severity. Instead, they began testing the rasp's ability to dig into what appeared to be an ice layer a few centimeters below the surface. But the engineers' conclusion was that the short circuit was bad enough that the next use of an oven might be the instrument's last test.

On sol 36, NASA headquarters intervened in the mission and barred Peter Smith from using TEGA on anything else until the team got a sample of ice into it—not soil, not icy soil, but ice. The requirement came from the NASA administrator via Ed Weiler, and it wasn't negotiable. It was very difficult to do, though, and it took the team until sol 60 to even attempt to deliver an icy sample. Project scientist Leslie Tamppari called this an "interface" sample, from a soil layer directly above the ice sheet but one likely to still contain water ice.[56] The initial delivery attempt failed—the icy sample stuck in the scoop—and Arvidson and Sabahi organized yet another tiger team. The lander's arm had a grinding tool known as the rasp, and the rasp's designers remembered that in their testing back on Earth, lights that were being used for videotaping had also caused their icy soil samples to stick to the scoop. They decided to do the sampling as early in the morning as possible, let the sample sit on the surface for a few hours so ice in the outer layer would sublimate away, then scoop up the sample while keeping the scoop out of the Sun until the sample reached TEGA.[57] That somewhat complicated process worked, and they got their sample into TEGA on sol 64 and cooked it. The data stream back from TEGA provided the direct confirmation of water ice at the landing site the science team had been hoping for, releasing the team from NASA's interdiction.[58] TEGA itself survived the testing, too, essentially eliminating the team's concerns about it.

A few days later, Phoenix became subject of a strange eruption of media coverage, due to an *Aviation Week and Space Technology* article that claimed the project had briefed the White House on major new findings "concerning the 'potential for life' on Mars."[59] The reporter, veteran Craig Covault, had apparently heard and misunderstood some discussions among the MECA team about a find that they were still arguing about internally. Some of the MECA team members thought the instrument had found a salt called perchlorate. But TEGA, which should also have found perchlorate's constituent atoms, had not. Perchlorate is toxic to humans (and thus relevant to human habitability), but bacteria can coexist with it. By itself, perchlorate's existence has little bearing on the presence or absence of life. The argument over whether perchlorate was really there or not was taking place within the Phoenix team at the same time that the confirmation of water ice was happening; when Covault did not hear the controversial "find" announced at the press conference on sol 65, he apparently thought the news was being suppressed for political reasons.

Perchlorate was not evidence of life, yet that's how his article was instantly misread across the Internet.[60] Covault had specifically stated that Phoenix had not discovered life on Mars, but that did not matter. The hint at a cover-up of martian life was too juicy for Web media to ignore. Smith forcefully denied the existence of the alleged White House briefing to a Space.com reporter, but the controversy whipped up by Covault's story ultimately forced the team to hold another press conference to debunk the "life on Mars" aspect of the story by explaining the disagreement between the two instruments, how it could happen, and what they were doing to figure out the true state of the soil chemistry.[61]

All of this occurred while the science and operations teams were making two transitions. The perchlorate press conference was on sol 70; sol 76 was to be the day the project shifted back from Mars to Earth time, and sol 90, the end of the primary mission phase, was the day the project shifted to distributed operations. The engineering support team moved from Arizona back to JPL and Lockheed's Denver facility, and most of the scientists returned to their home institutions. The overall pace of operations slowed, too. In part, this was because of the return to Earth time, making the planning cycle more challenging for the team, but primarily it was due to the rapidly waning martian Sun. Phoenix's power state was already declining and would fall more quickly after sol 90, reflecting the return of martian fall. The lander might be able to continue operating the meteorology experiments until perhaps sol 150, but the arm, MECA, and especially TEGA would be useable less and less frequently.

The Phoenix team ultimately got five surface samples into TEGA, four during the primary mission and only one during the extended mission, on sol 136. And the sol 136 sample was not fully analyzed before the lander failed on sol 152.[62] Similarly, the wet chemistry lab component of MECA was only able to analyze samples in three of its four cells; apparently, the sample delivered by the scoop to one of the cells got stuck in the cell's inlet, blocking it.

Phoenix's end came as a result of the predicted decline in power available from the waning Sun, combined with a dust storm and the buildup of water-ice clouds above the landing site in late October. The lander put itself into low-power safe mode on October 27, and for a few more days Goldstein's engineers were able to get engineering data, and some science data from the meteorology instruments, back via Mars Odyssey. They hoped that shutting down some of the heaters for the robot arm, TEGA, and other functions would enable them to gain a couple more weeks of meteorology data. But November 2 was the last time Phoenix communicated with any of the orbiters; NASA officials decided to stop listening for it December 1, which would have been Phoenix's sol 183.[63]

After the martian Sun rose again and the carbon dioxide ice had retreated from the site in 2010, JPL tried to reach Phoenix without success. The Mars Reconnaissance Orbiter was able to get a clear image of Phoenix in May, though, showing that one, and perhaps both, of the solar panels had been torn off.[64] They had never been designed to withstand the weight of the dry ice that had certainly built up on them during the long winter and had probably simply broken away.

What Hath Phoenix Wrought?

Both MER and Phoenix missions added to scientists' conviction that in its remote past Mars had been very wet. One interesting difference seemed to be that the MER data seemed to reveal a very acidic watery past for Mars, while Phoenix found the opposite. The MECA results for Phoenix's far northern perch showed a slightly alkaline chemistry, a little more in line with the needs of most life on Earth. Neither mission provided much evidence that liquid water persists on the current surface of Mars; Phoenix saw what most scientists expected, that water ice sublimates straight into the vapor phase without melting first, and only the possibility of brines remained to tantalize those wishing for a watery Mars.

The Phoenix project's trials with their inherited hardware also validated the sense that many JPL engineers had after the Mars 1998 losses, that the principal technical failing with the faster-better-cheaper era had been inadequate testing. Any complex design will inevitably have errors, and the key to success in the space business was thorough testing. That had been Tony Spear's argument back in the Mars Pathfinder days, and George Pace, whose lander Phoenix had originally been, echoed him many years later. The 2001 lander's test program had been "success oriented," developed based on the belief that the Polar Lander design was sound and would work; when it failed, that test program became inadequate. That and the missing documentary record led directly to Phoenix's troubles. The engineering teams working for Goldstein and his Lockheed counterpart, Ed Sedivy, had much more corrective work to do than they had anticipated.

The Phoenix project cost $402 million, not including the original cost of the 2001 lander. Because the 2001 lander was only part of Surveyor 2001 (2001 Mars Odyssey was the other component of that project), its cost cannot be known with great accuracy. Roger Gibbs, the Surveyor 2001 spacecraft manager, estimated the lander's price tag was $50 million; adding that to Phoenix's price gives a fairer cost of $452 million. Even that understates Phoenix's true cost, though, since most of the design cost was carried by Polar Lander. But it's impossible to adequately account for the value of that design work relative to Polar Lander's overall contract cost. For comparison, Mars Exploration Rover's cost was $810 million through the end of its primary mission. But that was for two rovers. When JPL had been asked to cost the second rover back in 2000, the basis had been that the second vehicle would increase the project's cost by a third, making a hypothetical single-rover MER mission about $540 million. Thus the Phoenix mission's cost was about midway between MER and Mars Pathfinder, whose inflation-adjusted cost was $354 million.[65]

Some, but not all, of the cost difference is attributable to Pathfinder's relative paucity of scientific instrumentation and the associated procurement, integration, and science team costs. Some is in testing. For example, Tom Rivellini did 15 full-scale airbag test drops for Mars Pathfinder, while MER did 52.[66] Phoenix did not have airbags, but its radar team ultimately did 73 test drops.[67] Testing rigor, and cost, expanded after Polar Lander's demise. The remainder of the cost differential is likely in documentation. The MER project was not able to recover Pathfinder's blueprints for its "build to print" goal; Dara Sabahi could not recover Pathfinder's parachute data either. Similarly, Phoenix could not recover some of the lander test data from Polar Lander and Surveyor 2001. All this had to be re-created, and that cost money. One of the tenets of the faster-better-cheaper argument was reduced documentation; the consequence was that future missions could not reuse designs without also re-creating the original test data, raising their own costs. Reduced documentation was a short-term savings with a substantial longer-term cost.

Phoenix's scientific results are, as of this writing, unclear. Hecht's team confirmed the perchlorate find, and in addition to the confirmation of water ice at the site from TEGA, Bill Boynton's team found that the site contained calcium carbonate, which they attributed to formation in thin films of liquid water.[68] The perchlorate finding threw into question the old Viking landers' results, which had not found organic material despite a constant influx of organic material on meteors. Like TEGA, the Viking landers' mass spectrometers had heated their samples, and perchlorate would have combusted with any organics, destroying them. Another result Tamppari found memorable was the first snowfall seen on Mars. It was photographed on the meteorology experiment's mirror, which reflected images of a wind telltale provided by Denmark.

The short-lived nature of the mission meant that there was little immediate funding for analysis of the Phoenix data; when the project ended, so did the science team's funding. Arvidson, who had been involved in Mars missions since Viking, commented: "It was crazy. I'd never been involved in a mission in which the science funding ended before the papers were published, so you really couldn't pay for page charges because the contract was over."[69] Tamppari explained later that NASA's rules required that the project's scientists provide all their data to the Planetary Data System within six months after the end of the mission, and that became their immediate focus. They then had to wait six months after uploading the data before submitting proposals for funding under the agency's research and analysis program so that other scientists could also write proposals on the data. Then there was another year's wait before selection and receipt of funding. That delay was intended to help more scientists utilize that data, but it meant fewer immediate results from the mission. Scientists could first get funding for analysis of Phoenix's data in 2010.[70] It will be a few more years before the project's findings are really understood.

Replanning the Mars Program, Again

Phoenix's short, exciting life paralleled the onset of another planning cycle for the Mars program, though unlike that undertaken in 2000, this "cycle" would drag on for years without resolution. The Mars science community, via its Mars Exploration Program Analysis Group, remained dedicated to the dream of sample return. During 2009 and 2010, this dedication was validated by the National Academy of Sciences's Committee on Planetary and Lunar Exploration's decadal survey, chaired by Steve Squyres. Published in 2011, Squyres's survey declared Mars sample return to be the highest-priority flagship-class planetary mission for the 2013–2022 period.[71]

Yet these were poor years in which to restart sample return planning. The worst banking crisis since the Great Depression resulted in a focus on budget cutting, and NASA was not spared. The Vision for Space Exploration program also ran far over budget, and the Obama administration, faced with shrinking tax revenues and growing costs, reopened the question of whether to develop a new launch vehicle. Portions of the aerospace industry had never been happy with the program's focus on development of a shuttle-derived, government owned and operated launch vehicle. The United States already had commercial launch vehicles with sufficient lift capacity to serve the International Space Station with cargo, and which could be improved to carry astronauts as well; the builders of those rockets, understandably, wanted the government's business. They found inroads into the new administration, and the administration empaneled (yet another) blue ribbon review of NASA's exploration policy. That panel concluded the Vision for Space Exploration's launch vehicle development needed about $59 billion over the FY 2010 budget guidelines it had been given by the White House, or $159 billion over the next decade. Completion on schedule would need, in essence, an increase of the NASA budget of about $6 billion per year.[72] That was not in the cards.

In 2010, the administration formally terminated the Vision for Space Exploration program, having already begun transitioning to a new effort, which it calls the Commercial Crew Development program. This was an expansion of another Bush administration effort, commercial cargo development, which sought to replace the shuttle's cargo-carrying capacity to the International Space Station with commercially procured launch vehicles. The administration's intended budget for the program was about $800 million per year to foster the development of crew-capable launchers, with much of the rest of the previous program's funding going to technology development, including heavy lift propulsion technology.[73] Congress, however, insisted on a parallel "heavy lift" launch vehicle development—and mandated it without funding it. In fact, the enacted fiscal year 2012 budget cut NASA by about $700 million overall. The funds for the heavy lift development came from reducing the funds available to the commercial crew program to about half of what the administration had requested.[74]

The administration's FY 2011 and 2012 budget submissions had projected NASA being flat-funded at $18.5 billion overall, with its Science Directorate also being flat funded for five years at just over $5 billion for the next five years. Much of the cut in 2012 came from funds that had once financed shuttle operations, so the agency's bet that it could retain those funds for its commercial crew development and heavy lifter development did not pan out. In order to expand the funds available to the commercial crew effort back to what it had intended in 2011, it again had to raid other accounts. The Science Directorate's contribution to the commercial crew development was about $162 million.

That $162 million contribution, though, hid the true damage to planetary science, which was cut $309 million.[75] The remaining funds largely went to fund an enormous cost overrun on the James Webb Space Telescope, which was approved for development in 2004 for $2.5 billion but will exceed $8 billion.[76] The $309 million cut to the planetary science budget amounted to a slice of about 20 percent. Mars exploration received the largest cut in absolute dollars, of $226.2 million or 40 percent of its previous budget; the smaller lunar science program begun in the Bush administration was cut by half, and will be zeroed out in 2015 unless Congress intervenes; and outer planets science was cut by about a third.[77]

These external pressures on the Mars program were mirrored by internal problems. Mars Phoenix's cost overrun paled compared to its successor, the Mars Science Laboratory, which, like Mars Observer two decades before, saw its launch

NASA Mars Program budget

Year	Millions of U.S.$	Year	Millions of U.S.$
1994	20.7320	2005	783.9461
1995	82.6254	2006	725.1960
1996	152.4078	2007	685.0571
1997	120.2400	2008	743.3464
1998	190.5024	2009	370.3808
1999	295.7780	2010	442.1438
2000	317.9664	2011	547.4000
2001	538.1635	2012	587.0000
2002	559.7025	2013*	360.8000
2003	597.6800	2014*	227.7000
2004	702.6950	2015*	188.7000

Source: For 1994–2001, adapted from NASA budgets at www.hq.nasa.gov/office/hqlibrary/find/newnasadoc.htm; for 2001–2011, from NASA budgets at www.nasa.gov/news/budget/index.html#.U7IHUiTDoQt; and for 2012–2015, from NASA operating plan submission, August 2013, www.nasa.gov/sites/default/files/632709main_NASA_FY13_Budget-Science-Planetary-508.pdf.

Note: All figures are in real year dollars.

*Estimated.

delayed from 2009 to 2011. The proximate cause of the delay was the inability of the vendor responsible for the "actuators," the electric motors and gears that drove the vehicle, to deliver on schedule. The consequence was a cost increase from $1.6 billion to $2.3 billion.[78] Most of this came from inside the Mars program itself: the sample return technology development program, known as "Safe on Mars," was eliminated and the funds available for a planned mission in 2016 were reduced, leading NASA to seek a partnership with the European Space Agency for a joint mission in that year.[79] The second Mars Scout mission was also delayed two years, in part due to conflicts of interest within the selection committee, and its funds became available. But funds for the next outer planet mission were cut, too. All of these transfers were supposed to be made up for in future years, but the external cuts imposed in 2011 and 2012 made some of those paybacks impossible.

The consequence of all this was programmatic turmoil. After the selection in September 2008 of the second Mars Scout mission, Mars Atmosphere and Volatile Evolution (MAVEN), and the first Mars mission to be managed by Goddard Space Flight Center in Maryland, the Scout program was suspended indefinitely. During 2009 and 2010, NASA and European Space Agency officials negotiated and funded a pair of joint "ExoMars" missions to be carried out in 2016 and 2018; NASA abruptly withdrew from them in 2012, and ESA turned to Russia instead.[80]

Reflecting the dismay among Mars scientists at this surprising turn of events, the chairman of the Mars Exploration Program Analysis Group wrote that his colleagues were

> shocked by the severe reduction in the President's proposed budget profile for NASA's Planetary Science Division, even though its programs have dramatically advanced our understanding of the Solar System and been so compelling to the public . . . The cuts threaten the very existence of the Mars Exploration Program, which has been one of the crown jewels of the Agency's planetary exploration. This occurs at a time when the significant discoveries of the last 15 years were about to inform and enable missions to Mars that will advance dramatically our quest of discovering life beyond the Earth, pave the way for human exploration, energize future scientists, inspire the public, and enhance U.S. prestige in space exploration.[81]

But the damage was already done. Neither protests from the scientific community nor from the equally dismayed Planetary Society had any significant effect on the decision.[82] Thus, after MAVEN's November 2013 launch, NASA's Mars Exploration Program, originally built around a demand for two launches every other year, will likely not have a launch until at least 2020. The next approved mission to Mars, InSIGHT, will instead be funded by the Discovery program (as Mars Pathfinder had been). And it will use a variant of Lockheed's lander, not JPL's Mars Science Laboratory–derived Skycrane.

The Phoenix project's success at reengineering Polar Lander was not mirrored by success in reengineering a sustainable Mars exploration program.

Conclusion

If the Mars Phoenix mission represents the final end of NASA's faster-better-cheaper saga, it also represents the completion of the Mars program's "follow the water" strategy. Mars's water has been definitively found, and via the radars on Mars Express and Mars Reconnaissance Orbiter, Mars's shallow subsurface ice has also been fairly well mapped globally. While questions about the planet's hydrologic past will continue to be asked—the heatflow measurements to be made by InSIGHT will hopefully address questions about whether Mars sustains deep subsurface reservoirs of liquid water, for example—scientific interest appears to be migrating back toward the Viking project's old focus on a search for life. Or, at least, a search for the remnants of ancient life. Water is only one of the needs of Earthlike life; living things on Earth also need nitrogen compounds (or produce them as waste products), and they've not yet been found on Mars.

It's important to understand, finally, that the Mars exploration program's sinking fortunes did not reflect a similar shrinkage of public enthusiasm for Mars. Quite the opposite was true. The very same years that the Mars program was being cut witnessed continued expansion in public interest—and in the offices of the Obama administration's NASA administrator.

Conclusion

I think NASA is becoming more and more risk averse.

—*Thomas R. Gavin, August 2011*

In February 2012, just after the announcement that the Mars program budget was being slashed again and the joint NASA / European Space Agency ExoMars missions cancelled, NASA administrator Charles Bolden held a town hall meeting at JPL. A former astronaut and Marine Corps general, he insisted that the Mars program would continue, and he teared up as he expounded on his intent to ensure that his granddaughters could be among the first to walk on Mars. "I would doubt very seriously if there's anyone in here who's more passionate than I am about going to Mars," he said.[1] Echoing his predecessor Dan Goldin's efforts to levy human exploration requirements on the Mars Surveyor program, after terminating the 2016 and 2018 missions, Bolden demanded that the Science Mission Directorate incorporate human exploration requirements in its post-2013 missions. This integration of human program requirements was necessary to provide the scientific support astronauts needed for these Mars expeditions, which he hoped would take place in the 2030s—2032 being the best of the launch periods in that decade for very large payloads. Still another planning group was formed later in 2012 to try to rework the Science Mission Directorate's goals to incorporate the demands of human missions to Mars, and in its report the group again promoted sample return as the primary goal of Mars scientific exploration.[2]

As the vagaries of the last two decades of Mars exploration reveal, scientists' desires were not the only driver of Mars exploration. Rather, there were many intersecting influences. Scientists established the intellectual framework, laid out in studies sponsored by the National Academy of Sciences and sometimes in NASA internal publications like the April 1995 strategy from the Office of Exobiology, and scientists developed the "follow the water" strategy that proved such a useful tool for explicating a rationale for continued Mars missions.[3] Scientists also drove Mars science via the proposal process, as more and more of what NASA did became subject to competition during the 1990s. But NASA's human exploration program intervened repeatedly during this period, demanding payload elements that, ultimately, neither the human exploration program nor the planetary science program could support.

Another major driver was Dan Goldin. In addition to pushing for a human mission to Mars after 2004, Goldin was responsible for establishing Mars exploration as a program of its own. He and Wes Huntress, his associate administrator for science for most of the 1990s, were also the engines behind faster, better, cheaper, which profoundly influenced the way Mars exploration was carried out during the 1990s. Engineering demands also influenced strategy; witness the engineering justification for the HiRISE camera for the Mars Reconnaissance Orbiter, the dependence of landing site selection on engineering safety criteria, and the desires of robotics engineers to promote the state of their own art. Last, NASA's need for public interest (and perhaps more to the point, its fear of public derision) influenced its choices. While it did not select any mission solely on the basis of public relations potential, the agency's leaders certainly did consider public reaction. And they were bothered when missions that they thought would have widespread appeal—Mars Global Surveyor's first camera to Mars in nearly a generation—got ignored. Scientists' desires were one driver of Mars exploration strategy, but not the only one.

Still another influence on the Mars program was foreign competition. Prior to the Cold War's end, competition with the USSR provided a framework to which demands for planetary exploration as part of a prestige race could be staked. But after the Soviet Union's collapse, this prestige narrative declined in effectiveness and was no longer able to recruit funding for NASA, as evinced by declining budgets throughout the 1990s. Instead, NASA pursued cooperative ventures with Russia, and then increasingly with the European Space Agency, as Russian planetary exploration effectively ended with the failure of its Mars '96 mission. This shift in focus enabled a more systematic exploration of Mars and offered at least the potential of expanded resources for Mars science by drawing on the treasuries of both the United States and European Union.

The conflicting demands placed on the Mars program by scientists, the human program, funding levels, foreign relations, and NASA leadership required JPL's engineers to engage in various forms of social engineering. They had to maintain effective communications within their own projects, with the science teams attached to their projects, among their own technical divisions, and with NASA management. They were not always successful. Climate Orbiter's loss resulted from a series of miscommunications and communications not acted upon; the Mars Surveyor 2001 project's collapse and reformation in April 1998 was a product of the inability of the project's management, as well as JPL's management, to communicate to NASA officials that they had imposed an impossible combination of requirements and resources. Recovering from the Polar Lander and Climate Orbiter losses involved restructuring the lines of communication between JPL's Mars office and NASA headquarters; inside JPL, one aspect of recovery was reimposing the requirement for formal tracking of spacecraft incidents and follow-up.

At the level of individual missions, scientists' ability to steer the course of research was stronger, but still constrained. Sometimes the constraints were institutional in nature—JPL or NASA imposing a "hands-off" contract regime to prevent scientists from interacting with instrument builders to limit change orders, and thus cost increases. Other times they were engineering constraints; on Mars Explorer Rover, the project's ability to land in certain desirable places depended on the engineers' ability to adapt the entry, descent, and landing system; once down, the ability to move, make measurements, and so forth depended on power available over and above the amount needed to keep the rovers "alive."

But the dependency of the scientists on the engineering teams was not one-way. Neither aerobraking nor the Pathfinder and MER EDL developments could have taken place successfully without scientists providing their expertise to the engineering teams. Scientists had to be willing to work with the engineering teams from the very beginning of the projects, and to be willing to make the time investment to understand what information the engineers needed. In his lessons-learned presentation at the end of the MER project, Mark Adler referred to the subset of scientists who were willing to work with engineers and able to communicate effectively with them as "engineering-friendly scientists." In parallel, JPL's systems engineering division deliberately seeks science-friendly engineers, knowing that not all members of its workforce are interested in the same things scientists are.

Legacies of Faster, Better, Cheaper

The loss of the Mars Surveyor 1998 missions brought the faster-better-cheaper era to a close. In his 2001 analysis, McCurdy argued that the strategies most effective at reducing costs were lessening mission complexity and keeping the project teams small.[4] Smaller teams could make decisions more quickly and needed to maintain fewer communications channels, allowing less formality. Less complex missions had fewer technical interfaces to manage, requiring less systems engineering involvement. In reality, of course, these two strategies are intertwined. A less complex mission needs fewer people to design, build, and test it, even if no other efforts to reduce manpower are taken.

Yet the faster-better-cheaper period's focus on reducing mission complexity did not last long, at least not within the Mars program. The successes of Mars Pathfinder and Mars Global Surveyor were taken as evidence within NASA that greater, not less, complexity was still possible on small budgets. Polar Lander, in particular, was scientifically more ambitious than Pathfinder had been, and it had less money. The original Surveyor 2001 lander mission, with its Athena rover and human exploration program instruments, took a long step toward Viking-level mission complexity. It had to be scaled back severely to get down to a budget that was twice Polar Lander's. In part because NASA managers felt the need to push the technological envelope, to make each future mission more exciting than the last, and in part because they needed to support Goldin's desire to gain a White House–level decision on sending astronauts to Mars in 2004, they abandoned one of the central demands of faster, better, cheaper.

The rising cost of Mars exploration after 2000, as the MER, MRO, and Phoenix stories illustrate, is attributable to two things: increasing complexity and increasing risk aversion. Mission complexity is a function of scientific ambition; the more ambitious the scientific goals, and the larger the payload, the more things cost. Risk aversion is illustrated by the return to explicit design and testing standards, reinforcement of problem-reporting systems, and the deployment of design principles, flight project practices, and the certificate of flight readiness process. By 2011, that risk aversion had grown strong enough to cause some of its originators to begin questioning it. JPL hosted a meeting of NASA's center directors on August 15, 2011, and Tom Gavin, who had re-created and enforced JPL's flight project practices and design principles as its director for flight projects during the later 2000s, stood up and commented that NASA had become more and more risk averse. To him, this countered another trend within NASA during that decade, commercialization of human spaceflight.[5] Imposing NASA's requirements and standards would result in raising costs, frustrating the very goal of commercialization. The Marshall Space Flight Center and Glenn Research Center directors agreed with him. The agency needed a new cultural revolution to free itself from what its own leaders were beginning to see as excessive technical conservatism.

At JPL, former Pathfinder spacecraft manager Brian Muirhead, by 2010 JPL's chief engineer, was trying to use his certificate of flight readiness process to cut back the overgrowth of reviews for the same reason. Reviews cost time and money, and many did not obviously contribute to mission success. He saw the demand for reviews as a trust issue. "Nobody trusts anybody. So we've got to try to get past that because if nobody trusts anybody, then the costs and everything else we're trying to do here to improve affordability [are] not going to get better."[6] He believed imposing sufficient rigor in the certification process could reduce the demand for reviews throughout the project life cycle by increasing trust in the final product.

One of Goldin's goals during the faster-better-cheaper era had been the elimination of "flagship" class missions, with Cassini-Huygens his poster child for projects run amok. Yet accomplishing programmatic goals, especially Mars sample return, requires missions of flagship scale, or at the very least a means of using a succession of smaller missions to advance the technologies needed by the programmatic goals. That was Tony Spear's criticism of Donna Shirley's Mars program; she had been prevented from crafting a program that would produce the technological needs for sample return by headquarters officials' demands that the Surveyor missions be competed. (To be fair, that demand flowed not just from headquarters—Cornell University's Steve Squyres and JPL's Dan McCleese were both supporters of competition, and reports from the National Research Council's Space Studies Board routinely argued for the protection of NASA's Discovery, Explorer, and New Frontiers programs, all of which are competed.) This tension was only resolved during the 2000s as the rapid expansion of the Mars program budget enabled the financing of both "strategic" missions like MRO and MER and the competed mission line, Mars Scout. Shrinking budgets after 2008 led to the suspension of the Mars Scout missions, so the tension was resolved again in the other direction.

The danger of returning to a strategic-mission-only program is what McCleese and Squyres warned about in the 1990s: "Our approach in the past has always been to begin from top-level science objectives, and then construct a program that meets those objectives. Time and again, this objectives-driven approach has failed, for the simple reason that it produces inflexible and ultimately unaffordable programs."[7] As I write this, what's left of the NASA Mars exploration program is the drive for sample return, a community-driven, top-down strategy that has to date proven unaffordable—repeatedly.

Science, Exploration, and Technological Novelty

Over the years I've been working on this book, I've had several conversations with historian of science Naomi Oreskes about one clear aspect of the Mars exploration story: novelty. Throughout the foregoing narrative, we have seen a demand for novelty from Mars specialists—not scientific novelty, but technological. Scientists sought long-range geological rovers long before the technology existed, and that validated the desire of JPL engineers to work toward that capability. Scientists advocated for converting the Mars Smart Lander into a mission capable of producing "breakthrough science," which meant new instrument technology on the giant rover. Echoes of this desire for technological novelty can be seen in Goldin's demand for airbag landings versus "old" Viking landing rockets, in NASA's repeated decisions to oversubscribe a spacecraft's power, mass, and telecommunications capabilities in order to squeeze the latest instrument technology aboard, and in ever-higher imaging resolutions that tax Deep Space Network capacity.

Scientists and engineers aren't the only source of this demand. The American public seems conditioned to expect technological novelty in Mars missions (and perhaps more broadly). NASA struggles to gain attention for Mars orbiters, even though they provide the majority of the scientific return, but has no troubles getting people interested in landers—even landers like Pathfinder, with minimal scientific capability. Landers are technological spectacle; public interest dissipates rapidly after touchdown, even if, like MER, the missions actually wear on for years. NASA is publically funded, of course, and its leaders seem to believe it needs these spectacles, and the public attention they garner, in order to stay funded.

It may seem natural that scientists seek continuous technological novelty, but this is a feature of late-twentieth-century science, not of science itself. One of the predecessor disciplines to planetary science, geology, became the premier science in the United States in the nineteenth century not because of continuous innovation but through routine application of instruments that changed very little. Their innovation was in rigorous mapping of both horizontal and vertical dimensions, of careful identification of rock units in unexplored regions, and of imagining how these discontinuous point measurements might be unified into a coherent understanding of continental evolution. A Mars program organized on those lines could send MER-class machines to Mars for the next several decades, building up knowledge of the planet's many terrains, rather than the current program's drive toward shipping a sample back from one spot. This hypothetical Mars program would produce a continuing stream of new scientific knowledge about Mars but without any new technology at all. Yet I cannot imagine such a Mars program being funded.

Where did this demand for technological novelty come from in American space science? Part of the answer is NASA itself, which is primarily an engineering agency. New technology is what engineers do. But the trend toward technological novelty seems to exist outside NASA's own realm. In Oreskes's history of plate tectonics, she contends that what finally made the shocking idea of moving tectonic plates acceptable in American geology was a novel form of measurement, the mapping of magnetic regions on the ocean floor. That was the Defense Department's funds at work, not NASA's, but those funds flowed to serve a technological need. In the case of sea floor mapping, the effort was to improve submarine and missile guidance. So perhaps the origin of this transformation lies at the end of World War II, when American science first gained sustained, large-scale public funding. Initially that funding flowed from the Office of Naval Research, but that source was soon eclipsed by other military funds, by the National Science Foundation in the 1950s, and, of course, by NASA. With the exception of the NSF, each of these funding sources sought technological achievement through scientific research. Perhaps the close, long-running relationship among scientists, engineers, and science managers (who are often, though not always, engineers), that this funding arrangement required explains the increasing adoption of technological novelty as a scientific goal in the late twentieth century.

What are the effects of the drive for novelty? It biases research programs toward those questions that can only be answered by new technology, and away from those that don't need it. Like the demand for community-driven flagship missions, it increases costs by ensuring one cannot use a "production line" for spacecraft—unlike communications satellites, nearly every planetary spacecraft is unique. And it leads to increasing sophistication of instrumentation, moving the ability to make instruments out of university departments and into industrial contractors and NASA centers.

Nevertheless, at least in the American context, technological novelty creates public interest. The technospectacles of Mars landings were tremendously popular. Thus, the desire for technological novelty has a powerful influence over scientific plans and over public reception. But spectacle has not guaranteed sustained political support.

Simulation and Mars Exploration

One of the many transitions that occurred at JPL in the 1990s was the tremendous increase in the use of simulation in the spacecraft design process. There are varieties of simulation, of course. The Viking project had given Mars Pathfinder the gift of a parachute design that had been tested in wind tunnels—a form of mechanical simulation—and in Earth's atmosphere, for both inflation stability at very low densities and high velocities, and for load at high densities. Nonetheless, Pathfinder and MER both carried out their own parachute strength and stability tests; in MER's case, going back into the wind tunnel after failure of their strength test methodology. The same kind of mechanical simulation, of course, was done in the airbag testing.

Another important use of simulation has been for training. What those at JPL call "operational readiness tests" are simulations designed to accomplish several things. In most cases, they serve as tests of the spacecraft, though the MER spacecraft were in flight by the time the team could perform them. ORTs also serve as tests of ground systems and as tests of software for both spacecraft and ground systems, and they are essential for training spacecraft operators. Because JPL generally builds unique spacecraft for each mission, the institution cannot use generic training for operations very effectively. Instead, it has developed a tradition of using the flight vehicles as part of training simulations to ensure adequate realism—the exception being MER, again, which had to employ its testbeds for training.

Digital simulation also became a major part of spacecraft design during the 1990s. Mars Pathfinder certainly wasn't the first use of digital simulation in robotic spacecraft design; trajectory design by JPL's navigation group has been done with the aid of digital computers since the 1960s, while the Viking project used a very limited set of Monte Carlo simulations to constrain certain requirements, such as fuel tank size for the landers, after traditional parametric studies left them with too much uncertainty. The great advantage of a large Monte Carlo simulation set run on a high-fidelity model has been its ability to define a rigorous probability envelope. The probability envelope allowed designers to know the likelihood that, for example, they'd exceed the limitations of their airbags—that is, as long as they could prove that the model was an accurate representation of the actual hardware. It wasn't always, as the Polar Lander story makes clear.

Years after the mission, Lockheed's Steve Jolly observed that the reason Pathfinder was a great success was that it had been designed to be tested.[8] Unlike the Polar Lander and Viking spacecraft, for which no reasonably valid descent test could be carried out on Earth, every piece of Pathfinder's descent sequence could be tested. As we saw, though, Rob Manning and Sam Thurman worked long and hard at figuring out how to structure their test program. They and Dara Sabahi were forced by their peers to conceive a test program that would permit comparing test data to model data in a very systematic way to validate all parts of the model. Even given the evident testability of the vehicles, in reality, their digital models still weren't perfect simulacra of the spacecraft.[9] They could not be, of course. They were simplifications. But such simplifications were still useful.

The need to convince themselves and their reviewers that they really understood how their spacecraft would behave caused the designers to think carefully about how every piece of their complex EDL system should work, what data could be extracted about its performance from various kinds of tests, and what tests were therefore most useful. Digital simulations could help identify which aspects of a project's vehicles were most sensitive, and should receive the most thorough testing. And simulation also helped identify tests that were less valuable: those that did not produce useful information that could be compared to the model output. In short, the Pathfinder team's desire to employ digital simulation demanded that they develop a rigorous understanding of their vehicle in order to test it, and to develop a set of tests that further informed that understanding. Their efforts to employ digital simulation didn't replace testing—simulation helped them think more clearly about testing, and they carried that forward into their future efforts.

Learning across Projects

As the above suggests, engineers in this story learned from their simulation efforts and carried that knowledge with them from one project to the next. Digital simulation was not the only area of learning that we have seen. Parachutes come up again and again in the narrative, with Pathfinder modifying the Viking parachute, George Pace's Surveyor 2001 team finding out from Pioneer Aerospace that the Pathfinder parachute had far less margin than it should have had, and undertaking a redevelopment program to strengthen it. That work, in turn, fed into the MER parachute development despite the cancellation of the 2001 lander. Similarly, when NASA forced JPL to rebid the Mars Reconnaissance Orbiter contract in 2000, Lockheed's engineers drew on the knowledge they'd gained from their prior Mars orbiters to design a more reliable vehicle.

But knowledge transfer was hardly smooth. John McNamee's engineers did not discover the Pathfinder parachute error and their build-to-print version of an early Pathfinder design was an incorrect choice. When Loren Zumwalt, the thermal engineer Lockheed assigned to the Phoenix lander project, began digging into Polar Lander's thermal design history, he became convinced that the design needed rework even beyond that imposed by changed mission requirements. And as we saw with the Phoenix project overall, design knowledge had to be reconstructed through retest due to lack of documentation.

Migration of knowledge was enabled by JPL's retention of the same subcontractors across long periods of time. This dependence of knowledge transfer on human continuity sets up a tension with the desire for low costs, though, since contractors ultimately know that their expertise can command higher fees—and so do NASA and JPL. Their options become to choose a higher-cost contractor with greater experience or a new, lower-cost contractor that might have a better, or at least newer, idea but lacks the experience to deliver a reliable product. The lower-cost route leads to higher risk, which for Mars exploration became unacceptable.

There is a third option, but one that requires a longer time horizon. JPL can choose a new contractor but invest a larger fraction of its own engineering resources in educating that contractor's engineers. In essence, that's what it did with Mars Observer, though unsuccessfully, and with Mars Phoenix, more successfully. In the short term, that isn't less expensive, but introducing new competition might lower costs for future missions. Whether this strategy is successful at lowering future costs over the long run depends on sufficient overall business to sustain multiple specialist contractors.

This dependence of knowledge transfer on human experience is known within both the history of technology and larger engineering communities. It's why, for example, Glenn Cunningham insisted on keeping George Pace as his spacecraft manager for Mars Global Surveyor even after criticism of Pace's leadership during the Mars Observer project. He lacked the time and money to redevelop expertise in a new manager. But that knowledge is forever in conflict with the desire to cut costs by refusing to pay for experience and, of course, with the longstanding desire of people to assign blame and punishment.

Humans, Machines, and Planetary Exploration

Despite the apparent primacy of machines, robotic exploration is a human enterprise. The robots are simply agents we dispatch to do various things that we can't do ourselves, in places where we cannot go. We can equip our exploring machines with instruments whose capabilities far surpass those possessed by people unaided by technology. But people are involved throughout—as planners, as managers, as designers, as researchers, as testers, as operators, as spectators. There are simply no *astronauts*.

Historian Stephen Pyne has argued that humanity has entered a Third Great Age of Discovery.[10] In previous ages, Europeans had voyaged into the larger Earth community, "discovering" (and trading with, conquering, or colonizing) the rest of Earth's cultures and peoples. While they were promoted, funded, and celebrated as eras of geographic discovery, what they actually were, to him, were eras of cultural discovery. The Third Age, beginning with the effort to map Antarctica and the deep oceans, and expanding outward into the solar system, is fundamentally different. "If the Third Age has propelled exploration beyond the ethnocentric realm of Western discovery, it has also thrust it beyond the sphere of the human and, with regard to space, perhaps beyond the provenance of life. No one will live off the land on Deimos, go native on Titan, absorb the art of Venus, the mythology of Uranus, the religious precepts of Mars, or the literature of Ceres. There will be no one to talk to except ourselves."[11] Human explorers were essential to the previous great ages of discovery because only we are capable of communicating culture. Humans had to be involved in "first contact." But in Pyne's Third Age, there is no first contact. There is no one at all.

Without the need for cultural communication, there is also no need for humans to leave Earth to explore. Our agents can do that for us. The incredible deployment of powerful computers and equally powerful display technology (incredible, at least, to this child of the mid-1960s) has made it possible to bring their explorations back not just to JPL but to everyone with an Internet connection—anywhere. In the first two ages of discovery, the exploring experience was limited to a lucky few (lucky, at least, if they survived it). The "public experience" of exploration in those ages came through works of art—paintings, lithographs, and books. The U.S. Congress created Yellowstone National Park in 1872 based on the sketches and watercolors of an exploring expedition artist named Thomas Moran.

But the public experience no longer needs to be quite so remote, filtered through the perceptions of other humans. Exploring machines can enable everyone to participate, if only virtually. The Third Age of Exploration is democratizing the experience. There are downloadable software tools that allow individuals to operate simulated rovers on real martian terrain—or, at least, within the imagery of real martian terrain. That emergent capability enables Web-connected members of the public to engage in virtual exploration in a way that was impossible in preceding decades. Human mediators are no longer necessary.

A potential counterargument is that only humans can communicate meaning, and that the astronaut-mediators served that purpose. Their disappearance, or, rather, their replacement by machines, then might explain the lack of support for space exploration among the contemporary public. But that argument is inconsistent with the evolution of NASA's public image. During the Apollo Moon program, when astronauts were national heroes, NASA had broad but shallow support. Those polled supported Apollo but did not think it worth the cost.[12] In the 2000s, after two decades of astronauts being stuck in Earth orbit while robots roamed the solar system, support for NASA's funding was consistently higher than it had been in Apollo's heyday.[13] It's simply a myth that public support for NASA has declined since the Moon landings. While funding for NASA has declined relative to the federal budget, public support for it has not. And, of course, the human operators of robotic explorers are just as capable of communicating meaning as astronauts had been. The cultural popularity of Pathfinder's engineers makes that clear.

There may still be reasons for humans to leave Earth—nationalist competition could compel it politically. Perhaps advocates of colonization will someday succeed in securing the resources of one of the Earth's wealthy nations to try to colonize the Moon or Mars. Moon or asteroid mining to retrieve space resources for use on Earth is a perennial justification for human presence in space. But exploration itself no longer requires humans to go anywhere.

Epilogue

On August 6, 2012, the Mars Science Laboratory, née Mars Smart Lander, mission placed its rover, now named Curiosity, on the surface of Mars. Its so-called Skycrane landing method (once called "rover on a rope" at JPL) worked flawlessly, placing the rover 2.4 kilometers from its target near Aeolis Mons.[1] Designed and built as the technological pathfinder for a Mars sample return mission later in the decade, like Mars Pathfinder 15 years before, it was already an orphan. It quickly found the signs of ancient water that it had been dispatched to look for, though to the true believers in current life on Mars, it delivered disappointment: even a year later, it had found no methane at all.[2]

NASA's next surface mission to Mars, the Discovery-program-funded InSight, will not use the Skycrane. It will use a clone of the Mars Phoenix mission's Lockheed-built lander to perform a pair of long-desired geophysical measurements on Mars, a heat flow measurement and a search for "marsquakes."[3] Beyond InSight, the Mars program has not yet produced a firm plan. The leaders of NASA and JPL are planning a "2020 rover" that they intend to continue pursuing "astrobiologically relevant" ancient terranes and possibly collect samples for retrieval by some future mission.[4] It is supposed to be built upon the Skycrane methodology, using leftover hardware and a "build to print" management style to try to keep the cost around $1.5 billion.[5] But the fiscal instability that has plagued NASA over the past decade leaves me skeptical that the mission will advance much beyond the PowerPoint stage.

Public enthusiasm for Mars has not waned with NASA's funding. Far from it. The past few years have seen a number of private initiatives to send not merely machines, but human explorers, to Mars. The "MarsOne" initiative, announced in 2012 by a Netherlands-based nonprofit organization, seeks to send another clone of Lockheed's lander to Mars in 2018 (with an undefined payload, as yet), and then establish a human colony there in 2025.[6] More than 200,000 people applied to go on its one-way trip to Mars; the four-person expedition will, apparently, be the subject of a "reality TV" show.[7] It is deploying one of the new tools of the Internet era, crowdsourcing, to raise at least some of its funds, though it did not reach an initial fundraising goal of $400,000 via this mechanism.[8]

A former JPL mission designer, Dennis Tito, who left the Lab in 1972 to found a very successful investment management company, also announced a humans-to-Mars expedition, called Inspiration Mars. Tito plans to send a man and a woman—married, past child-bearing age—on a round-trip flyby of Mars. His initial dramatic announcement of a privately funded 2018 mission during the 2013 IEEE Aerospace Conference in Big Sky, Montana, was met with skepticism by the assembled engineers. The standing-room-only crowd rapidly thinned as it became clear his mass numbers didn't add up. A year later, he was asking for NASA funding and use of its new, as yet unflown, Space Launch System.[9] This placed his mission on a schedule of no earlier than 2021.

Perhaps the most interesting example of Mars enthusiasm is Elon Musk, billionaire founder of the electronic payment company PayPal. Musk left PayPal and in 2002 founded a new company to produce less expensive launch vehicles, called SpaceX; located at Jack Northrop Field in Hawthorne, California, SpaceX has to date been the most successful of the participants in NASA's Commercial Orbital Transportation Services competition, with a total of four demonstration and cargo flights to the International Space Station. It is also a participant in the agency's Commercial Crew Development program, though it has not yet flown the crew capsule Musk unveiled on May 29, 2014.[10] Musk has said he wants to develop a human colony on Mars by the late 2020s, using a methane-fueled rocket that's still on his drawing board.[11]

Each of these martian hopefuls gives a different reason for his enthusiasm for Mars. Tito wants to inspire the next generation of explorers; Musk to make humans a multiplanet species—in case we wipe ourselves out on Earth. Science is not their goal. Instead, echoes of the American frontier abound in their words. "America is a nation of explorers. I'd like to see that we're expanding the frontier and moving things forward. Space is the final frontier, and we have to make progress," Musk told Alex Knapp of *Forbes*.[12]

"Second creation on the Moon, on Mars, or further out in space demands more complex technology, but the underlying story remains the classic nineteenth-century narrative," historian David E. Nye wrote in 2003.[13] Call Mars third creation, and we'll have it. Technology—really big, industrial, if not planetary-scale, technology—will be deployed to transform Mars into a livable world, in this narrative. And here's where the root conflict between martian explorers and Mars science lies: scientists want a pristine Mars, uncontaminated by Earth. How else can we sort out how Mars came to be? But the first human on Mars destroys that. Humans carry biomes with us, outside and inside; we can't sterilize ourselves the way JPL is required to sterilize its Mars landers. Thus we cannot hope to prevent contamination of Mars by our own microbes. The Mars scientists want to study won't exist anymore. Some other Mars will. The disconnect between NASA's human exploration program and its science program isn't merely one of dollars, roles, and missions. It's more fundamental—human exploration inevitably destroys the world scientists find so fascinating.[14]

Despite the current outpouring of Mars enthusiasm, I consider it unlikely that humans will reach Mars in my lifetime (at 48, I'm five years older than Elon Musk). As the Mars One funding campaign suggests, people want to go to Mars, but not badly enough to pay the fare. And despite Musk's achievements, no one yet knows how to land an Apollo capsule-sized payload on Mars. As we witnessed in this book, simply scaling things up doesn't always work for Mars. Nor has anyone ever created a self-sustaining biosphere independent of Earth's, and that's what a Mars colony has to be. Highly complex things that can't be done on Earth today are unlikely to be done on Mars in a mere 15 years or so. To make the obvious comparisons, JPL's Mars Science Laboratory took 11 years to reach the launch pad; if Musk gets his first astronaut to the International Space Station in 2015, then it will have taken him 13 years to get a human to low Earth orbit.

So I don't expect to witness First Landing. But I hope to see Mars science continue in the interim.

NASA Organization and Mars Exploration

For the purposes of Mars exploration, NASA has two distinct, interested communities. In this book, I've called them the science program and the human program. Mostly, but not entirely, these align with two entities on NASA's organization charts. NASA's human exploration program resided within what was originally known as the Office of Manned Space Flight ("Code M" to insiders), and from the 1970s to the late 1980s as the Office of Space Flight. In the mid-1990s, this office became the Space Flight Enterprise. The short-lived Human Exploration and Development of Space Enterprise was an unsuccessful renaming of the Space Flight Enterprise.

In 2004, the Space Flight Enterprise became the Space Operations Mission Directorate, and a new Exploration Systems Mission Directorate was created. (Also, use of the handy "Code M" label was banned.) The Space Operations Mission Directorate handled the day-to-day operation of the International Space Station and space shuttle, while the Exploration Systems Mission Directorate handled research and development on new launch vehicle and other human-exploration relevant technology. The two were merged in 2011 as the Human Exploration and Operations Mission Directorate.

NASA's science program is the main theme of this book. This program was originally housed in the Office of Space Science (or "Code S"); it became the Office of Space Science and Applications, lost the Applications portion in the 1970s, changed its name to Space Science Enterprise in the 1990s, and in the 2004 reorganization became the Science Mission Directorate. NASA's Mars Exploration Program was created as a unit of the Space Science Enterprise, and it remains a component of the Science Mission Directorate.

The key point of this recitation of organizational changes is this: NASA's human exploration and space science programs have always been institutionally separate. That separateness is further emphasized by the reality that they are also geographically separate. The Science Mission Directorate operates two centers, Goddard Space Flight Center in Maryland and JPL, while the old Office of Space Flight and its successor organizations operate Johnson Space Center in Texas, Marshall Space Flight Center in Alabama, Stennis Space Center in Mississippi, and Kennedy Space Center in Florida. (Since this book is about Mars exploration, I omit NASA's third

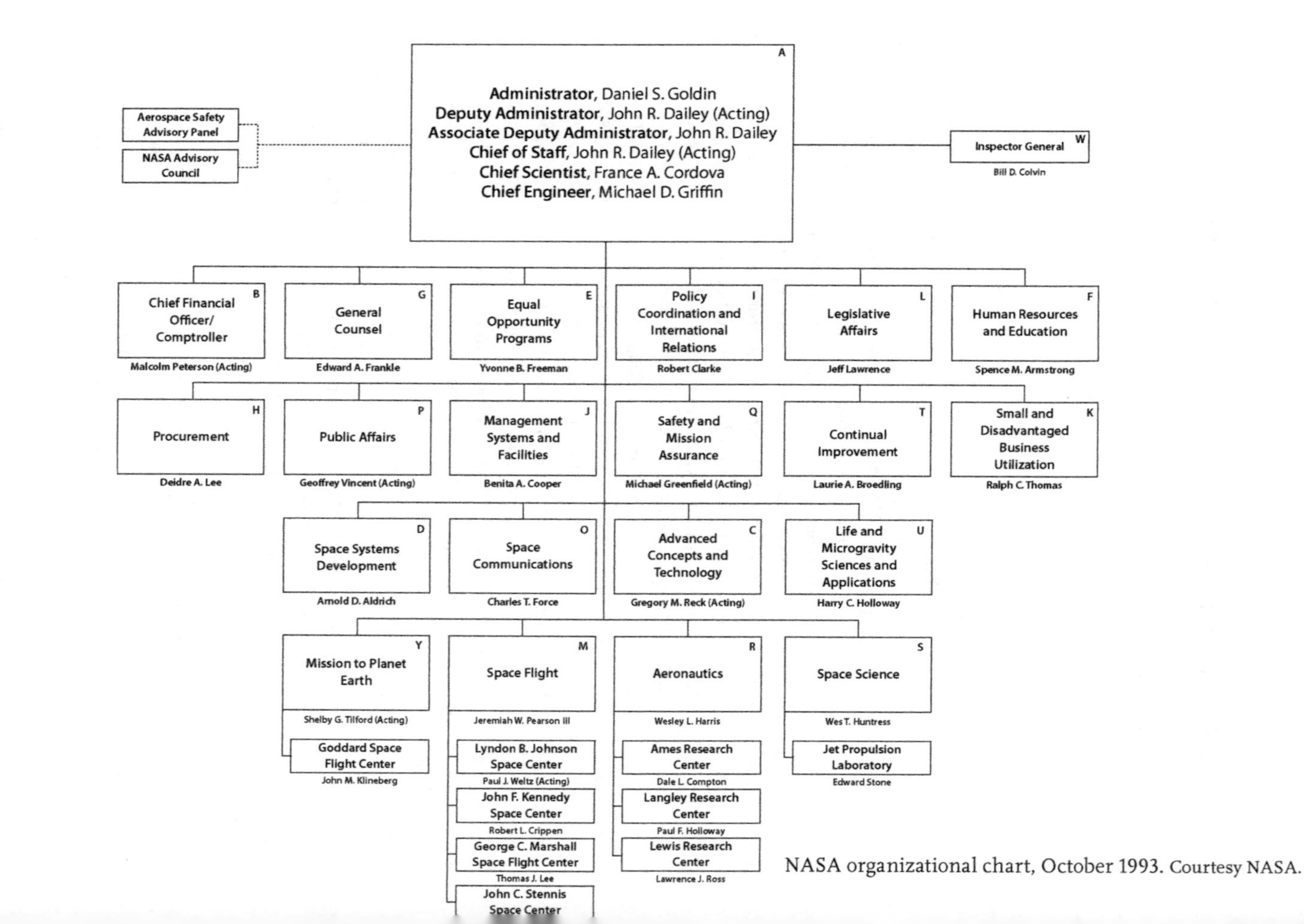

NASA organizational chart, October 1993. Courtesy NASA.

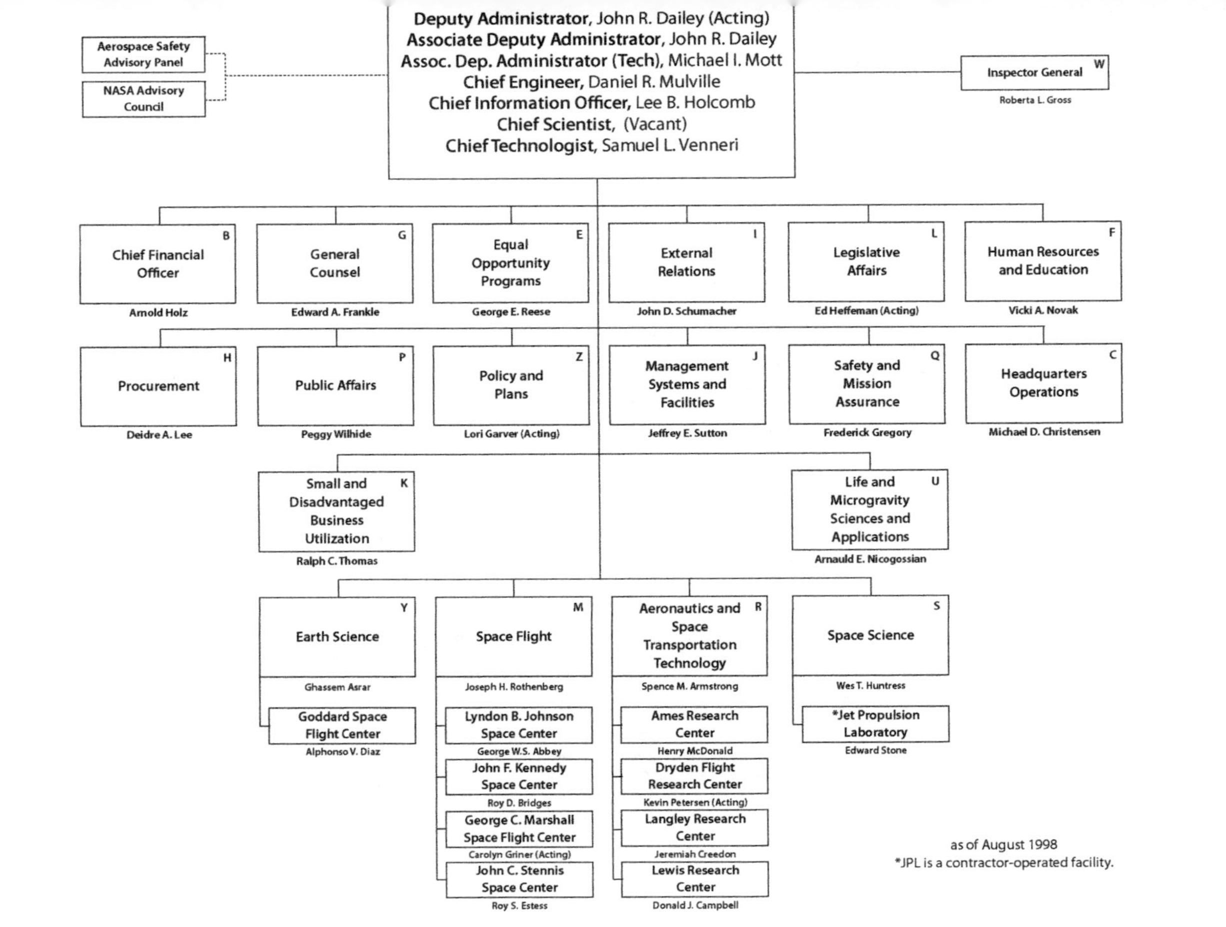

NASA organizational chart, August 1998. Courtesy NASA.

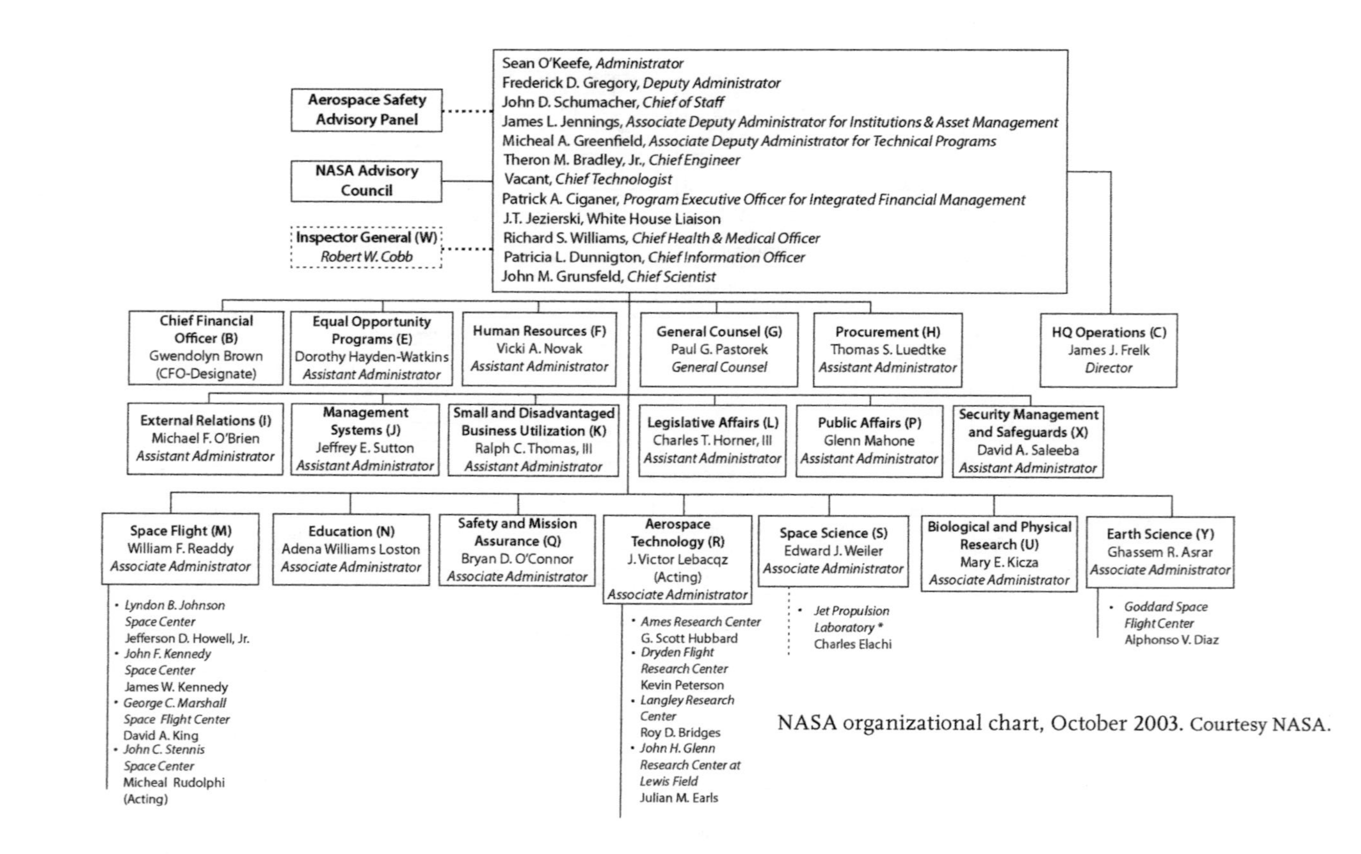

NASA organizational chart, October 2003. Courtesy NASA.

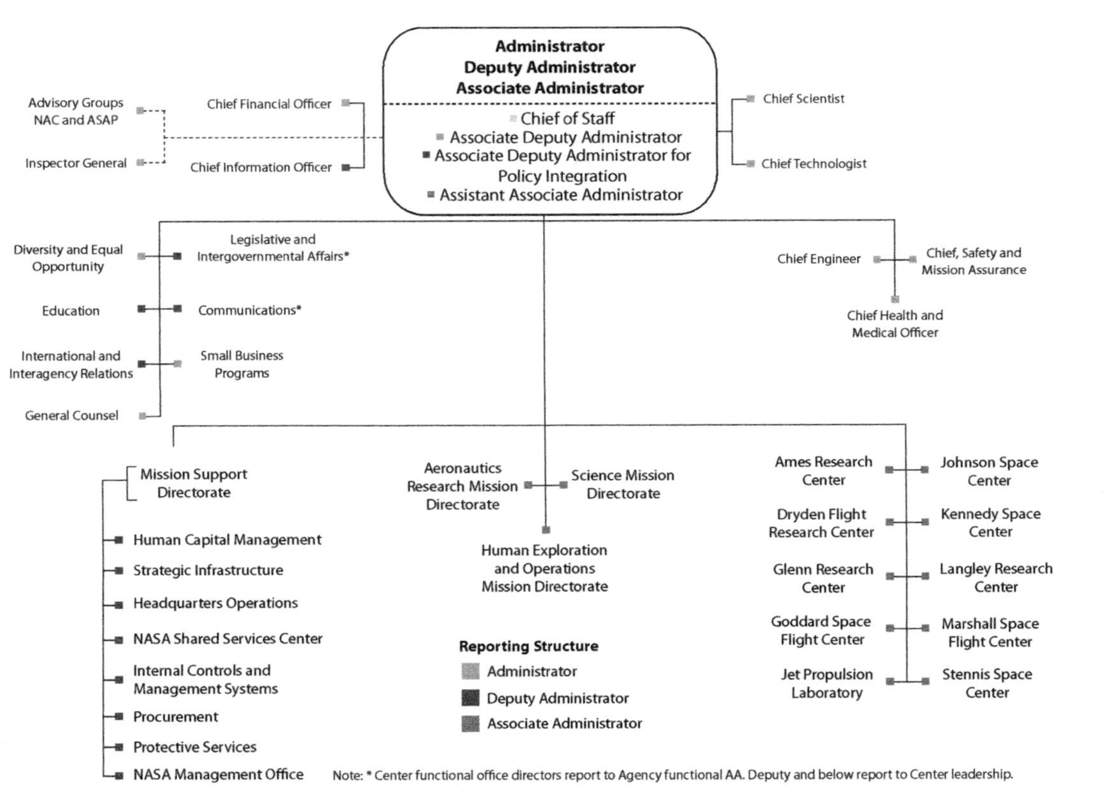

NASA organizational chart, September 2011. Courtesy NASA.

main component, Aeronautics and Technology, which operates four additional centers.)

Large, geographically and institutionally distinct organizations develop separate goals, agendas, and cultures. Thus it's a mistake to treat NASA as if, to borrow the great words of Ricardo Montalban in *Star Trek II: The Wrath of Khan* (1982), it were "one big happy fleet!"

NOTES

Note on Sources

This book starts in the "paper era," when sources were on paper and microfilm, and ends in the "digital era," when sources are largely "born digital" (and still microfilm). The consequence is that the kinds and locations of sources—even within JPL's fence—change dramatically, so the notes in this book are at times complex and confusing.

The first few projects I examine—Mars Observer, Mars Global Surveyor, and Mars Pathfinder—had established project libraries and document retention systems in the traditional manner: on paper and on microfilm. JPL's paper archives are organized in a manner familiar to most historians, using collection numbers and names. One important Mars Observer collection, for instance, is JPLA 33, "Mars Observer Project Records, 1983–1990." Throughout the notes, these paper archive collections are referred to by their collection number.

JPL also maintains a technical document library, known as Engineering Document Services (EDS). Documents preserved in that library have a quasi-unique "D-number," which I provide for sources in the notes. These are quasi-unique because some D-numbered documents are serials, though most are not. Thus, Mars Observer's Project Management Reports were D-1104, with an additional "issue number" and a date (e.g., Project Management Report, EDS D-1104, issue 49, 22 October 1987). Most EDS documents are not serials, however, and the D-number is a unique identifier. Often, these documents can be obtained from the NASA Technical Reports Server, described below.

JPL started its transition to digital documentation toward the end of the Pathfinder mission (and during the much longer Global Surveyor mission); by the Mars Exploration Rover mission of 2000, JPL had largely converted to digital documents (though the EDS documents continue to be microfilmed, for the time being). The nature of the depositories thus change during the Pathfinder and Global Surveyor missions, and so do the types of documents available in the repositories. Further, these and the Mars Surveyor 1998 missions took place under the faster-better-cheaper management paradigm of reduced documentation. Little still exists at JPL from the Surveyor 1998 missions in particular.

I partly compensated for this loss by obtaining documents directly from participants in these projects. The collections I gleaned were either "born digital" or were scanned by me into Adobe portable document format (pdf); they are all organized into a digital collection named Historian's Mars Exploration Collection (referred to as HMEC in the notes). In every case, I preserved the structure of the material I obtained. Thus, from Daniel J. McCleese, I obtained paper records in boxes from a temporary storage facility and filed their digital images in the HMEC collection by box number. From Robert M. Manning and Sam Thurman, I received entire hard drives of documents, and I placed their copies in HMEC in the same file structure the two engineers had used. HMEC references, therefore, do not share a common structure. The collection and file names are the most useful means for finding information within this artificial collection.

Beginning with the Mars Exploration Rover project, JPL implemented a digital document management system called "DocuShare." In that system, each project sets up its own secure library, organized by a project information manager. These libraries are searchable, though not organized with any consistency. I have elected to reference these documents by author, document name, date, library name (e.g., Mars Exploration Rover project library), and file name. While missions are still active, these libraries are closed to the public; when a mission ends, its library is transferred to the JPL Archives and falls under their published access rules.

The Laboratory also maintains vast audio-visual collections that exist on film, videotape of a variety of kinds, and digital files. Nearly all of this has been digitized, though the original cataloging format has been retained. In these notes, AVC refers to material that was originally on tape of some form, and SRC to a "born digital" file. The AVC library is organized by year, hence the name convention is AVC-year-number (e.g., AVC-1997-183). The SRC library is organized by number alone. One can access these collections either through a request to JPL's Media Relations office or the JPL Library and Archives Reference desk.

Another source frequently mentioned in the notes is the NASA Technical Reports Server (NTRS). An online, searchable database of technical material, it was once open to the public. However, much of its content has been closed since I started this work.

Introduction

1. A good recent survey of Mars in science and fiction is Markley, *Dying Planet*; also see Dick, *Biological Universe.*

2. See, e.g., National Commission on Space, *Pioneering the Space Frontier,* 140–42.

3. Robert Zubrin is the most vocal promoter of this view; see Zubrin, "Significance of the Martian Frontier." Also see Nye, *America as Second Creation,* 289.

4. Ezell and Ezell, *On Mars.*

5. Conway, *Atmospheric Science at NASA*, chap. 4.

6. A very nice popular summary of the water controversy is Morton, *Mapping Mars,* 162–72; for a detailed scientific discussion, see Carr, *Water on Mars.*

7. Horowitz, *To Utopia and Back,* 101–42; Dick and Strick, *Living Universe,* 80–101.

8. Snyder, "NASA and Planetary Exploration," 263–300; Westwick, *Into the Black,* 42–58, Murray, *Journey into Space,* 253–86.

9. Huntress and Marov, *Soviet Robots in the Solar System,* 367–86.

10. JPL's teething pains in space exploration are well documented in Koppes, *JPL and the American Space Program,* and Hall, *Lunar Impact.*

11. See Koppes, *JPL and the American Space Program,* and Westwick, *Into the Black,* for overviews of JPL history.

12. McCurdy, *Faster, Better, Cheaper*, provides a detailed overview of this period.

14. Mindell, *Digital Apollo.*

15. Ibid., 268–71.

Chapter 1 • Planetary Observers, Mars Observer

1. Committee on Planetary and Lunar Exploration, *Strategy for Exploration of the Inner Planets*, 43–52, quote at 49.

2. Ibid., 52–53.

3. Solar System Exploration Committee, "Planetary Exploration through the Year 2000: An Augmented Program," NASA TM-89235, part 2 (Washington, DC, 1986) (hereafter SSEC pt. 2), p. 29.

4. Geoffrey Briggs, interview with author, 31 March 2006.

5. Solar System Exploration Committee, "Planetary Exploration through the Year 2000: A Core Program," NASA TM-89235, part 1 (Washington, DC, 1983) (hereafter SSEC pt. 1), p. 69–70.

6. SSEC pt. 1, p. 70; Briggs interview.

7. SSEC pt. 1, p. 74.

8. SSEC pt. 1, p. 29.

9. SSEC pt. 2, p. 37.

10. SSEC pt. 1, pp. 97–99.

11. McCleese remembers this as having been called the Mars Water Mission; see Daniel J. McCleese, interview with author, 15 February 2006.

12. Talon Gold developed a complex in-space tracking and pointing system for ballistic missile defense. See Westwick, *Into the Black*, 104.

13. Charles Polk, "Mars Observer Project History," December 1990, EDS D-8095, p. 18.

14. Mars Observer Proposal Information Package Announcement of Opportunity OSSA-2-85 Volume V: Science Working Group Report (also EDS D-2244 vol. 5), 3 May 1985, p. 2.

15. Morton, *Mapping Mars*, 168–69.

16. Polk, "Mars Observer Project History," 19.

17. Ibid., 18; William I. Purdy, interview with author, 1 May 2006; George Pace, interview with author, 12 January 2006.

18. Purdy interview.

19. Mars Observer Project Science Group Meeting #1, April 1986, EDS D-3245, pp. 24–33.

20. "JPL Selects Subcontractors for Mars Observer," *Aviation Week and Space Technology*, 31 March 1986, p. 23.

21. Purdy interview; Pace interview; Polk, "Mars Observer Project History," 22–23.

22. Mars Observer Project Science Group Meeting #1.

23. Pace interview.

24. Thomas Thorpe, interview with author, 22 September 2005.

25. Polk, "Mars Observer Project History," 39.

26. Ibid., 52.

27. Ibid., 51–52.

28. Ibid., 34.

29. Ibid., 53.

30. Purdy interview.

31. Polk, "Mars Observer Project History," 55.

32. Final cost of the Titan III was $157 million. Mars Observer Project Management Report, EDS D-1104 issue 74, 30 November 1989, folder 31, JPLA 33.

33. Murray, *Journey into Space,* 347–48.

34. Ibid., 332–34.

35. Morton, *Mapping Mars,* 243–47.

36. National Commission on Space, *Pioneering the Space Frontier,* 3, 65–73; on the frontier thesis in space history, see Launius, "Perceptions of Apollo," 129–39; see also Limerick, "Imagined Frontiers," 249–61.

37. Polk, "Mars Observer Project History," 48–49, discusses Lunar Geophysical Orbiter. For more on the mission and spares issues, see Purdy interview and Pace interview.

38. Polk, "Mars Observer Project History," 50; W. I. Purdy Jr. to W. E. Giberson, 12 June 1987, "Re: Significant Events during the Week Ending 6/12/87: Mars Observer," folder 6, JPLA 33.

39. Mars Observer as Precursor for the Mars Rover Sample Return Mission, 6 June 1987, report MHC/7-10-1987, appendix 14 to Minutes of the Mars Rover Sample Return Science Working Group Meeting #2, 29–30 June 1987, MRSR_SWG_Jun1987.pdf, Mars Rover Sample Return Science Working Group records, HMEC; Polk, "Mars Observer Project History," 57–58, 62.

40. Minutes of the Mars Observer Project Science Group Sixth Meeting: Attachment A, 10 February 1988, quoted in Polk, "Mars Observer Project History," 76.

41. Polk, "Mars Observer Project History," 76.

42. Purdy interview.

43. D. H. Albaugh to W. E. Giberson, "Re: Significant Events during the Week Ending 6/30/88," 30 June 1988, Mars Observer, folder 6, JPLA 33.

44. Polk, "Mars Observer Project History," 80.

45. Albaugh to Giberson, 9 September 1988; George Pace, personal communication, 26 September 2010.

46. Huntress and Marov, *Soviet Robots in the Solar System,* 389.

47. Polk, "Mars Observer Project History," 80.

48. Wilford, *Mars Beckons,* 158–70; Edelson and McLucas. "United States and Soviet Planetary Exploration."

49. Huntress and Marov, *Soviet Robots in the Solar System,* 367–86.

50. Mars Observer Program Operating Plan 88-1 Preview: Presentation to NASA Headquarters, 29 January 1988, folder 53, JPLA 265.

51. Pace interview, 12 January 2006; also see Mars Observer Program Operating Plan 88-1 Preview.

52. George Pace, interview with author, 26 April 2011; W. E. Giberson to W. I. Purdy Jr., "Re: Significant Events during the Week Ending 8/7/87," 7 August 1987, file 6, JPLA 33; D. H. Albaugh to W. E. Giberson, "Re: Significant Events during the Week Ending 1/8/88," 8 January 1988, file 6, JPLA 33.

53. Mars Observer Project Management Report, 23 August 1991, folder 118, JPLA 265; Mars Observer Fall 1991 Quarterly Review, 18 November 1991, folder 120, JPLA 265.

54. William Boynton, interview with author, 8 May 2006. See also Mars Observer Project Management Report, EDS D-1104, issue 74, 30 November 1989, folder 31, JPLA 33, and

Mars Observer Project Management Report, EDS D-1104 issue 75, 21 December 1989, folder 32, JPLA 33.

55. Boynton interview.

56. Mario Acuna, interview with author, 10 September 2007; Mars Observer Project Management Report, EDS D-1104 issue 76, 26 January 1990, folder 33, JPLA 33.

57. Pace personal communication, 21 September 2010.

58. Mars Observer Project Management Report, EDS D-1167-1 issue 97, 22 November 1991, folder 121, JPLA 265; Mars Observer Project Management Report, 20 December 1991, EDS D-1167-1 issue 98, folder 123, JPLA 265.

59. Mars Observer Project Management Report, EDS D-1167-1 issue 99, 24 January 1992, folder 124, JPLA 265; Pace interview, 12 January 2006.

60. Mars Observer Project Management Report, 26 March 1992, folder 130, JPLA 265; Pace interview, 12 January 2006.

61. Mars Observer Project Management Report, 25 September 1992, EDS D-1167 issue 106, folder 139, JPLA 265.

62. Viking Orbiter cost from McCurdy, *Faster, Better, Cheaper,* 63; inflation factors from NASA, Aeronautics and Space Report of the President, FY 2007, p. 133, available at http://history.nasa.gov/presrep.htm. These figures do not include the cost of the launch vehicles.

Chapter 2 • Politics and Engineering on the Martian Frontier

1. Ronald Reagan, "Remarks at Edwards Air Force Base, California, on Completion of the Fourth Mission of the Space Shuttle Columbia," 4 July 1982, in *The American Presidency Project*, ed. Gerhard Peters and John T. Woolley, www.presidency.ucsb.edu/ws/?pid=42704.

2. Ride, *Leadership and America's Future in Space,* 33.

3. Committee on Planetary and Lunar Exploration, *Strategy for Exploration of the Inner Planets,* 50.

4. The other was Mark Craig of Johnson Space Center, which was responsible for the ascent vehicles, landers, and sample acquisition technology for the study.

5. Minutes of the Sixth Meeting of the Mars Rover Sample Return Science Working Group, 8–9 April 1988, MRSR_SWG_Apr1988.pdf, Mars Rover Sample Return Science Working Group records, HMEC, p. 1.

6. Shirley was also known by her married name, Donna Shirley Pivirotto.

7. Notes and Handouts from the Third Mars Rover Sample Return Science Working Group Meeting, 6–7 August 1987, MRSR_SWG_Aug1987.pdf, Mars Rover Sample Return Science Working Group records, HMEC, pp. 34–56.

8. Ibid.

9. Ibid. Shirley tells this story slightly differently in her memoir; see Shirley, *Managing Martians,* 122–23.

10. Notes and Handouts from the Fourth Mars Rover Sample Return Science Working Group, 23–24 November 1987, MRSR_SWG_Nov1987.pdf, Mars Rover Sample Return Science Working Group records, HMEC, pp. 4, 106–10.

11. Notes and Handouts from the Fifth Mars Rover Sample Return Science Working Group, 11–12 January 1988, MRSR_SWG_Jan1988.pdf, Mars Rover Sample Return Science Working Group records, HMEC, p. 2.

12. Notes and Handouts from the Fifth Mars Rover Sample Return Science Working Group, 11–12 January 1988, MRSR_SWG_Jan1988.pdf, Mars Rover Sample Return Working Group records, HMEC, pp. 3–4.

13. Notes and Handouts of the Seventh Mars Rover Sample Return Science Working Group Meeting, 20–21 June 1988, MRSR_SWG_Jun1988.pdf, Mars Rover Sample Return Science Working Group records, HMEC.

14. Malin et al., "Design and Development of the Mars Observer Camera."

15. Notes and Handouts of the Eleventh Meeting of the Mars Rover Sample Return Science Working Group, 23–24 February 1989, MRSR_SWG_Feb1989.pdf, Mars Rover Sample Return Science Working Group records, HMEC, pp. 51–55.

16. Hogan, *Mars Wars*, 41–42.

17. Ibid., 46–48; Craig Covault, "Manned Lunar Base, Mars Initiative Raised in Secret White House Review," *Aviation Week & Space Technology*, 17 July 1989, 24–26.

18. George H. W. Bush, "Remarks on the 20th Anniversary of the Apollo 11 Moon Landing," 20 July 1989, Public Papers of the President, http://bushlibrary.tamu.edu/research/papers/1989/89072000.html (accessed 3 October 2006).

19. Hogan, *Mars Wars*, 53–54.

20. Glenn Cunningham, interview with author, 11 July 2006; Code El Mid-Term Review of JPL Support of 90-Day Study [briefing package], 15 September 1989, courtesy of Cunningham.

21. Code El Mid-Term Review of JPL Support of 90-Day Study, p. B-6.

22. Ibid., C-1–C-16.

23. Notes and Handouts from the Twelfth Mars Rover Sample Return Science Working Group Meeting, 8–9 May 1989, MRSR_SWG_May1989.pdf, Mars Rover Sample Return Science Working Group records, HMEC, p. 4.

24. Code El Mid-Term Review of JPL Support of 90-Day Study, E-1–E-34.

25. Ibid., 3–11; Hogan, *Mars Wars*, 58.

26. Hogan, *Mars Wars*, 60. Nuclear thermal propulsion operates by heating a propellant gas in a fission reactor directly, providing a high specific thrust. Nuclear electric propulsion differs by ionizing a propellant gas electrically instead of heating it.

27. Ibid., 61. Most copies of the 90-Day Study had the cost estimates removed due to their political sensitivity, but the NASA History Office preserves a copy with them.

28. Ibid., 63.

29. Ibid., 65, 75–81.

30. In his masterpiece of political history, Walter McDougall discusses the political power of space technology's image making: McDougall, . . . *the Heavens and the Earth*, 301–6; the notion of Apollo as technospectacle I draw from Nye, *American Technological Sublime*, 225–45.

31. Hogan, *Mars Wars*, 110–12, 124–25, 128–35.

32. Notes and Handouts from the Third Mars Science Working Group Meeting, 23–24 August 1990, EDS D-14194, pp. 1, 156–202.

33. Ibid., 1. Also see Notes and Handouts from the First MESUR Science Definition Team Meeting, 22 February 1991, EDS D-13436, pp. 1–12.

34. Mishkin, *Sojourner*, 61–63.

35. Golombek personal communication, 3 January 2011.

36. Wesley T. Huntress, interview with author, 17 March 2008.

37. Wes Huntress, interview with Peter Westwick, 8 July 2003, JPL Oral Histories; also see Roy, "Origin of the Smaller, Faster, Cheaper Approach."

38. Roy, "Origin of the Smaller, Faster, Cheaper Approach."

39. Notes and Handouts from the Sixth Mars Science Working Group Meeting, 29–30 October 1991, EDS D-15432, pp. 3–4.

40. Ibid., pp. 3–4; Appendix M, U.

41. Ibid., pp. 3–4; Appendix R, S.

42. Michael H. Carr to Wes Huntress, 31 October 1991, appendix U to Notes and Handouts from the Sixth Mars Science Working Group Meeting.

43. Scott Hubbard, interview with author, 11 December 2006.

44. Dale Compton, interview with author, 18 August 2010.

45. L. A. Fisk to Edward C. Stone, 28 October 1991, Stone-MESUR Decision-1.docx, Wesley T. Huntress materials, HMEC. Thanks to Michael Neufeld for sharing this source.

46. Huntress interview, 8 July 2003; Huntress interview, May 2008; Tony Spear, "Discovery Presentation," 17 December 1991, EDS D-15857.

47. Tony Spear, interview with Peter Westwick, 12 February 2004, JPL Oral Histories. Spear remembers both February and April as having been when this happened; Huntress's recommendation to Len Fisk to carry out the SLIM demonstration is in Huntress to Fisk, 3 March 1992, PostFY1992strategy.docx, Huntress materials, HMEC.

48. Spear interview, 12 February 2004.

49. Matt Golombek, interview with author, 24 August 2006; Notes and Handouts from the Sixth MESUR Science Definition Team Meeting, 5–6 November 1992, EDS D-13074, pp. 9–11.

50. Steven W. Squyres to Carl B. Pilcher, 12 November 1992, enclosure to Notes and Handouts from the Sixth MESUR Science Definition Team Meeting.

51. Huntress and Marov, *Soviet Robots in the Solar System*, 389–90.

52. MESUR Retreat at Descano Gardens, 23 March 1992, EDS D-13219, pp. 157–59.

53. L. Van Warren to Distribution, "A Meeting with the NASA Administrator," 30 December 1992, folder 79, JPLA 267.

54. Brian Muirhead to R. Rhoads Stephenson, Re: Summary Comments on MESUR Pathfinder Review 11/1–4/92, IOM 3520-BKM-92-043, 10 November 1992, in MESUR Pathfinder Review Board Reports, EDS D-13126.

55. Notes and Handouts from the Sixth MESUR Science Definition Team Meeting, 5–6 November 1992, EDS D-13074, p. 8.

56. Muirhead to Stephenson, 10 November 1992.

57. Sixth MESUR Science Definition Team Meeting, pp. 15–16, 4, 16.

58. Ibid., p. 6.

59. Ibid., pp. 15–16, 4, 16.

60. Matt Golombek, interview with author, 8 June 2006; MESUR Pathfinder, Announcement of Opportunity: Image Investigation for the MESUR Pathfinder Mission, 22 December 1992, EDS D-13253.

61. Muirhead, *High Velocity Leadership*, 15–18, 30–31.

62. Robert M. Manning, interview with author, 14 June 2006.

63. Anthony Spear, interview with author, 31 October 2006.

64. Pathfinder Design, Implementation, and Cost Review, 20–23 July 2003, EDS D-13187, pp. 412–19; Muirhead, *High Velocity Leadership*, 21–22; Rivellini interview with author, 23 August 2006.

65. Tomasso Rivellini, interview with author, 23 August 2006.

66. Muirhead, *High Velocity Leadership*, 21–22; Pathfinder Design, Implementation, and Cost Review, pp. 260–65.

67. Muirhead, *High Velocity Leadership*, 21–22.

68. Ibid.; Pathfinder Design, Implementation, and Cost Review, pp. 260–65.

69. Pathfinder Design, Implementation, and Cost Review, p. 220.

70. This budget was also subject to some controversy. John Casani bet the subsystem manager, David Lehman, a case of fine red wine that he couldn't stay within this budget. Casani lost and paid up at a party at Lehman's house.

71. MESUR Pathfinder Design, Implementation, and Cost Review: Review Board Report, 19–23 July 1993, EDS D-13054, pp. 13, 56.

72. Ibid., p. 45.

73. Manning personal communication, 27 February 2011.

Chapter 3 • Attack of the Great Galactic Ghoul

1. Westwick, *Into the Black*, 258.

2. Ibid., 230.

3. Ibid., 262; John Casani, interview with Peter Westwick, 22 July 2004, JPL Oral Histories.

4. Westwick, *Into the Black*, 246–47.

5. Daniel S. Goldin, speech to the National Academy of Engineering, Washington, DC, 7 October 1993, Speeches of NASA Administrator Daniel S. Goldin, 1989–2001, NASA Historical Reference Collection.

6. See Westwick, *Into the Black*, 224–25 and 246–50, for more detail on JPL-Caltech-NASA relationship in the early 1990s.

7. Report of the Mars Observer Mission Failure Investigation Board, vol. 1, 31 December 1993, NASA Historical Reference Collection, p. D-3.

8. Ibid., pp. D-3, D-5.

9. George Pace, interview with author, 12 January 2006.

10. Report of the Mars Observer Mission Failure Investigation Board, vol. 1, B-4.

11. Ibid., D-2–D-3.

12. George Pace, personal communication, 16 September 2010.

13. William Boynton, interview with author, 8 May 2006; Philip Christensen, interview with author, 21 November 2006.

14. Westwick, *Into the Black*, 258.

15. Glenn Cunningham, interview with author, 1 August 2005.

16. Ibid.; Mars Global Surveyor POP 94-1 review, 3 May 1994, folder 93, JPLA 258.

17. Mars Observer Project Science Group 25th Meeting Minutes and attachment A, 6 October 1993, folder 680, JPLA 244.

18. Ibid.

19. Glenn Cunningham, personal communication, 9 December 2010.

20. Mars Observer Project Science Group 26th Meeting Minutes, 4–5 November 1993, folder 681, JPLA 244.

21. Christensen interview.

22. Mars Global Surveyor Spacecraft Contract (summary table), in Mars Global Surveyor POP 94-1 review. The RFP had required a contract cost cap of $54 million, which included $51.2 million for the spacecraft and $2.8 million for Martin Marietta's component of operations development. The cap did not include the contractor's fee.

23. Cunningham interview.

24. Ibid.

25. Notes and Handouts from the Mars Science Working Group Meeting, 7–8 February 1994, EDS D-14394, p. 7.

26. D. J. McCleese, "Mars Together: Long Range Plans for U.S. Exploration of Mars," June 1994; John B. McNamee, "Mars Together, Potential Mission Options 1998–2007, 6 June 1994," both in Mars_together_1994b.pdf, McCleese materials, HMEC; Notes and Handouts from the 11th Mars Science Working Group Meeting, Washington, DC, 27–28 October 1994, EDS D-15434, p. 2.

27. Notes and Handouts from the Mars Science Working Group Meeting, p. 7.

28. Ibid., p. 22.

29. Ibid., p. 8.

30. Ibid., p. 9.

31. Notes and Handouts from the 10th Mars Science Working Group Meeting, 13 May 1994, EDS D-15433, p. 11.

32. David A. Paige et al., "Mars Polar Pathfinder," 24 October 1994, Polar_Pathfinder_proposal.pdf, HMEC.

33. Mars Exploration Program Office Stone Report, Stone_reports_1994.pdf, 23 September 1994, Shirley materials, HMEC.

34. Mars Exploration Program Office Stone Report, 2 September 1994, and Mars Exploration Program Office Stone Report, 23 September 1994, both in Stone_reports_1994.pdf, Shirley materials, HMEC; John B. McNamee, interview with author, 5 February 2007.

35. John B. McNamee, "Re: July SL Monthly—Mars Surface Exploration Pre-Project," 25 July 1994, Stone_reports_1994.pdf, Shirley materials, HMEC.

36. Mars Exploration Program Office Stone Report as of 4 November 1994, Stone Report as of 11 November 1994, and Stone Report as of 30 December 1994, all in Stone_reports_1994.pdf, Shirley materials, HMEC.

37. William L Piotrowski to Donna L. Shirley, "Re: '98 Mars Orbiter Instrument Selection Process," 10 February 1995, Stone_reports_1994.pdf; D. Shirley, Notes from a Meeting with Code SL, 7 March 1995, Shirley materials, Stone_report_1995b.pdf; Mars Exploration Program Office Stone Report as of 12 July 1995, Stone_reports_1995b.pdf, all in Shirley materials, HMEC.

38. Mars Exploration Program Office Stone Report as of 14 April 1995, stone_reports_1995a.pdf, Shirley materials, HMEC.

39. Mars Exploration Program Office Stone Report as of 31 March 1995, Shirley materials, HMEC. Martin Marietta was in the process of merging with Lockheed Corporation, forming Lockheed Martin Corporation.

40. McNamee interview.

41. John McNamee to Joe Boyce, 26 June 1995, stone_reports_1995c.pdf, Shirley materials, HMEC.

42. Jurgen Rahe to John McNamee, 2 November 1995, stone_reports_1995b.pdf, Shirley materials, HMEC; David A. Paige, "Mars Volatiles and Climate Surveyor Integrated Payload Proposal for the Mars Surveyor Program '98 Lander," vol. 1, 4 August 1995, MVACS_Proposal_UCLA.pdf, folder 1995, MCO/MPL documents, HMEC.

43. Mars Exploration Program Office Stone Report as of 15 December 1995, Stone_reports_1995a, Shirley materials, HMEC; Rahe to McNamee, 2 November 1995.

44. Sarah Gavit, interview with author, 1 February 2007.

45. Sarah A. Gavit, Deep Space 2 Final Report, Project Overview, n.d., EDS D-19978.

46. 1998 Mars Surveyor Lander and Orbiter Project: Quarterly Report to NASA Code SL, 6 December 1995; Sarah Gavit oral history, 1 February 2007; Memorandum of Agreement between the Mars Exploration and New Millennium Program Offices for the Accommodation of the Mars Microprobes on the Mars 1998 Lander, 16 February 1996, in Mars Surveyor 1998 Preliminary Design Review, 4 March 1996, EDS D-13700, pp. 33–36.

47. Donna Shirley, "Mars Exploration Program Strategy" [presentation], 17 November 1995, and Mars Exploration Program Office Stone Report as of 21 July 1995, both in Stone_reports_1995a.pdf, Shirley materials, HMEC; McNamee interview.

48. Boston, Ivanov, and McKay, "On the Possibility of Chemosynthetic Ecosystems."

49. Exobiology Program Office, "An Exobiological Strategy for Mars Exploration," April 1995, NASA SP-530, NTRS, p. 1.

50. Ibid., p. 2.

51. Geoffrey Briggs to Dan McCleese, 6 February 1995, MELTSWG_1995d.pdf, McCleese materials, HMEC.

52. Carl Pilcher to Geoff Briggs et.al., "Re: Your notes on MELTSWG meeting," 7 July 1995, Pilcher_notes_MELTSWG.pdf, McCleese materials, HMEC.

53. Geoffrey Briggs, "Chairman's notes following MELTSWG meeting 20/21 April 1995," 8 May 1995, MELTSWG_1995d.pdf, McCleese materials, HMEC.

54. D. J. McCleese to Donna Shirley, n.d.; R. S. Saunders, "MELTSWG Meeting at JSC," 29 August 1995, both in stone_reports_1995c.pdf, Shirley materials, HMEC.

55. Manning, personal communication, 7 March 2011.

Chapter 4 • Engineering for Uncertainty

1. Muirhead, *High Velocity Leadership*, 156.

2. Robert M. Manning, interview with author, 20 December 2006.

3. Ibid.

4. Robert D. Braun, interview with author, 18 April 2007.

5. MPF Entry, Descent, and Landing System Peer Review, 1–2 September 1994, EDS D-13079.

6. Robert M. Manning, interview with author, 14 June 2006.

7. Ibid.

8. Sam Thurman, interview with author, 21 July 2006.

9. Brian Muirhead, interview with author, 16 October 2006.

10. Mars Pathfinder Critical Design Review, 6 September 1994, EDS D-13190, p. 129.

11. Mars Pathfinder Critical Design Review, 7–8 September 1994, EDS D-13190 and D-13191.

12. Muirhead, *High Velocity Leadership*, 91.

13. Muirhead interview.

14. Quote from Robert M. Manning to Erik Conway, e-mail communication, 12 April 2007; also see Manning interview, 14 June 2006.

15. Manning interview, 14 June 2006.

16. James S. Martin Jr. et al., "Mars Pathfinder Independent Readiness Review Report," 1 March 1995, EDS D-13070.

17. Ibid.

18. Tommaso Rivellini, interview with author, 23 August 1996.

19. Ibid.

20. Ibid.

21. Tommaso Rivellini, interview with author, 2 October 2006.

22. Tommaso Rivellini, notebook #1316, pp. 54 and 57, courtesy Tom Rivellini; Rivellini interview, 2 October 2006; ILC Dover, "Mars Pathfinder Flight Airbag Subsystem, Full Scale Development Test Report," n.d., accession A08841, JPLA.

23. Sam Thurman, interview with author, 14 September 2006.

24. Rob Manning, Sam Thurman, and Dara Sabahi, "Entry, Descent, Landing System Status," 30 January 1996, EDS D-14221, p. 8.

25. Muirhead, *High Velocity Leadership*, 150–51, 159–60.

26. Mishkin, *Sojourner*, 132–33; NASA news release 95-112, "NASA Names First Rover to Explore the Surface of Mars," 14 July 1995, www.nasa.gov/home/hqnews/1995/95-112.txt.

27. Mishkin, *Sojourner*, 171–72.

28. Muirhead, *High Velocity Leadership*, 172; and Mars Pathfinder System Level Solar Thermal Vacuum Test Cruise Configuration Test Report, 22 November 1996, EDS D-14025.

29. Mishkin, *Sojourner*, 190.

30. Muirhead, *High Velocity Leadership*, 177–78.

31. M. Golombek, ed., "Mars Pathfinder Landing Site Workshop," 18–19 April 1994, Lunar and Planetary Institute Technical Report no. 94-04, p. 1, also available as NASA CR-196745, NTRS. This ellipse defined the 90% probability envelope (or 2 sigma).

32. Minutes of the Second Meeting of the Mars Pathfinder Project Science Group, 9–10 June 1994, pp. 12–13, courtesy Matt Golombek.

33. Michael H. Carr to Jim Martin, 3 March 1995, enclosure to Notes and Handouts from the Fourth Mars Pathfinder Project Science Group Meeting, 18–19 May 1995, courtesy Matt Golombek.

34. Ken Edgett, interview with author, 10 July 2007.

35. Landing Site Workshop II, 30 September 1995, EDS D-13313, pp. 17–20.

36. Ibid., p. 10.

37. Ibid., p. 13.

38. Michael Carr to Jim Martin, 24 November 1995, enclosure to "Notes and Handouts from the Fifth Mars Pathfinder Project Science Group Meeting and Mars Pathfinder Landing Site Certification Meeting," 15–16 November 1995, courtesy Matthew Golombek.

39. Mars Pathfinder Landing Site Review with NASA Headquarters, 1 March 1996, EDS D-13430.

40. Glenn E. Cunningham, interview with author, 1 August 2005; George Pace, interview with author, 12 January 2006.

41. Peter C. Theisinger, interview with author, 1 March 2006.

42. Richard Zurek, "Report of the MGS Aerobraking Meeting, Orbit-to-Orbit Variability in Atmospheric Density," 31 March 1995, attachment to Mars Global Surveyor Aerobraking Margin Decision Meeting, 31 March 1995, EDS D-18314; Theisinger interview.

43. G. E. Cunningham to Distribution, "Re: Aerobraking Margin," 3 April 1995, folder 95, JPLA 258; Aerobraking Margin Decision Meeting, 31 March 1995.

44. Mary Kaye Olsen to Mars Global Surveyor Project Manager, "Re: Mars Global Surveyor Mission Operations," 31 May 1995, folder 165, JPLA 258; Mars Global Surveyor Mission System Critical Design Review Day 1, 26 September 1995, EDS D-15899.

45. Mishkin, *Sojourner*, 223–26; Muirhead, *High Velocity Leadership*, 194–96.

46. Mars Pathfinder "Lessons Learned" Presentations, 17 April 1997, EDS D-14531.

Chapter 5 • Mars Mania

1. Sawyer, *Rock from Mars*, 89; Matt Golombek, personal communication, 29 December 2011.

2. Sawyer, *Rock from Mars*, 115–33; 165–67.

3. Advanced Projects Design Team, "Mars Sample Return Benchmark Study," 16–21 April 1996, TeamX_MSR_Apr96.pdf, McCleese materials, HMEC.

4. D. J. McCleese, "Science and Political Considerations for Mars Sample Return," 1996, MSR_May96.pdf; "Mars 2005 Science and Implementation Committee Report to Complex," 1996, MSR_Complex_Jun96.pdf, McCleese materials, HMEC.

5. McCleese, "Mars 2005 Science and Implementation Committee Report"; "Sample Return Science and Implementation Committee Mission Engineering/Technology Key Issues," 23–24 May 1996, MSR_engineering_May96.pdf, McCleese materials, HMEC.

6. McCleese, "Mars 2005 Science and Implementation Committee Report" and "Sample Return Science and Implementation Committee."

7. Mars Exploration Directorate Stone Report as of 9 August 1996, stone_reports_1996b.pdf; Mars Exploration Directorate Stone Report as of 16 August 1996, stone_reports_1996a.pdf, Shirley materials, HMEC.

8. Daniel J. McCleese personal communication, 3 March 2011.

9. COMPLEX, "Scientific Assessment of NASA's Mars Sample Return Mission Options" [unpublished letter review], 3 December 1996, COMPLEX_review_1996.pdf, McCleese materials, HMEC, p. 2. Emphasis in original.

10. Mars Exploration Directorate Stone Report as of 7 February 1997, stone_reports_1996a.pdf, Shirley materials, HMEC; NASA FY 1999 Budget Briefing, ftp://ftp.hq.nasa.gov/pub/pao/budget/1999/99budget_summary.pdf.

11. Rob Manning, interview with author, 23 May 2007.

12. Mishkin, *Sojourner*, 238–39.

13. Ibid., 249–54; Muirhead, *High Velocity Leadership*, 200–203; Matt Golombek, interview with author, 28 August 2007.

14. Mishkin, *Sojourner*, 249–54; Muirhead, *High Velocity Leadership*, 200–203; Golombek interview.

15. Vaughn, "Mars Pathfinder Landing Inspires Thousands."

16. He was also politically controversial. In conservative circles he was seen as traitorous for his friendliness with Soviet scientists and for his advocacy of nuclear disarmament. This also played a role in his exclusion from elite scientific circles.

17. Paul G. Backes, Kam S. Tso, and Gregory K. Tharp, "Mars Pathfinder Mission Internet-Based Operations Using WITS," 1998 IEEE International Conference on Robotics and Automation, May 1998; Paul Backes, interview with author, 30 September 2009.

18. Rob Manning, interview with author, 14 June 2006.

19. Mars Pathfinder Landing Day Commentary Coverage, 4 July 1997, AVC-1997-183, part 2 of 22, JPLA.

20. Ibid.

21. "Hello, Earth to Mars: Pathfinder Restores a Sense of Wonder about Space," *Los Angeles Times*, 8 July 1997. See also "Spellbound by Pathfinder," *Pasadena Star-News*, 10 July 1997; John Noble Wilford, "A New Breed of Scientists Studying Mars Takes Control," *New York Times*, 14 July 1997; and Peter N. Spotts, "Trek to Mars Introduces NASA's Next Generation," *Christian Science Monitor*, 9 July 1997.

22. James Gorman, "Mars Mission Separates the Men from the Toys," *New York Times*, 13 July 1997.

23. Summarized from Mishkin, *Sojourner*, 265–77.

24. M. P. Golombek et.al., "Overview of the Mars Pathfinder Mission and Assessment of Landing Site Predictions," *Science* 278 (5 December 1997), plate 6, 1741–1742.

25. Mishkin, *Sojourner*, 282–84.

26. Golombek et al., "Overview of the Mars Pathfinder Mission"; Smith et al., "Results from the Mars Pathfinder Camera"; Matijevic et al., "Characterization of the Martian Surface Deposits"; Rieder et al., "Chemical Composition of Martian Soil and Rocks."

27. Bruce A. Smith, "Global Surveyor Nears Mars," *Aviation Week and Space Technology*, 8 September 1997; quote from Robin Lloyd, "Mars Craft Passes Crucial Fuel-Tank Test," *Pasadena Star-News*, 10 September 1997.

28. M. Dan Johnston, interview with author, 4 April 2007.

29. Michael Dornheim, "Mars Atmosphere Thicker Than Expected," *Aviation Week and Space Technology*, 29 September 1997.

30. Acuna et al., "Magnetic Field and Plasma Observations."

31. Glenn E. Cunningham, "Mars Global Surveyor Aerobraking Status Report," 17 October 1997, folder 135, JPLA 258.

32. Bud McAnally, interview with author, 16 November 2006.

33. Robert Braun, interview with author, 18 April 2007; George Pace, personal communication, 7 October 2010.

34. Charles Whetsel, interview with author, 25 September 2007.

35. Johnston interview.

36. This was orbit PO15.

37. Frank Palluconi, "Mars Global Surveyor Project Science Group Fourteenth Meeting Minutes," 31 October 1997, EDS D-11752 issue 13, p. 5.

38. Cunningham, "Mars Global Surveyor Aerobraking"; Bruce A. Smith, "Solar Panel Problem Triggers Major MGS Assessment," *Aviation Week and Space Technology* (20 October 1997), 25–27.

39. Cunningham, "Mars Global Surveyor Aerobraking"; Smith, "Solar Panel Problem Triggers Major MGS Assessment."

40. Palluconi, "Mars Global Surveyor Project Science Group Fourteenth Meeting Minutes," pp. 5–6.

41. Glenn E. Cunningham, "Mars Global Surveyor Aerobraking Status Report," 28 October 1997, folder 137, JPLA 258; Cunningham presentation to Huntress, 28 October 1997, folder 138, JPLA 258.

42. Palluconi, "Mars Global Surveyor Project Science Group Fourteenth Meeting Minutes," pp. 7–10.

43. Ibid., pp. 5–9.

44. Charles Whetsel to Glenn Cunningham et al., 23 January 1998, folder 158, JPLA 258.

45. M. H. Acuna, interview with author, 10 September 2007.

46. Smith, "Solar Panel Problem Triggers Major MGS Assessment," 25.

47. Frank Palluconi, "Mars Global Surveyor Project Science Group Fifteenth Meeting Minutes," 16 December 1997, EDS D-11752, issue 14, pp. 5–6.

48. Peter C. Theisinger, interview with author, 28 January 2005.

Chapter 6 • The Faster-Better-Cheaper Future

1. Jurgen Rahe to Donna L. Shirley, 31 January 1996, Stone_reports_1996b, Shirley materials, HMEC.

2. John B. McNamee, "1998 Mars Surveyor Lander and Orbiter Project Preliminary Design Review, March 1996" [presentation], EDS D-13700, p. 38.

3. Arden Albee and Frank Carr to Norm Haynes, "Review Board Report for the 1998 Mars Project Preliminary Design Review," 29 March 1996, Stone_reports_1996b.pdf, Shirley materials, HMEC.

4. Ibid.

5. Mars Exploration Directorate Stone Report as of 19 April 1996, Stone_reports_1996b.pdf, Shirley materials, HMEC.

6. Ibid.

7. John B. McNamee to Jurgen Rahe, "Re: Microphone Accommodation on the 1998 Mars Surveyor Lander," 17 June 1996, Stone_reports_1996a.pdf, Shirley materials, HMEC.

8. Lou Friedman to Donna Shirley, 17 August 1996, Stone_reports_1996a.pdf, Shirley materials, HMEC.

9. John B. McNamee, interview with author, 15 January 2008.

10. Sam Thurman, interview with author, 15 June 2007.

11. Mars Surveyor Program 1998 Lander and Orbiter Flight Systems Critical Design Review, 22 January 1997, EDS D-14569 book 2, p. 79.

12. Ibid., pp. 79, 84, 87–88.

13. Ibid., p. 80.

14. Ibid., p. 174.

15. Ibid., pp. 120–23.

16. Greg Heinsohn, "Lander History," 23 April 1998, lander_history_23april1998.pdf, McCleese materials, HMEC.

17. Norm R. Haynes to Dr. E. C. Stone, "Re: Mars Exploration Directorate Significant Events Week ending 10 January 1997," 1_13_97.doc, Shirley materials, HMEC. Also see Portree, *Humans to Mars*, 94.

18. See Portree, *Humans to Mars*, pp. 89–90; Zubrin, Baker, and Gwynne, "Mars Direct."

19. Norm R. Haynes to E. C. Stone, "Mars Exploration Directorate Significant Events Week Ending March 21, 1997," IOM MED-SIG-97-020, 3_21_97.doc, Shirley materials, HMEC.

20. See Mars Surveyor Program Announcement of Opportunity, 2001 Orbiter Mission Proposal Information Package (Final), 30 June 1997, EDS D-25348, p. 1-1.

21. See Mars Surveyor Program Announcement of Opportunity, 2001 Lander Mission Proposal Information Package (Final), 30 June 1997, EDS D-25347, p. 3-26.

22. "Science Team and Instruments Selected for Mars Surveyor 2001 Missions," 6 November 1997, NASA news release 97-260, www.jpl.nasa.gov/news/releases/97/2001team.html.

23. Philip R. Christensen et al., "The Thermal Emission Imaging System (THEMIS) for the Mars 2001 Odyssey Mission," *Space Science Reviews* 110 (January/February 2004): 85–130.

24. Program Scientist, Mars Pathfinder and Mars Surveyor '98 [Boyce] to Director, "Re: Science on U.S. Mars Missions," 12 September 1997, SciFunding_BoyceLtr.pdf, McCleese materials, HMEC; Gail Robinson presentation in Mars Surveyor 2001 Project Quarterly Status Review, 12 February 1998, EDS D-15990-01, p. 17.

25. Jake Matijevic, interview with author, 11 July 2007.

26. Mars 2001 Project Science Group, "Mars 2001 PSG #2 Minutes at JPL," 24 February 1998, EDS D-15672 PSG #2B.

27. Robinson presentation in Mars Surveyor 2001 Project Quarterly Status Review, p. 17.

28. Mars '01 Assessment (draft), 24 April 1998, draft_msp01_reviewnotes.pdf, McCleese materials, HMEC.

29. For example, the 2001 lander propulsion system switched to a pressure-regulated system to gain extra landed mass capability, had 25% larger, "ultraflexible" solar arrays to support a more power-hungry payload, and used Pathfinder's larger 2.65 meter aeroshell vice the 2.4 meter Surveyor 1998 aeroshell to provide more packaging volume. See MSP '01 Lander Baseline Design Review, 10 December 1997, EDS D-15927-A.

30. Mars '01 Assessment (draft #2), 24 April 1998, Draft2_MSP-1_Review.pdf, McCleese materials, HMEC.

31. Ibid.

32. McCleese notes from MED Retreat, 26 May 1998, MED_notes_Feb1998.pdf, McCleese materials, HMEC; Mars Exploration Directorate Stone Report as of 22 May 1998, Stone_reports_1998.pdf, Shirley materials, HMEC.

33. Mars Exploration Directorate Stone Report as of 8 May 1998, Stone_reports_1998.pdf, Shirley materials, HMEC; also see Squyres, *Roving Mars*, 44–47.

34. Squyres, *Roving Mars*, 47–49; Mars Exploration Directorate Stone Report as of 5 June 1998, Stone_reports_1996a.pdf, Shirley materials, HMEC.

35. Mars Expeditions Strategy Group, June 1998 Meeting Report, Shirley materials, MESG Meeting Minutes 6_98.doc, HMEC.

36. George Pace, interview with author, 23 January 2007; Roger Gibbs, interview with author, 22 April 2008.

37. Wesley T. Huntress, interview with author, 9 April 2008.

38. "Huntress Announces His Departure from NASA," 18 February 1998, NASA Release 98-31, www.nasa.gov/home/hqnews/1998/98-031.txt; "Weiler Named Associate

Administrator for Space Science," 16 November 1998, NASA Release 98-204, www.nasa .gov/home/hqnews/1998/98-204.txt.

39. Mars Exploration Directorate Stone Report as of May 22, 1998, Stone_reports_1998 .pdf, Shirley materials, HMEC.

40. Mars 2001 Project Science Group Meeting #4 minutes, 17 August 1998, EDS D-15672, PSG #4B, p. 6.

41. Dan McCleese to Mars '01 Project Office, 6 August 1998, IOM_CurieRover_06081998 .pdf, McCleese materials, HMEC; Squyres, *Roving Mars*, 49–50.

42. William J. O'Neil, "Final Mars Architecture Team Meeting: Project and Mission Overview," 2 March 1999 [presentation, 20070110111742911.pdf, MSR 2003 Architecture materials, HMEC], p. 28.

43. Mars Expeditions Strategy Group June 1998 Meeting Report, Stone_reports_1998 .pdf, Shirley materials, HMEC.

44. McCleese to Files, 20 October 1995, MSR_notes_oct95.pdf, McCleese materials, HMEC.

45. DS2 Final Report, November 2000, EDS D-19978, section X: System Integration and Test, p. X-12; Tom Rivellini, interview with author, 7 January 2008.

46. DS2 Final Report, Section M: Telecommunications, p. M-23, and Section I: Mechanical, p. I-26.

47. Mars Climate Orbiter pre-ship review, 14 August 1998, EDS D-16068, p. 20.

48. Mars Polar Lander Pre-Ship Readiness Review, 15 September 1998, EDS D-16198, p. 26.

49. Ibid.

50. Ibid., p. 96.

51. Thurman interview.

52. Sam W. Thurman, "Re: Mars Climate Orbiter Mission Status 12/13/98," MCO_13Dec98.pdf, Thurman materials, HMEC.

53. Sam W. Thurman, "Re: Mars Polar Lander / Mars Climate Orbiter Mission Status 01/06/99," Status_6Jan99.pdf, Thurman materials, HMEC.

54. 1998 Mars Surveyor Climate Orbiter and Polar Lander Final Report, 29 January 1999, MSP 98 Final Report.ppt, Thurman materials, HMEC, p. 40.

55. Ibid., p. 47.

56. Ibid.; Report on the Loss of the Mars Polar Lander and Deep Space 2 Missions, 22 March 2000, EDS D-18709, p. 17.

Chapter 7 • Revenge of the Great Galactic Ghoul

1. "NASA on the Ropes—Annus Horribilis," *Economist*, 1 April 2000, p. 74.

2. Donna Shirley, interview with author, 21 February 2006.

3. S. W. Thurman and R. W. Zurek, "Mars Polar Lander / Deep Space-2 Landing Site Selection Briefing," 24 August 1999, MS'98 Site Cert (NASA HQ).pdf, and "Mars Polar Lander / Deep Space-2 Landing Site Evaluation/Selection Overview," 4 August 1999, MS'98 Site Review Aug99.pdf, both in Thurman materials, HMEC.

4. Richard Zurek, interview with author, 26 June 2007.

5. Sam W. Thurman to Richard A. Cook, "Re: MCO/MPL Mission Status, September 1, 1999," status_1Sep99.pdf, Thurman materials, HMEC.

6. Pete Theisinger, interview with author, 31 May 2007.

7. Mars Program Independent Assessment Team Report, 14 March 2000, MPIAT_report 2000.pdf, HMEC, p. 51.

8. George Pace, personal communication.

9. Sam Thurman to Richard A. Cook, "Re: MCO/MPL Mission Status 04-Mar-99," status_4Mar99.pdf, Thurman materials, HMEC.

10. Sam Thurman to Richard A. Cook, "Re: MCO/MPL Mission Status 18-Mar-99," status_18Mar99.pdf, Thurman materials, HMEC.

11. JPL Special Review Board, "Report on the Loss of the Mars Climate Orbiter Mission," 11 November 1999, EDS D-18441, pp. 43–44.

12. Sam Thurman to Richard A. Cook, "Re: MCO/MPL Mission Status 04-Aug-99," status_4Aug99.pdf; Thurman to Cook, "Re: MCO/MPL Mission Status 28-Jul-99," status_28Jul99.pdf, both in Thurman materials, HMEC; Steve Jolly, interview with author, 24 July 2007, and interview with author, 1 October 2007.

13. The final report on the loss of Climate Orbiter covers the navigation issue in detail. See JPL Special Review Board, "Report on the Loss of the Mars Climate Orbiter Mission," 11 November 1999, EDS D-18441.

14. Thurman to Cook et al., "Re: MCO/MPL Mission Status, 08-Sept-99," Status_8Sept99 .pdf, Thurman materials, HMEC.

15. JPL Special Review Board, "Report on the Loss of the Mars Climate Orbiter Mission," p. 47.

16. Ibid., p. 21.

17. Cook, Thurman, and McNamee, "Mars Climate Orbiter Mars Orbit Insertion, presentation to Edward C. Stone," 22 September 1999, MCO MOI (Stone Brief).ppt, Thurman materials, HMEC.

18. Steven D. Jolly to Kenny Starnes et al., "Re: Final MCO Status 9_23_99," 24 September 1999, MCO_status_23Sept99.pdf, Thurman materials, HMEC.

19. Ibid.; Mars Climate Orbiter Insertion Event, 23 September 1999, AVC-1999-147, JPLA.

20. Jolly to Starnes et al., "Re: Final MCO Status 9_23_99."

21. Mars Climate Orbiter Press Briefing, 23 September 1999, AVC-1999-148, JPLA.

22. Jolly to Starnes et al., "Re: Final MCO Status 9_23_99."

23. Office of the Director to All Personnel, "Re: Probable Loss of Mars Climate Orbiter Mission," 23 September 1999, AllPers_23Sept99.pdf, Thurman materials, HMEC.

24. Jolly to Starnes et al., "Re: Final MCO Status 9_23_99."

25. Sam Thurman to Richard Cook et al., "Re: Mars Climate Orbiter End of Mission Status Report," 29 September 1999, MCO_status_final_29Sep99.pdf, Thurman materials, HMEC.

26. Frank Jordan, interview with author, 3 April 2008.

27. Rob Manning, interview with author, 21 October 2007; Manning, personal communication, September 12, 2013.

28. Jolly interview, 24 July 2007.

29. Jordan interview.

30. Norm Haynes, "MCO Peer Review Team Report," n.d., MCO Investigation Material, NASA Historical Reference Collection. The report was presented at a Climate Orbiter review at headquarters, 18–19 October 1999. Frank Jordan's memory disagrees with the dates but corresponds to the events: see Jordan interview.

31. Haynes, "MCO Peer Review Team Report."

32. Mars Climate Orbiter Mishap Investigation Board Phase I Report, 10 November 1999, MCO_report_phase_1.pdf; Mars Climate Orbiter Mishap Investigation Board, "Report on Project Management in NASA," 13 March 2000, MCO_MIB_Report.pdf, Manning materials, HMEC.

33. Report on the Loss of the Mars Climate Orbiter Mission, EDS D-18441, 11 November 1999, pp. 38–40.

34. Another excellent discussion of these events is Edward A. Euler, Steven D. Jolly, and H. H. "Lad" Curtis, "The Failures of the Mars Climate Orbiter and Mars Polar Lander: A Perspective from the People Involved," AAS Paper #01-074, presented at the 24th Annual AAS Guidance and Control Conference, Breckinridge, CO, 31 January–4 February 2001.

35. Report on the Loss of the Mars Climate Orbiter Mission, pp. 37–40.

36. Jordan interview.

37. JPL Special Review Board, "Report on the Loss of the Mars Polar Lander and Deep Space 2 Missions," 22 March 2000, EDS D-18709, pp. 33–34.

38. Ibid., pp. 34–35.

39. Mars Climate Orbiter Mishap Investigation Board, "Report on Project Management in NASA," 13 March 2000, MCO_MIB_Report.pdf, Manning materials, HMEC, p. 32.

40. JPL Special Review Board, "Report on the Loss of the Mars Polar Lander and Deep Space 2," pp. 90–91.

41. Morton, *Mapping Mars*, 35–36.

42. Mars Polar Lander / DS-2 Mission Status Briefing 1:30PM, 3 December 1999, AVC-1999-191, part 1, JPLA.

43. Mars Polar Lander / DS-2 Commentary 7:30 PM, 3 December 1999, AVC-1999-193, part 1 of 2, JPLA.

44. Mars Polar Lander / DS2 Mission Status Briefing 10:40 p.m., AVC-1999-192, part 1, JPLA.

45. Mars Polar Lander / DS2 Mission Status Briefing, 9 December 1999, AVC-1999-196, JPLA.

46. Edward C. Stone, "Loss of Mars Climate Orbiter" [presentation to the CIT Board of Trustees], 7 December 1999, folder 806, JPLA 259.

47. Edward C. Stone, speech delivered at town hall meeting, 10 December 1999, folder 807, JPLA 259; "Aftermath of MPL/DS2 and the Challenges Ahead," town hall meeting, 10 December 1999, AVC-1999-207, JPLA.

48. John B. McNamee to Edward C. Stone et al., "Re: Why no Downlink Capability during EDL for MPL," 8 December 1999, whynodownlink_8Dec99.pdf, Thurman materials, HMEC.

49. Bruce Moomaw and Cameron Park, "Was Polar Lander Doomed by Fatal Design Flaw?" *SpaceDaily*, 16 February 2000.

50. David Olschansky, "Flight System Response Introduction," in Mars Surveyor Program 2001 Lander Return to Flight Review, 28 February 2000, EDS D-18778, pp. III-A, 14.

51. Parker Stafford, interview with author, 30 July 2007.

52. JPL Special Review Board, "Report on the Loss of the Mars Polar Lander and Deep Space 2," Missions," 22 March 2000, EDS D-18709, p. 6, 20.

53. Ibid., p. 21.

54. Sarah Gavit, interview with author, 1 February 2007.

55. Mars Program Independent Assessment Team, 14 March 2000, MPIAT_report 2000.pdf, HMEC, pp. 25–29.

56. Ibid., p. 34.

57. Ibid., p. 49.

58. Ibid., p. 11.

59. "Don't Blame NASA Alone for Mars Mission Failures," *Nature*, 6 April 2000, 527.

60. Daniel S. Goldin, "When the Best Must Do Even Better: Remarks at the Jet Propulsion Laboratory, 29 March 2000," e000861.pdf, Speeches of NASA administrator Daniel S. Goldin, 1989–2001, NASA Historical Reference Collection.

61. Morton, *Mapping Mars*, 68–69.

62. James R. Asker, Robert Wall, and William B. Scott, "NASA Ponders a New Path to Mars," *Aviation Week and Space Technology*, 13 December 1999, 35; Michael A. Dornheim, "Mars Reassessment to Goldin in March," *Aviation Week and Space Technology*, 20–27 December 1999.

63. Westwick, *Into the Black*, 284.

Chapter 8 • Recovery and Reform

1. Edward C. Stone speech delivered at "Town Hall Meeting," December 10, 1999, folder 807, JPLA 259.

2. Ibid.

3. Edward C. Stone, interview with A. Blaine Baggett, 20 March 2001, SRC-000229, JPLA.

4. Larry Dumas, "Human Factors: Presentation to the Mars Program Independent Assessment Team, 2 February 2000," folder 818, JPLA 259.

5. Stone to Edward J. Weiler, 7 March 2000, folder 101, JPLA 259.

6. Thomas R. Gavin, interview with author, 13 September 2005.

7. Scott Hubbard, interview with author, 6 December 2006.

8. Chris Jones, interview with author, 15 June 2005.

9. Firouz Naderi, interview with A. Blaine Baggett, 5 May 2005, JPL Oral Histories.

10. Squyres, *Roving Mars*, 71–72.

11. Ibid., 73.

12. Ibid., 73.

13. Mark Adler, interview with author, 2 May 2008; Daniel J. McCleese, personal communication, 3 March 2011.

14. Adler interview; Rob Manning, interview with author, 22 October 2007.

15. Manning interview.

16. Mark Adler et al., "Mars Geologist Pathfinder '03," 26 April 2000, mgp-casani -000426.ppt, Manning materials, HMEC.

17. Ibid.

18. Ibid., p. 50.

19. J. Matijevic, "Actions, Notes, Advisories, and Results from the evaluation of the '2003 Mars Geologist Pathfinder (MGP)' proposal," n.d., Review___426.doc, Manning materials, HMEC.

20. Ibid.

21. Summarized from Mars Surveyor Program 2001 Lander Return to Flight Review, 28 February 2000, EDS D-18778; George Pace, "Mars 2003 Lander Mission Concept Review," 3 May 2000, EDS D-19429.

22. Richard W. Zurek, "Mars Surveyor Orbiter Science," 6 July 2000, MSO_PMSR.ppt, Zurek materials, HMEC; Lockheed Martin, "Mars Surveyor Orbiter Final Report," 15 September 2000, JPL subcontractor report 9950-1660.

23. Joy Crisp, "Science Objectives, Payload, and Requirements," in Preliminary Mission Systems Design and Cost Review, 10 July 2000, PMSDCR_D1_All1.pdf, MER Project Library.

24. James Garvin, interview with author, 25 March 2009.

25. Squyres, *Roving Mars*, 77.

26. Adler interview; Squyres, *Roving Mars*, 77.

27. Peter C. Theisinger, interview with author, 28 January 2005.

28. Zurek, "Mars Surveyor Orbiter Science."

29. Squyres, *Roving Mars*, 74.

30. Glenn Cunningham, interview with author, 1 August 2005.

31. Squyres, *Roving Mars*, 90; JPL Task Plan for Mars Exploration Rover-2003, Phase B, 7 August 2000, file: "second '03 rover option," box 202, Goldin Papers, National Archives.

32. Hubbard, *Exploring Mars*, 86.

33. "Martian Gamble," *Science* 289, no. 5479 (28 July 2000): 521.

34. In fact, JPL practice had been to build two "flight model" spacecraft and a "proof test model" that was identical. The proof test model was assembled and tested first, to practice the complex procedures.

35. Mars Exploration Rover Dual Lander Feasibility Review, 7 August 2000, dual_lander_review_080700.pdf, MER Project Library, pp. 65–66, 111.

36. Gentry Lee to Scott Hubbard, 7 August 2000; David Swenson to Thomas R. Gavin et al., "Re: MER, Material faxed to Dr. E. Weiler on 8/8/00," 8 August 2000, both in file: "second '03 rover option," box 202, Goldin Papers.

37. Barry Goldstein, interview with author, 22 January 2009; Larry Dumas, interview [video], SRC-1175, JPLA.

38. Scott Hubbard to Edward J. Weiler, 8 August 2000; Associate Administrator for Space Science to Administrator, "Re: Second Rover for 2003 Mars Mission Option," 8 August 2000, both in file: "second '03 rover option," box 202, Goldin Papers; see also Andrew Lawler and John MacNeil, "Plan for Two Rovers Squeezes NASA Budget," *Science* 289, no. 5482 (18 August 2000): 1119.

39. Squyres gives a slightly different gloss on these events in *Roving Mars*, 89–92.

40. Dara Sabahi, interview with author, 28 July 2008; Tom Rivellini, interview with author, 14 November 2008.

41. Tom Rivellini, "From Legs to Wheels," n.d. [ca. December 2005], History of Sky Crane for PMSR.ppt, Rivellini materials, HMEC; Rivellini interview.

42. Sabahi interview; Rivellini interview.

43. Sabahi interview.

44. Ibid.

45. Rivellini interview.

46. Tom Rivellini, "Robust Landing ATD Task Overview, January 2001," Overview.ppt, Rivellini materials, HMEC; Rivellini interview; Sabahi interview.

47. Hubbard, Naderi, and Garvin, "Following the Water."

48. James Graf, interview with author, 12 January 2009.

49. Roger Gibbs, interview with author, 22 April 2008.

50. Ibid.

51. Ibid.

52. Mars '01 Project Red Team Report to the Governing Program Management Council, 29 June 2000, MSP01_redteam_06292000.pdf, 2001 Mars Odyssey Project Library.

53. 2001 Mars Odyssey Mission Readiness Review, 12 March 2001, EDS D-20361, p. 6.

54. Gibbs oral history, 22 April 2008; Mars '01 Orbiter Propulsion Risk Evaluation Technical Interchange Meeting, 15 August 2000, EDS D-19544.

55. David A. Spencer, interview with author, 21 October 2007.

56. Philip Christensen, interview with author, 21 June 2008.

57. William Boynton, interview with author, 8 May 2006.

58. George Pace, personal communication, September 2010.

59. Boynton interview; Boynton et al., "Distribution of Hydrogen in the Near Surface of Mars." The Viking-era hypothesis was stated in Farmer and Doms, "Global Seasonal Variation of Water Vapor on Mars."

60. Westwick, *Into the Black*, 286.

Chapter 9 • Margins on the Final Frontier

1. James Garvin, interview with author, 25 March 2009.

2. Peter C. Theisinger, interview with author, 11 March 2009; Preliminary Mission Systems Design and Cost Review [presentation], 10–11 July 2000, PMSDCR_D1_ALL1 .pdf, Manning MER materials, HMEC.

3. Joel Krajewski, interview with author, 17 April 2009; J. Matijevic, Keith Novak, and Chuck Phillips, "Surface Power / Thermal Status Update," 19 April 2001, EDS D-31305.

4. Krajewski interview.

5. J. Krajewski et al., "LEM Accommodation and Primary Battery Sizing Trade," 11 August 2000, Flight System Design Team Meeting, EDS D-31281.

6. Redesign summarized in R. Lindemann, "Rover Design Modifications, MER Re-Baseline Peer Review," 12 December 2000, EDS D-10170.

7. Rob Manning, "MER EDL Chronology Rev. B.," 7 July 2004, p. 5, MER_EDL_ChronologyRevB.ppt, Manning MER materials, HMEC.

8. Theisinger interview.

9. Report of the Mars Exploration Rover Project and Flight System Preliminary Design Review, 30 October 2000, pp. 4–7, MER_PDR1-301000.doc.pdf; Program Executive, MER-2003 Project (Lavery) to Associate Administrator for Space Science (Weiler), "Re: Mars Exploration Rover-2003 Mission Success Criteria," 31 January 2001, D_Lavery_01-31_2001.pdf, both in Mars Exploration Rover (MER) Project Library, JPL.

10. Report of the Mars Exploration Rover Project and Flight System Preliminary Design Review, p. 7.

11. See Rob Manning and Scott Doudrick, "Flight System Requirements, Preliminary Design Review," 17–19 October 2000, EDS D-2160, p. 14.

12. Richard Cook, "Flight System Design Update, MER Delta Preliminary Design Review," 31 January 2001, EDS D-20161, pp. 103–7; J. Matijevic, Novak, and Phillips, "Surface Power / Thermal Status Update."

13. MER Airbag Subsystem Peer Review, 4 April 2001, MER document 420-6197, MER420-6197.pdf, MER Project Library, p. 29.

14. Cook, "Flight System Design Update," p. 16; Manning, "MER EDL Chronology Rev. B.," p. 4.

15. Cook, "Flight System Design Update," p. 15.

16. Ibid., pp. 31–33; Gun-Shing Chen, "Lander/Rover Redesign for Mass Reduction, MER Re-Baseline Peer Review," 12 December 2000, EDS D-20170.

17. Mass Reduction for the Cruise Stage, MER Re-baseline Review, 12 December 2000, EDS D-20170; Cook, "Flight System Design Update," pp. 11–13.

18. Cook, "Flight System Design Update," p. 61.

19. This was Kobie Boykins.

20. Squyres, *Roving Mars*, 110.

21. Ibid., 105–7.

22. Keith Novak, interview with author, 23 April 2009; Cook, "Flight System Design Update," p. 107.

23. "New Security Measures at JPL," *Mars Explorer Times* 1, no. 11 (25 September, 2001), MER Project Library.

24. Dara Sabahi, interview with author, 28 July 2008.

25. Rob Manning, interview with author, 3 October 2013, HMEC.

26. Wayne Lee, interview with author, 28 July 2008.

27. Ibid.; ILC Dover, "Mars Exploration Rover Airbag Characterization Test Report," March 2001, DM 0 (Margin) Test Report.doc, MER Project Library.

28. Cook, "Flight System Design Update," pp. 18–21.

29. David Kass, interview with author, 23 July 2009; Kass et al., "Analysis of Atmospheric Mesoscale Models for Entry, Descent, and Landing."

30. Cook, "Flight System Design Update," pp. 18–21.

31. Manning, "MER EDL Chronology Rev B."; Wayne Lee, "EDL Systems Implementation Status, Project Critical Design Review," 22 August 2001, EDS D-20622, p. 6.

32. ILC Dover, "Mars Exploration Rover Airbag System: Development Model 1 Test Report," May 2001, DM 1 Test Report.doc, MER Project Library.

33. ILC Dover, "Mars Exploration Rover Airbag System: Development Model 3 Test Report," October 2001, DM 3 Test Report.doc, MER Project Library.

34. Manning, "MER EDL Chronology Rev B."

35. Kass interview; Kass et al., "Analysis of Atmospheric Mesoscale Models for Entry, Descent, and Landing."

36. Manning, "MER EDL Chronology Rev B."; Squyres, *Roving Mars*, 127–29.

37. Quoted in Squyres, *Roving Mars*, 127.

38. Andrew Johnson et al., "Feasibility of Image-Based Horizontal Velocity Estimation for MER EDL," 6 November 2001, horzVelFeasibility.ppt, Manning materials, HMEC.

39. Squyres, *Roving Mars*, 124.

40. Rob Manning, "MER EDL Chronology Rev B."; also see "EDL Team Airbag Qualification Status," MER EDL Subsystem Qualification Review, 4 December 2002, subsys_qual_02_airbags.ppt, Manning materials, HMEC.

41. Sabahi oral history, 28 July 2008.

42. ILC Dover, "Mars Exploration Rovers Airbag System Development Models 4 & 5 Test Report," n.d., DM4-5 Test Report.doc, MER Project Library.

43. EDL Team Airbag Qualification Status.

44. Manning, "MER EDL Chronology Rev B."

45. Review Board Report for DIMES System Requirements and Implementation Review, 11 April 2002, DIMES_rjt.doc, Manning materials, HMEC.

46. Andrew Johnson, interview with author, 21 July 2009.

47. Andrew Johnson et al., "Descent Image Motion Estimation System", n.d. RI_MER_DIMES_20041008.ppt, Manning materials, HMEC; Andrew Johnson et al., "Field Testing of the Mars Exploration Rovers Descent Image Motion Estimation System."

48. Robin Bruno, "MER PDS Structural Test Readiness Review," 2 April 2002, EDS D-31239.

49. LaRC et al., "MER Parachute Structural Test Summary May June 2002," 26 June 2002, EDS D-29715.

50. Lee interview.

51. Robin Bruno, interview with author, 18 June 2009.

52. Allen Witkowski, interview with author, 20 March 2012.

53. PDS Strength Recovery Review, 18 July 2002, EDS D-29714; EDL Team PDS Qualification Status, 4 December 2002, EDS D-25459.

54. Squyres, *Roving Mars*, 138.

55. Ibid., 139–40.

56. ILC Dover, "Mars Exploration Rovers Airbag System Development Model 6 Test Report," 14 August–9 October 2002, DM6DropTestReport.doc, MER Project Library.

57. Rob Manning, "EDL Simulation Results, Landing Site Peer Review," 26–27 March 2003, EDL_Simulation_Results.ppt, Manning materials, HMEC.

58. Kass interview; Rob Manning, interview with author, 5 August 2009; Manning, "EDL Simulation Results"; B. Gentry Lee and Daniel McCleese, "Mars Exploration Rover Landing Site Selection Peer Review Summary Report," 31 March 2003 (revised 5 April 2003), PeerReviewReportFinal.doc, Manning materials, HMEC.

59. Manning interview, 5 August 2009.

60. Miguel San Martin, "Dual-TIRS Algorithm Description and Performance, TIRS Utilization Review," 21 July 2003, Dual_20TIRS_20Description.ppt, MER Project Library; Manning interview, 5 August 2009.

61. NASA Independent Review Team Assessment of the Mars Exploration Rover Project's Mission System CDR and ATLO Readiness Reviews, 1 March 2002, EDS D-31906, p. 31.

62. Squyres, *Roving Mars*, 144–45.

63. Manning interview, 5 August 2009; Matt Wallace, interview with author, 26 March 2009; Matt Wallace, "ATLO Overview, MER ATLO Readiness Review," 23–25 January 2002, EDS D-22493.

64. Wallace, "ATLO Overview," p. 42.

65. Wallace oral history, 26 March 2009.

66. Peter C. Theisinger, "Project Status Report," 13 June 2002, MER doc. 420-6170, MER Project Library.

67. Theisinger oral history, 11 March 2009.

68. Gary M. Kinsella, "MER 1 System Thermal Test Report, Spacecraft Cruise Configuration," 20 March 2003, EDS D-24425.

69. Matt Wallace, "KSC Plans, MER Preship Review #3," 25 February 2003, EDS D-25521.

70. Benjamin Riggs to Distribution, IOM 352B-BR-03-IOM010, 16 November 2002, MER_MP_Report_-_11162002.doc, Manning materials, HMEC; MER 1/B Final Mass Properties Report, 17 June 2003, MER MP Report MER1 Final.doc, MER Project Library; Manning, MER EDL Chronology Rev B., 7 July 2004, Manning materials, HMEC.

71. Squyres, *Roving Mars*, 199–217.

72. "Girl with Dreams Names Mars Rovers 'Spirit' and 'Opportunity,'" 8 June 2003, News Release 2003-081, http://marsrover.nasa.gov/newsroom/pressreleases/20030608a.html.

73. Peter C. Theisinger, "Budget Discussion, MER Project CDR," 21–23 August 2001, Budget Discussion Final.ppt, MER Project Library, p. 15.

Chapter 10 • Sending a Spy Satellite to Mars

1. Hubbard, *Exploring Mars*, 136–37.

2. James Garvin, interview with author, 17 August 2010.

3. G. Scott Hubbard, James Garvin, and Edward J. Weiler, "Mars Reconnaissance Orbiter Formulation Authorization," 17 January 2001, EDS D-22733; J. Edward Weiler to Director, JPL, "2005 Mars Reconnaissance Orbiter Project Phase A to B Confirmation," 10 May 2002, EDS D-22734.

4. James Graf, interview with author, 12 January 2009.

5. Richard Zurek and Ronald Greely, co-chairs, "Report of the NASA Science Definition Team for the Mars Reconnaissance Orbiter," 19 February 2001, MRO_DSTreport.pdf, Mars Reconnaissance Orbiter (MRO) Project Library, JPL, p. 4.

6. Ibid., p. 3.

7. Ibid., pp. 9, 12–13.

8. Ibid., pp. 14–15.

9. Mars Reconnaissance Orbiter Announcement of Opportunity, 6 June 2001, NASA AO #01-OSS-02, www.worldcat.org/title/mars-reconnaissance-orbiter-2005-announcement-of-opportunity-soliciting-proposals-for-basic-research-for-period-ending-august-22-2001/oclc/53363201, p. 6-7.

10. Ibid., pp. 2-8, 2-9.

11. Ibid., p. 2-10.

12. Steve Jolly, interview with author, 19 July 2011.

13. Ibid.

14. Ibid.

15. Howard Jay Eisen, "MRO Project Mission System Review: Flight System Description and Implementation," 17–18 January 2002, 1PMSR_7_FS_HE.ppt, MRO Project Library, p. 19.

16. Steven Jolly, interview with author, 19 July 2011.

17. Graf interview.

18. Alfred S. McEwen, interview with author, 7 September 2010.

19. McEwen et al., "Mars Reconnaissance Orbiter's High Resolution Imaging Science Experiment."

20. McEwen interview; McEwen et al., "Mars Reconnaissance Orbiter's High Resolution Imaging Science Experiment"; "HiWish: Public Suggestion Page," www.uahirise.org/hiwish/.

21. Scott Murchie, interview with author, 5 August 2011.

22. Murchie et al., "Compact Reconnaissance Imaging Spectrometer for Mars Investigation."

23. Murchie interview.

24. Graf interview.

25. Tim Gasparrini, Steve Jolly, and Todd Bayer, "Flight System Part One: Overview, Preliminary Design Review," 22–23 July 2002, 1PDR_6_FS_1.ppt, MRO Project Library, p. 37.

26. David Kass, "MCS Rules and Guidelines, Mars Reconnaissance Orbiter Project Science Group meeting #7," 14 March 2004, 40412_0_MCS Kass Presentation.ppt, Zurek materials, HMEC.

27. Glenn Cunningham, "Report of the Mars Reconnaissance Orbiter Standing Review Board for the Preliminary Design Review Non-Advocate Review," 9 August 2002, 020908_pdr_results.pdf, MRO Project Library, p. JEG-37.

28. Graf interview.

29. R. Zurek for Jim Graf, "MRO Project Status Presented at the Project Science Group Meeting #5," 11 March 2003, PSG5graf_rwz.pdf, Zurek materials, HMEC.

30. Ibid.

31. Barry Goldstein, interview with author, 14 February 2011; Leslie Tamppari, interview with author, 12 July 2011.

32. Peter H. Smith, "Phoenix: 'Follow the Water,' Mars Scout Step 2 Concept Study Report," 15 May 2003, Phoenix_Proposal.pdf, Goldstein materials, HMEC.

33. Mars Phoenix had two significant foreign contributions that brought the total proposed cost to $341 million; ibid., p. 2.

34. Ibid., p. F-1.

35. "NASA's First Scout Mission Selected for 2007 Mars Launch," 4 August 2003, news release 03-256, www.nasa.gov/home/hqnews/2003/aug/HQ_03256_mars_scout.html.

36. NASA Mars Exploration Program Mars 2007 Smart Lander Mission Science Definition Team Report, 11 October 2001, MRO_SDTreport.pdf, MRO Project Library, Appendix 1, p. 25.

37. Ibid., p. 2.

38. Mars Science Laboratory Mission Project Science Integration Group Final Report, 6 June 2003, PSIG_Final_Full_Report4.pdf, MSL Materials, HMEC, quote from p. 7; C. W. Whetsel, "The End of the Beginning: Mars Science Laboratory Mission at PDR," 6 October 2006, MSL_Through_Confirmation_Whetsel.ppt, Rivellini MSL Materials, HMEC.

39. Pallet and Skycrane, March 2003, pal_vs_sky_short.ppt, Rivellini MSL materials, HMEC; Whetsel, "End of the Beginning."

40. Committee on Planetary and Lunar Exploration. *Assessment of Mars Science and Mission Priorities*, 4, 102–3, quote at 103.

41. Whetsel, "End of the Beginning"; Brian Muirhead, "Mars Science Laboratory Mission Concept Review, Flight System Trade Studies and Reference Design," 28–29 October 2003, MCR_Flight_system-FINAL.ppt, Rivellini MSL materials, HMEC.

42. Scott Murchie and Dennis Fort, "CRISM presentation at Project CDR," 21–23 May 2003, 1O_CDR_CRISM_sm.ppt, MRO Project Library.

43. Ibid.; Murchie interview; Fred Vescelus, "MRO Payload Standing Review Board Chairman Report," 21 May 2003, 1J_CDR_Payload_SRB_fv.ppt, MRO Project Library.

44. Brian K. Muirhead, interview with author, 10 August 2011.

45. Jim Graf, "Project Risk Summary, Audit #3," 15 June 2005, 10 Project Risk Summary.ppt, MRO Project Library.

46. Graf interview.

47. Mars Reconnaissance Orbiter Launch press kit, August 2005, p. 6, http://mars.jpl.nasa.gov/mro/files/mro/mro-launch.pdf.

Chapter 11 • Robotic Geologists on the Red Planet

1. NASA, "The Vision for Space Exploration," February 2004, http://history.nasa.gov/Vision_For_Space_Exploration.pdf (accessed 7 August 2013).

2. Ibid., p. 19.

3. "Europe Is Going to Mars," 11 June 1999, ESA news release 22-1999, http://sci.esa.int/mars-express/13295-pr-22-1999-europe-is-going-to-mars/.

4. Wright, Sims, and Pillinger, "Scientific Objectives of the Beagle 2 Lander"; John Noble Wilford, "Looking for a Little Life, 3 Visitors Descend on Mars," *New York Times*, 23 December 2003.

5. Patrick E. Tyler, "No Signal from Mars: British Disappointed, but Won't Give Up," *New York Times*, 26 December 2003.

6. Prasun N. Desai and Philip C. Knocke, "Mars Exploration Rovers Entry, Descent, and Landing Trajectory Analysis," 2004, AIAA/AAS Astrodynamics Specialist Conference and Exhibit, 16–19 August 2004, Providence, RI, published as AIAA paper 2004-5092.

7. Jason Willis, interview with author, 24 June 2009; R. Manning, "Overview: Pyro Backup Timer Restoration Review," 7 October 2003, EDS D-25549.

8. Manning, "Overview: Pyro Backup Timer Restoration Review"; James A. Donaldson and Glenn Reeves, "Proposal for Restoring the Backup Pyro Timer Capability on the MER Spacecraft," 7 October 2003, EDS D-25549; Willis interview.

9. Willis interview.

10. James A. Donaldson, "Avionics Subsystem Pyro Loss Anomaly Description (Z82923)," 12 January 2004, 0079737.ppt; Manning, "Pyro Timer Enable Timeline," n.d., pyrotimerenabletimeline.xls, both in Manning materials, HMEC.

11. Mark Adler and Rob Manning remember this conversation; Gostelow doesn't. See Mark Adler, interview with author, 21 September 2009; Kim Gostelow, interview with author, 26 August 2009.

12. Adler interview.

13. MER Spirit news briefing, 8:00 p.m., 4 January 2004, AVC-2004-016; MER Spirit post-landing briefing, 3 January 2003, AVC-2004-006, both in JPLA.

14. MER Spirit post-landing briefing, 3 January 2004.

15. Rob Manning, "Spirit EDL Status & Opportunity EDL Readiness, Independent Status Review," 16 January 2004, EDS D-27728.

16. Adam Steltzner et al., "Spirit Parachute Inflation Results and Opportunity Parachute Deployment Targeting," in Manning, "Spirit EDL Status & Opportunity EDL Readiness," p. 33.

17. MER Spirit news briefing, 11 January 2004, AVC-2004-028; MER Spirit news briefing, 12 January 2004, AVC-2004-029, both in JPLA; Squyres, *Roving Mars*, 253–57.

18. MER Opportunity post-landing press conference, 24 January 2004, AVC-2004-068, JPLA.

19. Ibid.

20. Squyres, *Roving Mars*, 294.

21. Ibid., 296–99.

22. Squyres et al., "Opportunity Rover's Athena Science Investigation."

23. "New Phase of Exploration Beginning for Mars Rovers," 26 March 2004, JPL news release 2004-92, www.jpl.nasa.gov/news/news.php?release=92.

24. A search of the Lexis-Nexis Academic Universe database for "Mars Pathfinder" garnered 533 newspaper stories for 1997, while a "Mars Exploration Rover" search gave 176 stories. See also Christine Russell, "Covering Controversial Science: Improving Reporting on Science and Public Policy," Joan Shorenstein Center on the Press, Politics and Public Policy, working paper #2006-4, Spring 2006.

25. MER Opportunity post-landing press conference.

26. Richard Cook, Joy Crisp, and Steve Squyres, "Mars Exploration Rover Extended Mission Proposal," 31 March 2004, EM1_03-31-2004.pdf, Mars Exploration Rover (MER) Project Library, JPL, p. 10; Squyres, *Roving Mars*, 331–34.

27. Keith Novak et al., "The Mars Exploration Rover Thermal Design: A Successful Case Study," presented at the 26th International Conference on Environmental Systems, Norfolk, VA, 16–20 July, courtesy Keith Novak.

28. Cook, Crisp, and Squyres, "Mars Exploration Rover Extended Mission Proposal," pp. 14–15.

29. James K. Erickson, Joy Crisp, and Steve Squyres, "Mars Exploration Rover Extended Mission Proposal," 6 August 2004, EM2_Proposal_2004-08-06, MER Project Library, pp. 12–13; John Callas, interview with author, 29 March 2009.

30. Squyres, *Roving Mars*, 331–49.

31. Matt Golombek, interview with author, 11 August 2010.

32. Squyres, *Roving Mars*, 331–49.

33. Ibid., 371; MER Analyst's Notebook, MER Opportunity/B, sol 315, http://an.rsl.wustl.edu/mer/.

34. "Movie Clip Shows Whirlwinds Carrying Dust on Mars," 21 April 2005, JPL news release 2005-061, www.nasa.gov/centers/jpl/newsmer=042105.html.

35. Biesiadecki, Leger, and Maimone, "Tradeoffs between Directed and Autonomous Driving."

36. Squyres, *Roving Mars*, 350–67.

37. James K. Erickson, "Mars Exploration Rover Extended Mission III Project Plan," 30 March 2005, Ext_3OpPlan51-1.doc; James K. Erickson et al., "Mars Exploration Rover Extended Mission Proposal," 6 August 2004, EM2_Proposal_2004-08096.pdf, both in MER Project Library.

38. "Rover Team Tests Mars Moves on Earth," 6 May 2005, JPL news release 2005-072, www.jpl.nasa.gov/news/news.php?release=2005-072; "NASA's Opportunity Rover Rolls Free on Mars," 6 June 2005, JPL news release 2005-095, www.jpl.nasa.gov/news/news.php?release=2005-095.

39. Erickson, "Mars Exploration Rover Extended Mission III," pp. 9–12; "NASA Mars Rovers Head for New Sites after Studying Layers," 12 April 2006, JPL news release 2006-054, www.jpl.nasa.gov/news/news.php?release=2006-054; JPL image PIA07506 shows the route toward Erebus, available at http://photojournal.jpl.nasa.gov; Golombek interview.

40. Janet Amelia Vertesi, "Seeing Like a Rover: Images in Interaction on the Mars Exploration Rover Mission" (Ph.D. thesis, Cornell University, May 2009).

41. "NASA Mars Rovers Head for New Sites"; Erickson, "Mars Exploration Rover Extended Mission III," pp. 7–8; John Callas, W. Bruce Banerdt, and Steven W. Squyres, "Mars Exploration Rover Fifth Extended Mission Proposal," April 2007, MER_EM5_Proposal_Final_04-2997.pdf, MER Project Library, p. 3.

42. Biesiadecki, Leger, and Maimone, "Tradeoffs between Directed and Autonomous Driving"; Biesiadecki et al., "Mars Exploration Rover Surface Operations."

43. Callas, Banerdt, and Squyres, "Mars Exploration Rover Fifth Extended Mission Proposal," p. 10.

44. "Mars Rovers Survive Severe Dust Storms, Ready for Next Objectives," 11 September 2007, JPL news release 2007-098, www.jpl.nasa.gov/news/news.php?release=2007-098; John Callas, Steven W. Squyres, and W. Bruce Banerdt, "Mars Exploration Rover Sixth Extended Mission Proposal," February 2008, MER_EM_6_2008.pdf, MER Project Library, p. 34.

45. Biesiadecki, Leger, and Maimone, "Tradeoffs between Directed and Autonomous Driving"; Callas, Banerdt, and Squyres, "Mars Exploration Rover Fifth Extended Mission Proposal."

46. Robyn D. Diegan to David B. Lavery, 9 March 2007, "Re: Letter Revision E to JPL Task Plan No. 69-8371 entitled 'Mars Exploration Rovers (MER) Project: Phase E, Attachment A,'" MER_EM4.pdf, MER Project Library.

47. James Graf, interview with author, 12 January 2009.

48. Alfred McEwen, "HiRISE Report, MRO PSG #6," 15 September 2003, PSG-9-14-2003-HiRISE.ppt, Zurek materials, HMEC.

49. Alfred McEwen, "HiRISE Issues and Science Results, MRO PSG #13," 11 April 2006, McEwen.ppt, Zurek Materials, HMEC.

50. "The Opportunity Rover at 'Victoria Crater,'" 6 October 2006, http://hiroc.lpl.arizona.edu/images/TRA/TRA_000873_1780/opportunity.html.

51. Ehlmann et al., "Orbital Identification of Carbonate-Bearing Rocks on Mars."

52. Holt et al., "Radar Sounding Evidence for Buried Glaciers."

53. Krasnopolsky, Maillard, and Owen, "Detection of Methane in the Martian Atmosphere"; Formisano et al., "Detection of Methane in the Atmosphere of Mars."

54. M. J. Mumma et al., "A Sensitive Search for Methane on Mars," Division for Planetary Sciences of the American Astronomical Society (DPS) 35th Meeting, Monterey, CA, 1–6 September 2003, Mars Atmosphere II Poster session, http://aas.org/archives/BAAS/v35n4/dps2003/388.htm; M. J. Mumma et al., "Detection and Mapping of Methane and Water on Mars," DPS 36th Meeting, Louisville, KY, 8–12 November 2004, Mars Atmosphere I: Methane and High Altitude, http://aas.org/archives/BAAS/v36n4/dps2004/445.htm.

55. Mumma et al., "Strong Release of Methane on Mars"; Encrenaz, "Search for Methane on Mars." The most explicit criticism of the methane claims is Zahnle, Freedman, and Catling, "Is There Methane on Mars?"

56. Formisano et al., "Detection of Methane in the Atmosphere of Mars."

57. Daniel J. McCleese, ed., "Mars Exploration Strategy, 2009–2020," n.d. http://mepag.jpl.nasa.gov/reports/3336_Mars_Exp_Strat.pdf (accessed 31 May 2011), p. 12.

58. NASA FY 2006 Budget Estimates, www.nasa.gov/about/budget/FY_2006/index .html (accessed 1 June 2011), p. SAE 2-20; NASA FY 2007 Budget Estimates, www.nasa.gov /about/budget/FY_2007/index.html (accessed 1 June 2011), pp. SAE SMD 2-5.

59. Garvin interview, 17 August 2010.

60. NASA FY 2012 Budget Estimates, www.nasa.gov/about/budget/FY_2012/index .html (accessed 25 January 2012), p. P-2; NASA FY 2010 Budget Estimates, www.nasa.gov /about/budget/FY_2010/index.html (accessed 25 January 2012), p. SCI-86.

61. Callas, Squyres, and Banerdt, "Mars Exploration Rover Sixth Extended Mission Proposal," pp. 13–23.

62. Callas interview.

63. Katherine Sanderson, "Mars Rovers under Strain from Cuts," *Nature*, 25 March 2008 (online ed.), doi:10.1038/news.2008.681.

64. Eric Hand, "NASA Science Chief Resigns," *Nature*, 26 March 2008, doi:10.1038 /news.2008.694.

65. Andrew Lawler, "NASA's Stern Quits over Mars Exploration Plans," *Science* 320 (4 April 2008): 31.

66. Fuk K. Li to Doug McCuistion, "Re: Plans for Extended Missions for Fiscal Year 2009," 16 October 2009, MER_EM_6_2008.pdf; Katrina H. Christian to David B. Lavery, "Re: Letter revision G to JPL Task Plan No. 69-8371 entitled 'Mars Exploration Rovers Project: Phase E,'" 31 March 2009, 69-8371G.pdf, both in MER Project Library, JPL.

67. *Free Spirit*, http://www.jpl.nasa.gov/freespirit/free-spirit-archive.cfm.

68. "Spirit May Have Begun Months-Long Hibernation," 31 March 2010, JPL news release 2010-106, www.jpl.nasa.gov/news/news.php?release=2010-106. A comprehensive set of updates on the effort to revive Spirit is archived at www.nasa.gov/mission_pages /mer/spirit-update.html.

69. "NASA Mars Rove Arrives at New Site on Martian Surface," 10 August 2011, JPL news release 2011-248, www.jpl.nasa.gov/news/news.php?release=2011-248.

70. R. Bonnefoy, chair, "Beagle 2 ESA/UK Commission of Inquiry," 5 April 2004; "Lessons Learnt from Beagle 2 and Plans to Implement Recommendations from the Commission of Inquiry," 24 May 2004, both at www.esa.int/ESA_in_your_country/Ireland/Lessons _learnt_from_Beagle_2_and_plans_to_implement_recommendations_from_the_Com mission_of_Inquiry.

Chapter 12 • Reengineering a Spacecraft, and a Program

1. Barry Goldstein, "Phoenix Termination Review," 25 January 2007, TR Final 1-25-07.ppt, Goldstein materials, HMEC, p. 62.

2. Loren Zumwalt, interview with author, 13 October 2011.

3. In "real year" dollars. Peter Smith, "Phoenix: 'Follow the Water,' Mars Scout Step 2 Concept Study Report," 15 May 2003, Phoenix_Proposal.pdf, Goldstein materials, HMEC, p. J-7; Phoenix Termination Review, 26 January 2007, TR Final 1-25-07.ppt, HMEC, p. 59.

4. Phoenix Quarterly Review, 4 May 2004, Phoenix Quarterly Presentation—2004-05. ppt, Mars Phoenix Project Library, JPL; Barry Goldstein, interview with author, 3 May 2007.

5. Robert Shotwell, interview with author, 23 June 2011.

6. Tim Priser and Rob Grover, "E01—Implementation and Overview, EDL Critical Design Review," 14 November 2005, edl_cdr_01_overview.ppt, Phoenix Project Library.

7. Gary Parks interview with author, 1 September 2011.

8. Phoenix Telecom Review Team, "Phoenix Project Telecom Subsystem Inheritance Review Out Brief," 10–12 October 2004, Out Brief PHX Telecom IR.ppt, Phoenix Project Library, p. 9.

9. Parks interview.

10. Barry Goldstein, Glenn Knosp, and Genji Arakaki, "Phoenix Quarterly Presentation," 3 May 2005, Phoenix Quarterly Presentation-2005-05.ppt, Phoenix Project Library; Wayne Lee, interview with author, 13 June 2011.

11. Gary Parks and Edward Sedivy, "G—Flight System, Phoenix Critical Design Review," 14–16 November 2005, G-Flight System_as_presented.ppt; Barry Goldstein, Glenn Knosp, and Genji Arakaki, "Phoenix Project Status Report," 1 November 2005, Phoenix Quarterly Presentation-2005-11.ppt; C. May and L. Zumwalt, "Flight System Thermal Analysis, Thermal Delta Critical Design Review," 29 November 2005, Delta-CDR_Thermal_Final.ppt, pp. 82–93, all in Phoenix Project Library.

12. Loren Zumwalt, interview with author, 13 October 2011; Shotwell interview.

13. Goldstein interview; Parks interview.

14. Goldstein, Knosp, and Arakaki, "Phoenix Quarterly Presentation," 3 May 2005, 2 August 2005, both in Phoenix Project Library.

15. Robert Shotwell, interview with author, 22 June 2011.

16. Shotwell interview, 23 June 2011; Dave Spencer, Glenn Knosp, and Genji Arakakai, "Phoenix Project Status Report," 9 January 2007, Phoenix PSR—1007-01.ppt, Phoenix Project Library.

17. C. Jones, "Mars Polar Lander Red Team Final Report," 23 November 1999, EDS D-1945.

18. Dara Sabahi, interview with author, 1 December 2011; Dara Sabahi, "Phoenix IRT Radar Review," 1 December 2006, Phoenix_Radar_IRT_Overview_120106.ppt, Phoenix Project Library.

19. Sabahi interview.

20. Sabahi, "Phoenix IRT Radar Review."

21. Scott Shaffer, interview with author, December 5, 2011.

22. Sabahi, "Phoenix IRT Radar Review."

23. Barry Goldstein, Glenn Knosp, and Genji Arakaki, "Phoenix Quarterly Report," 1 November 2006, Phoenix Quarterly Presentation—2006-11.ppt, Phoenix Project Library.

24. Sabahi, "Phoenix IRT Radar Review."

25. Fawwaz Ulaby, Chair, Independent Review Team, to Fuk Li, 12 December 2006, IRT Final Report.doc, Phoenix Project Library.

26. Chronological radar activity chart, Radar_Bubble_Chart_070517.ppt, 17 May 2007, courtesy Sabahi.

27. Goldstein, "Phoenix Termination Review," p. 59.

28. Ibid., p. 71.

29. The project estimated termination costs of $40.1 million but would save an estimated $36.5 million from cancelling the launch vehicle contract. Thus, termination would cost less than completing the mission but would not be "free." Ibid., pp. 73–74.

30. Barry Goldstein, interview with author, 17 November 2011.

31. SJS [Scott Shaffer], "Characterization and Open Issues, Phoenix IRT Review," PHX_Radar_IRT_openissues_20070517.pdf, 17 May 2007, Phoenix Project Library.

32. Barry Goldstein, "Project Status, Phoenix Project Flight Readiness Review," 27 June 2007, B—Project Status.ppt, Phoenix Project Library.

33. Parks interview; Malin Space Science Systems, "Mars Descent Imager (MARDI) Update," 12 November 2007, www.msss.com/msl/mardi/news/12Nov07/index.html (accessed 29 December 2011).

34. "NASA Mission to asteroid belt rescheduled for September Launch," 7 July 2007, news release 2007-075, www.jpl.nasa.gov/news/news.cfm?release=2007-075.

35. Barry Goldstein, Glenn Knosp, and Helenann Kwong-Fu, "Phoenix Project Status Report," 2 October 2007, Phoenix PSR—2007-10.ppt, Phoenix Project Library.

36. Peter Smith, PI, Mars Scout Step 2 Concept Study Report, "Phoenix: 'Follow the Water,'" 15 May 2003, Phoenix_Proposal.pdf, Goldstein materials, HMEC, p. E-9.

37. Arvidson et al., "Mars Exploration Program 2007 Phoenix Landing Site Selection and Characteristics."

38. Phoenix Mars Scout Mission Science Team Meeting #9, 20–21 June 2006, STM#9-Minutesv3.doc, Phoenix Science Team Meeting Minutes, HMEC; Phoenix Mars Scout Mission, Science Team Meeting #10, 16–17 October 2006, STM10MeetingMinutesv4.doc, Phoenix Science Team Meeting Minutes, HMEC, courtesy Deborah Bass.

39. Ray Arvidson, interview with author, 5 January 2012.

40. Golombek et al., "Size-Frequency Distributions of Rocks"; "THEMIS Helps Phoenix Land Safely on Mars," n.d., http://themis.asu.edu/news/themis-helps-phoenix-land-safely-mars; Ray Arvidson, interview with author, 5 January 2012.

41. Ray Arvidson et al., "Phoenix Landing Site Certification Review: Update on Hazard Maps," 27 March 2008, 2 Arvidson.pdf, Phoenix Project Library; Arvidson et al., "Mars Exploration Program 2007 Phoenix Landing Site Selection and Characteristics"; Golombek et al., "Size-Frequency Distributions of Rocks."

42. Arvidson et al., "Phoenix Landing Site Certification Review."

43. Tim Gasparrini, "Phoenix Final Touchdown Analysis, Site Certification Review," 27 March 2008, 3_Gasparrini_Rev1.pdf, Phoenix Project Library.

44. Joel Krajewski, "Surface Overview, Surface Phase Critical Events Readiness Review," 8 April 2008, D—Surface Overview.ppt, Phoenix Project Library.

45. "NASA Mars Lander Prepares to Move Arm," 27 May 2008, http://phoenix.lpl.arizona.edu/05_27_pr.php.

46. "Descent of the Phoenix Lander (PSP_008579_9020)," 27 May 2008, www.uahirise.org/phoenix-descent.php (accessed 3 January 2011).

47. Erik Baily et al., "Phoenix EDL Reconstruction," 7 October 2008, EDS D-66509.

48. James K. Erickson, interview with author, 10 January 2012.

49. "NASA Mars Lander Prepares to Move Arm."

50. "NASA's Phoenix Lander Robotic Arm Camera Sees Possible Ice," 30 May 2008, http://phoenix.lpl.arizona.edu/05_30_pr.php.

51. "Hard Substrate, Possibly Ice, Uncovered Under the Mars Lander," 31 May 2008, www.nasa.gov/mission_pages/phoenix/images/press/20080531.html.

52. "NASA'S Phoenix Mars Lander Checking Soil Properties," 7 June 2008, http://phoenix.lpl.arizona.edu/06_07_pr.php; "Phoenix Sifts for Samples, Continues Imaging

Landing Site," 8 June 2008, http://phoenix.lpl.arizona.edu/06_08_pr.php; "NASA's Phoenix Mars Lander Testing Sprinkle Technique," 9 June 2008, http://phoenix.lpl.arizona.edu/06_09_pr.php.

53. "NASA's Phoenix Lander Has An Oven Full Of Martian Soil," 11 June 2008, http://phoenix.lpl.arizona.edu/06_11_pr.php; "Phoenix Analyst's Notebook," Sol Summaries, http://an.rsl.wustl.edu/phx/solbrowser/default.aspx (accessed 9 May 2012).

54. "Bright Chunks at Phoenix Lander's Mars Site Must Have Been Ice," 19 June 2008, http://phoenix.lpl.arizona.edu/06_19_pr.php.

55. "Phoenix Returns Treasure Trove For Science," 26 June 2008, http://phoenix.lpl.arizona.edu/06_26_pr.php.

56. Leslie Tamppari, interview with author, 5 August 2011.

57. Arvidson interview, 6 January 2012.

58. Kessler, *Martian Summer*, 213–39; "NASA Spacecraft Confirms Martian Water, Mission Extended," 31 July 2008, http://phoenix.lpl.arizona.edu/07_31_pr.php; Kenneth Chang, "Test of Mars Soil Sample Confirms Presence of Ice," *New York Times*, 1 August 2008; Smith et al., "H2O at the Phoenix Landing Site."

59. Craig Covault, "White House Briefed on Potential for Mars Life," *Aviation Week and Space Technology*, 1 August 2008.

60. Covault clarified his own story three days later; see Covault, "Phoenix Data More Negative on Potential for Life," *Aviation Week and Space Technology*, 4 August 2011.

61. Andrea Thompson, "Scientist Says Buzz over Mars Life Is 'Bogus,'" *Space.com*, 4 August 2008, www.msnbc.msn.com/id/26012037/ns/technology_and_science-space/t/scientist-says-buzz-over-mars-life-bogus/ (accessed 29 December 2011); "Phoenix Mars Team Opens Window on Scientific Process," 5 August 2011, http://phoenix.lpl.arizona.edu/08_05_pr.php; Kessler, *Martian Summer*, 262–72.

62. Arvidson et al., "Results from the Mars Phoenix Lander Robotic Arm Experiment."

63. "Mars Phoenix Lander Finishes Successful Work on Red Planet," 10 November 2008, news release 08-284, www.nasa.gov/mission_pages/phoenix/news/phoenix-20081110.html; "NASA Finishes Listening for Phoenix Mars Lander," 1 December 2008, http://phoenix.lpl.arizona.edu/12_01_pr.php.

64. "Phoenix Mars Lander Is Silent, New Image Shows Damage," 24 May 2010, news release 2010-175, www.jpl.nasa.gov/news/news.cfm?release=2010-175.

65. Costs from Theisinger, "GPMC/Quarterly Presentation, Mars Exploration Rover," 6 May 2003, MER_GPMC_May_02_Final.pptx, MER Project Library, p. 69; George Pace, "Mars 2003 Lander Mission Concept Review," 3 May 2000, EDS D-19429, p. 22; Goldstein, "Phoenix Termination Review," p. 61.

66. Tommaso P. Rivellini, "Development Testing of the Mars Pathfinder Inflatable Landing System," presented at ASCE Meeting, Albuquerque, NM, 1–6 June 1996; Douglas S. Adams, Mars Exploration Rover Airbag Landing Loads Testing and Analysis," presented at the 45th AIAA Structures, Structural Dynamics, and Materials Conference, Palm Springs, CA, 19 April 2004.

67. Dara Sabahi, "Chronological Radar Activity Chart, May 2007 IRT," 17 May 2007, Radar_Bubble Chart_070517.ppt, Phoenix Project Library.

68. Hecht et al., "Detection of Perchlorate"; Boynton et al., "Evidence for Calcium Carbonate."

69. Arvidson interview, 6 January 2011.

70. Tamppari interview.

71. Committee on the Planetary Science Decadal Survey, *Vision and Voyages*.

72. Review of U.S. Human Spaceflight Plans Committee—Final Report, October 2009, www.nasa.gov/offices/hsf/meetings/10_22_pressconference.html, p. 84.

73. NASA FY 2011 Budget Estimates, www.nasa.gov/news/budget/2011.html, costs from p. iv.

74. NASA Office of the Inspector General, Office of Audits, "NASA's Challenges to Meeting Cost, Schedule, and Performance Goals," 27 September 2012, report no. IG-12-021, http://oig.nasa.gov/audits/reports/FY12/IG-12-021.pdf, p. 28.

75. NASA FY 2013 Budget Estimates, www.nasa.gov/news/budget/2013.html#.U5C QoBaXJW0, p. SC-3.

76. John Casani, chair, "James Webb Space Telescope Independent Comprehensive Review Panel Final Report," 29 October 2010, www.nasa.gov/pdf/499224main_JWST -ICRP_Report-FINAL.pdf.

77. NASA FY 2013 Budget Estimates, p. PS-1.

78. NASA Office of the Inspector General, Office of Audits, "NASA's Management of the Mars Science Laboratory Project," 8 June 2011, report no. IG-11-019, www.hq.nasa .gov/office/oig/hq/audits/reports/FY11/IG-11-019.pdf.

79. Doug McCuistion, "MEPAG," presented to MEPAG #20, Rosslyn, VA, 3–4 March 2009, http://mepag.jpl.nasa.gov/meeting/2009-03/02_MEPAG_McCuistion_Mar_09 .pdf.

80. Yudhijit Bhattacharjee, "Scientists Decry Cuts That Would Doom ExoMars Missions," *Science* 355 (24 February 2012): 900, doi:10.1126/science.335.6071.900.

81. David Des Marais, chair, to Jim Bell, 5 March 2012, http://mepag.jpl.nasa.gov /meeting/feb-12/MEPAG_mtg_2728Feb2012.pdf.

82. The NASA FY 2013 Operating Plan, released in August 2013, restored about $10 million of the $226 million cut. See FY 2013 Operating Plan for Public Law 113-6 Appropriations, 1 August 2013, www.nasa.gov/sites/default/files/632709main_NASA_FY13_Budget -Science-Planetary-508.pdf.

Conclusion

1. J. D. Velasco, "NASA Administrator Visits JPL in Wake of Budget Proposal That Cuts Mars Exploration," *Pasadena Star-News*, 21 February 2012, www.pasadenastarnews .com/ci_20025017; AVC-2012-038, JPLA.

2. Mars Program Planning Group, "Summary of the Final Report," 25 September 2012, www.nasa.gov/pdf/691580main_MPPG-Integrated-v13i-Summary%20Report-9-25-12.pdf.

3. U.S. Exobiology Program, *Exobiological Strategy for Mars Exploration*.

4. McCurdy, *Faster, Better, Cheaper*, 152–56.

5. NASA Center Directors Meeting at JPL, 16 August 2011, AVC-2011-167, JPLA.

6. Brian K. Muirhead, interview with author, 10 August 2011.

7. Notes and Handouts from the Mars Science Working Group Meeting, Pasadena, CA, 7–8 February 1994, EDS D-14394, p. 8.

8. Steve Jolly, interview with author, 24 July 2007.

9. The key issue is their adequacy as simplifications of the real vehicles. A classic discussion of this for scientific modeling is Cartwright, *How the Laws of Physics Lie.*

10. Pyne, "Seeking Newer Worlds" and *Voyager.*

11. Pyne, *Voyager,* 305.

12. Launius, "Public Opinion Polls and Perceptions of U.S. Human Spaceflight."

13. Jeffrey M. Jones, "Majority of Americans Say Space Program Costs Justified," 17 July 2009, www.gallup.com/poll/121736/majority-americans-say-space-program-costs-justified.aspx.

Epilogue

1. Kenneth Chang, "Curiosity Rover Lands Safely on Mars," *New York Times,* 6 August 2012.

2. Kenneth Chang, "Hope of Methane on Mars Fades," *New York Times,* 3 November 2012; Marc Kaufman, "Mars Rover Finds Ancient Streambed—Proof of Flowing Water," *National Geographic,* 27 March 2012.

3. "New NASA Mission to Take First Look Deep Inside Mars," 20 August 2012, http://insight.jpl.nasa.gov/newsdisplay.cfm?Subsite_News_ID=31164&SiteID=8 (accessed 3 December 2012).

4. J. F. Mustard et al., "Report of the Mars 2020 Science Definition Team," 1 July 2013, http://mepag.jpl.nasa.gov/reports/MEP/Mars_202_DST_Report_Final.pdf (accessed 30 July 2013).

5. G. Scott Hubbard, "Mars 2020 Is No Redo," 28 January 2014, www.planetary.org/blogs/guest-blogs/2014/0128-mars-2020-is-no-redo.html.

6. "Mars One Selects Lockheed Martin to Study First Private Unmanned Mission to Mars," 10 December 2013, Lockheed Martin news release, www.lockheedmartin.com/us/news/press-releases/2013/december/1210-ss-marsone.html.

7. Mark Carreau, "Restless Americans Look to Mars One for New Home," *Aviation Week,* 2 January 2014.

8. "Mars One's Indiegogo Campaign Reached $313,744, or 78%, of Its Goal by the Close of 9 February 2014," www.indiegogo.com/projects/mars-one-first-private-mars-mission-in-2018.

9. Dennis A. Tito et al., "Feasibility Analysis for a Manned Mars Free-Return Mission in 2018," IEEE Aerospace Conference, Big Sky, MT, 2–9 March 2013; Brian Vastag, "Dennis Tito's Mission to Mars: Launching in 2018 for the Children (and to Beat China)," *Washington Post,* 27 February 2013; Frank Morring Jr., "Lawmakers Skeptical of 2021 Human Mars Flyby Idea," *Aviation Week,* 27 February 2014.

10. Ken Kremer, "Elon Musk Premiers SpaceX Manned Dragon V2 Astronaut Transporter—First Photos," *Universe Today,* 29 May 2014, www.universetoday.com/112213/elon-musk-premiers-spacex-manned-dragon-v2-astronaut-transporter-1st-photos/.

11. Chris Anderson, "Elon Musk's Mission to Mars," *Wired,* 21 October 2012, www.wired.com/2012/10/ff-elon-musk-qa/all/; Tariq Malin, "Elon Musk says SpaceX Making 'Progress' toward Mars Colony," *Space.com,* 19 May 2014, www.space.com/25934-elon-musk-mars-colony-spacex-rockets.html; Sebastian Anthony, "SpaceX Says It Will Put Humans on Mars by 2026, Almost 10 years Ahead of NASA," *ExtremeTech,* 18 June 2014, www.extremetech.com/extreme/184640-spacex-says-it-will-put-humans-on-mars-by-2026-almost-10-years-ahead-of-nasa.

12. Alex Knapp, "SpaceX Billionaire Elon Musk on the Business and Future of Space Travel," *Forbes,* 23 April 2012, www.forbes.com/sites/alexknapp/2012/04/23/spacexs-elon-musk-on-the-business-and-future-of-space-travel/.

13. Nye, *America as Second Creation,* 289.

14. This conflict between science and exploration is one of the subtexts of Kim Stanley Robinson's magnificent Mars trilogy: *Red Mars, Green Mars,* and *Blue Mars.*

BIBLIOGRAPHY

Archival Sources

Jet Propulsion Laboratory, Pasadena, California
- Engineering Document Services (EDS)
- Historian's Mars Exploration Collection (HMEC)
- Jet Propulsion Laboratory Archives (JPLA)
- JPL Oral Histories
- Mars Exploration Rover (MER) Project Library
- Mars Phoenix Project Library
- Mars Reconnaissance Orbiter (MRO) Project Library

NASA History Office, NASA Headquarters, Washington, DC
- NASA Historical Reference Collection

NASA Scientific and Technical Information Program, www.sti.nasa.gov
- NASA Technical Reports Server (NTRS)

National Archives and Records Administration, College Park, Maryland
- Papers of Daniel S. Goldin, RG 255, Records of the National Aeronautics and Space Administration.

Published Sources

Acuna, M. H., J. E. P. Connerney, P. Wasilewski, R. P. Lin, K. A. Anderson, C. W. Carlson, J. McFadden, et al. "Magnetic Field and Plasma Observations at Mars: Initial Results of the Mars Global Surveyor Mission." *Science* 279, no. 5357 (13 March 1998): 1676–80. doi:10.1126/science.279.5357.1676.

Arvidson, R., D. Adams, G. Bonfiglio, P. Christensen, S. Cull, M. Golombek, J. Guinn, et al. "Mars Exploration Program 2007 Phoenix Landing Site Selection and Characteristics." *Journal of Geophysical Research—Planets* 113 (19 June 2008). doi:10.1029/2007JE003021.

Arvidson, R. E., R. G. Bonitz, M. L. Robinson, J. L. Carsten, R. A. Volpe, A. Trebi-Ollennu, M. T. Mellon, et al. "Results from the Mars Phoenix Lander Robotic Arm Experiment." *Journal of Geophysical Research—Planets* 114 (2 October 2009). doi:10.1029/2009JE003408.

Bergreen, Laurence. *Voyage to Mars: NASA's Search for Life beyond Earth.* New York: Riverhead Books, 2000.

Biesiadecki, J. J., E. T. Baumgartner, R. G. Bonitz, B. K. Cooper, F. R. Hartman, P. C. Leger, M. W. Maimone, et al. "Mars Exploration Rover Surface Operations: Driving Opportunity at Meridiani Planum." *IEEE Robotics and Automation Magazine* 13, no. 2 (2006): 63–71. doi:10.1109/MRA.2006.1638017.

Biesiadecki, J. J., P. C. Leger, and M. W. Maimone, "Tradeoffs between Directed and Autonomous Driving on the Mars Exploration Rovers," *International Journal of Robotics Research* 26, no. 1 (2007): 91–104, doi:10.1177/0278364907073777.

Boston, P. J., M. V. Ivanov, and C. P. McKay. "On the Possibility of Chemosynthetic Ecosystems in Subsurface Habitats." *Icarus* 95, no. 2 (February 1992): 300–308. doi: 10.1016/0019-1035(92)90045-9.

Boynton, W. V., W. C. Feldman, S. W. Squyres, T. H. Prettyman, J. Bruckner, L. G. Evans, R. C. Reedy, et al. "Distribution of Hydrogen in the Near Surface of Mars: Evidence for Subsurface Ice Deposits." *Science* 297, no. 5578 (5 July 2002): 81–85. doi:10.1126 /science.1073722.

Boynton, W. V., D. W. Ming, S. P. Kounaves, S. M. M. Young, R. E. Arvidson, M. H. Hecht, J. Hoffman, et al. "Evidence for Calcium Carbonate at the Mars Phoenix Landing Site." *Science* 325, no. 5936 (3 July 2009): 61–64. doi:10.1126/science.1172768.

Bromberg, Joan Lisa. *NASA and the Space Industry*. Baltimore: Johns Hopkins University Press, 1999.

Carr, Michael H. *Water on Mars*. New York: Oxford University Press, 1996.

Cartwright, Nancy. *How the Laws of Physics Lie*. New York: Oxford University Press, 1983.

Christensen, P. R., B. Jakosky, H. H. Kieffer, M. C. Malin, H. Y. McSween, K. Nealson, G. L. Mehall, et al. "The Thermal Emission Imaging System (THEMIS) for the Mars 2001 Odyssey Mission." *Space Science Reviews* 110, no. 1–2 (2004): 85–130. doi:10.1023/B: SPAC.0000021008.16305.94.

Committee on Planetary and Lunar Exploration. *Assessment of Mars Science and Mission Priorities*. Washington, DC: National Academies Press, 2003.

———. *Strategy for Exploration of the Inner Planets: 1977–1987*. Washington, DC: National Academy of Sciences, 1978.

Committee on the Planetary Science Decadal Survey. *Vision and Voyages for Planetary Science in the Decade 2013–2022*. Washington, DC: National Academies Press, 2011.

Committee to Review the Next Decade Mars Architecture. *Assessment of NASA's Mars Architecture, 2007–2016*. Washington, DC: National Academies Press, 2006.

Conway, Erik. *Atmospheric Science at NASA: A History*. Baltimore: Johns Hopkins University Press, 2008.

Dick, Steven J. *The Biological Universe: The Twentieth Century Extraterrestrial Life Debate and the Limits of Science*. New York: Cambridge University Press, 1996.

Dick, Steven J., and James E. Strick. *The Living Universe: NASA and the Development of Astrobiology*. New Brunswick, NJ: Rutgers University Press, 2004.

Edelson, B. I., and J. L. McLucas. "United States and Soviet Planetary Exploration—the Next Step Is Mars, Together." *Space Policy* 4, no. 4 (1988): 337–49.

Ehlmann, B. L., J. F. Mustard, S. L. Murchie, F. Poulet, J. L. Bishop, A. J. Brown, W. M. Calvin, et al. "Orbital Identification of Carbonate-Bearing Rocks on Mars." *Science* 322, no. 5909 (19 December 2008): 1828–32. doi:10.1126/science.1164759.

Encrenaz, Therese. "Search for Methane on Mars: Observations, Interpretation and Future Work." *Advances in Space Research* 42, no. 1 (1 July 2008): 1–5. doi:10.1016/j.asr.2007.01.069.

Ezell, Edward Clinton, and Linda Neuman Ezell. *On Mars: Exploration of the Red Planet, 1958–1978*. SP-4212. Washington, DC: NASA, 1984.

Farmer, C. B., and P. E. Doms. "Global Seasonal Variation of Water Vapor on Mars and the Implications for Permafrost." *Journal of Geophysical Research—Solid Earth* 84, no. B6 (1979): 2881–88. doi:10.1029/JB084iB06p02881.

Formisano, V., S. Atreya, T. Encrenaz, N. Ignatiev, and M. Giuranna. "Detection of Methane in the Atmosphere of Mars." *Science* 306, no. 5702 (3 December 2004): 1758–61. doi:10.1126/science.1101732.

Golombek, M. P., R. C. Anderson, J. R. Barnes, J. F. Bell, N. T. Bridges, D. T. Britt, J. Bruckner, et al. "Overview of the Mars Pathfinder Mission: Launch through Landing, Surface Operations, Data Sets, and Science Results." *Journal of Geophysical Research—Planets* 104, no. E4 (25 April 1999): 8523–53. doi:10.1029/98JE02554.

Golombek, M. P., R. A. Cook, T. Economou, W. M. Folkner, A. F. C. Haldemann, P. H. Kallemeyn, J. M. Knudsen, et al. "Overview of the Mars Pathfinder Mission and Assessment of Landing Site Predictions." *Science* 278, no. 5344 (5 December 1997): 1743–48. doi:10.1126/science.278.5344.1743.

Golombek, M. P., A. Huertas, J. Marlow, B. McGrane, C. Klein, M. Martinez, R. E. Arvidson, et al. "Size-Frequency Distributions of Rocks on the Northern Plains of Mars with Special Reference to Phoenix Landing Surfaces." *Journal of Geophysical Research—Planets* 113 (15 July 2008). doi:10.1029/2007JE003065.

Golombek, M. P., H. J. Moore, A. F. C. Haldemann, T. J. Parker, and J. T. Schofield. "Assessment of Mars Pathfinder Landing Site Predictions." *Journal of Geophysical Research—Planets* 104, no. E4 (25 April 1999): 8585–94. doi:10.1029/1998JE900015.

Hall, R. Cargill. *Lunar Impact: A History of Project Ranger*. SP-4210. Washington, DC: NASA, 1977.

Hartmann, William K. and Odell Raper. *The New Mars: The Discoveries of Mariner 9*. Washington, DC, NASA, 1974.

Hecht, M. H., S. P. Kounaves, R. C. Quinn, S. J. West, S. M. M. Young, D. W. Ming, D. C. Catling, et al. "Detection of Perchlorate and the Soluble Chemistry of Martian Soil at the Phoenix Lander Site." *Science* 325, no. 5936 (3 July 2009): 64–67. doi:10.1126/science.1172466.

Hogan, Thor. *Mars Wars: The Rise and Fall of the Space Exploration Initiative*. SP-2007-4410. Washington, DC: NASA, August 2007.

Holt, J. W., A. Safaeinili, J. J. Plaut, J. W. Head, R. J. Phillips, R. Seu, S. D. Kempf, et al. "Radar Sounding Evidence for Buried Glaciers in the Southern Mid-Latitudes of Mars." *Science* 322, no. 5905 (21 November 2008): 1235–38. doi:10.1126/science.1164246.

Horowitz, Norman H. *To Utopia and Back: The Search for Life in the Solar System*. New York: W. H. Freeman, 1986.

Hubbard, G. S., F. M. Naderi, and J. B. Garvin. "Following the Water: The New Program for Mars Exploration." *Acta Astronautica* 51, no. 1–9 (2002): 337–50.

Hubbard, Scott. *Exploring Mars: Chronicles from a Decade of Discovery*. Tucson: University of Arizona Press, 2011.

Huntress, Wesley T., Jr, and Mikhail Ya. Marov, *Soviet Robots in the Solar System: Mission Technologies and Discoveries*. Chichester, UK: Springer Praxis, 2011.

Johnson, A., R. Willson, J. Goguen, J. Alexander, and D. Meller. "Field Testing of the Mars Exploration Rover's Descent Image Motion Estimation System." *Proceedings of the 2005 IEEE International Conference on Robotics and Automation* (April 2005): 4463–69. doi:10.1109/ROBOT.2005.1570807.

Kass, D. M., J. T. Schofield, T. I. Michaels, S. C. R. Rafkin, M. I. Richardson, and A. D. Toigo. "Analysis of Atmospheric Mesoscale Models for Entry, Descent, and Landing." *Journal of Geophysical Research—Planets* 108, no. E12 (1 December 2003): 8090. doi:10.1029/2003JE002065.

Kessler, Andrew. *Martian Summer: Robot Arms, Cowboy Spacemen, and My 90 Days with the Phoenix Mars Mission*. NY: Pegasus Books, 2011.

Koppes, Clayton R. *JPL and the American Space Program: A History of the Jet Propulsion Laboratory*. New Haven, CT: Yale University Press, 1982.

Krasnopolsky, V. A., J. P. Maillard, and T. C. Owen. "Detection of Methane in the Martian Atmosphere: Evidence for Life?" *Icarus* 172, no. 2 (December 2004): 537–47. doi:10.1016/j.icarus.2004.07.004.

Launius, Roger D. "Perceptions of Apollo: Myth, Nostalgia, Memory, or All of the Above." *Space Policy* 21 (2005): 129–39.

———. "Public Opinion Polls and Perceptions of U.S. Human Spaceflight." *Space Policy* 19 (2003): 163–75.

Launius, Roger D., and Howard McCurdy. *Robots in Space: Technology, Evolution, and Interplanetary Travel*. Baltimore: Johns Hopkins University Press, 2008.

Limerick, Patricia Nelson. "Imagined Frontiers: Westward Expansion and the Future of the Space Program." In *Space Policy Alternatives*, ed. Radford Byerly. Boulder, CO: Westview Press, 1992.

Malin, M. C., G. E. Danielson, M. A. Ravine, and T. A. Soulanille. "Design and Development of the Mars Observer Camera." *International Journal of Imaging Systems and Technology* 3, no. 2 (1991): 76–91.

Markley, Robert. *Dying Planet: Mars in Science and the Imagination*. Durham, NC: Duke University Press, 2005.

Matijevic, J. R., J. Crisp, D. B. Bickler, R. S. Banes, B. K. Cooper, H. J. Eisen, J. Gensler, et al. "Characterization of the Martian Surface Deposits by the Mars Pathfinder Rover, Sojourner." *Science* 278, no. 5344 (5 December 1997): 1765–68.

McCurdy, Howard E. *Faster, Better, Cheaper: Low Cost Innovation in the U.S. Space Program*. Baltimore: Johns Hopkins University Press, 2001.

McDougall, Walter. *. . . the Heavens and the Earth*. New York: Basic Books, 1985.

McEwen, A. S., E. M. Eliason, J. W. Bergstrom, N. T. Bridges, C. J. Hansen, W. A. Delamere, J. A. Grant, et al. "Mars Reconnaissance Orbiter's High Resolution Imaging Science Experiment (HiRISE)." *Journal of Geophysical Research—Planets* 112, no. E5 (17 May 2007). doi:10.1029/2005JE002605.

Mindell, David A. *Digital Apollo: Human and Machine in Spaceflight*. Cambridge, MA: MIT Press, 2008.

Ming, D. W., R. V. Morris, R. Woida, B. Sutter, H. V. Lauer, C. Shinohara, D. C. Golden, et al. "Mars 2007 Phoenix Scout Mission Organic Free Blank: Method to Distinguish

Mars Organics from Terrestrial Organics." *Journal of Geophysical Research—Planets* 113 (29 October 2008). doi:10.1029/2007JE003061.

Mishkin, Andrew. *Sojourner: An Insider's View of the Mars Pathfinder Mission.* New York: Berkeley Books, 2003.

Morton, Oliver. *Mapping Mars: Science, Imagination, and the Birth of a World.* New York: Picador, 2002.

Muirhead, Brian K. *High Velocity Leadership.* New York: HarperCollins, 1999.

Mumma, M. J., G. L. Villanueva, R. E. Novak, T. Hewagama, B. P. Bonev, M. A. DiSanti, A. M. Mandell, and M. D. Smith. "Strong Release of Methane on Mars in Northern Summer 2003." *Science* 323, no. 5917 (20 February 2009): 1041–45. doi:10.1126/science.1165243.

Murchie, S. L., F. P. Seelos, C. D. Hash, D. C. Humm, E. Malaret, J. A. McGovern, T. H. Choo, et al. "Compact Reconnaissance Imaging Spectrometer for Mars Investigation and Data Set from the Mars Reconnaissance Orbiter's Primary Science Phase." *Journal of Geophysical Research—Planets* 114 (1 October 2009). doi:10.1029/2009JE003344.

Murray, Bruce C. *Journey into Space: The First Thirty Years of Planetary Exploration.* New York: W. W. Norton, 1989.

National Commission on Space. *Pioneering the Space Frontier.* New York: Bantam Books, 1986.

Nye, David. *America as Second Creation: Technology and Narratives of New Beginnings.* Cambridge, MA: MIT Press, 2003.

Portree, David S. *Humans to Mars: Fifty Years of Mission Planning, 1950–2000.* SP-2001-4521. Washington, DC: NASA, 2001.

Pyne, Stephen J. "Seeking Newer Worlds: An Historical Context for Space Exploration." In *Critical Issues in the History of Space Flight,* ed. Steven J. Dick and Roger D. Launius, 7–35. Washington, DC: NASA, 2006.

———. *Voyager: Seeking Newer Worlds in the Third Great Age of Discovery.* New York: Viking, 2010.

Ride, Sally K. *Leadership and America's Future in Space: A Report to the Administrator.* Washington, DC: NASA, August 1987.

Rieder, R., T. Economou, H. Wanke, A. Turkevich, J. Crisp, J. Bruckner, G. Dreibus, and H. Y. McSween. "The Chemical Composition of Martian Soil and Rocks Returned by the Mobile Alpha Proton X-ray Spectrometer: Preliminary Results from the X-ray Mode." *Science* 278, no. 5344 (5 December 1997): 1771–74. doi:10.1126/science.278.5344.1771.

Roy, Stephanie A. "The Origin of the Smaller, Faster, Cheaper Approach in NASA's Solar System Exploration Program." *Space Policy* 14, no. 3 (August 1998): 153–71. doi:10.1016/S0265-9646(98)00021-6.

Sawyer, Kathy. *The Rock from Mars: A Detective Story on Two Planets.* New York: Random House, 2006.

Shirley, Donna. *Managing Martians.* NY: Broadway Books, 1998.

Smith, P. H., J. F. Bell, N. T. Bridges, D. T. Britt, L. Gaddis, R. Greeley, H. U. Keller, et al. "Results from the Mars Pathfinder Camera." *Science* 278, no. 5344 (5 December 1997): 1758–65. doi:10.1126/science.278.5344.1758.

Smith, P. H., L. K. Tamppari, R. E. Arvidson, D. Bass, D. Blaney, W. V. Boynton, A. Carswell, et al. "H2O at the Phoenix Landing Site." *Science* 325, no. 5936 (3 July 2009): 58–61. doi:10.1126/science.1172339.

Smith, P. H., L. Tamppari, R. E. Arvidson, D. Bass, D. Blaney, W. Boynton, A. Carswell, et al. "Introduction to Special Section on the Phoenix Mission: Landing Site Characterization Experiments, Mission Overviews, and Expected Science." *Journal of Geophysical Research: Planets* 113, no. E3 (2008). doi:10.1029/2008JE003083.

Snyder, Amy Paige. "NASA and Planetary Exploration." In *Exploring the Unknown, Selected Documents in the History of the U.S. Civil Space Program, vol. 5: Exploring the Cosmos*, 263–300. SP-2001-4407. Washington, DC: NASA, 2001.

Squyres, Steven. *Roving Mars: Spirit, Opportunity, and the Exploration of the Red Planet.* New York: Hyperion Books, 2005.

Squyres, S. W., R. E. Arvidson, E. T. Baumgartner, J. F. Bell, P. R. Christensen, S. Gorevan, K. E. Herkenhoff et al. "Athena Mars Rover Science Investigation." *Journal of Geophysical Research: Planets* 108, no. E12 (2003). doi:10.1029/2004JE002121.

Squyres, S. W., R. E. Arvidson, J. F. Bell III, J. Brückner, N. A. Cabrol, W. Calvin, M. H. Carr, P. R. Christensen, et al. "The Opportunity Rover's Athena Science Investigation at Meridiani Planum, Mars." *Science* 306, no. 5702 (3 December 2004): 1698–1703. doi:10.1126/science.1106171.

U.S. Exobiology Office. *An Exobiological Strategy for Mars Exploration.* SP-530. Washington, DC: NASA, April 1995.

Vaughn, Jennifer. "Mars Pathfinder Landing Inspires Thousands Celebrating the Adventure of Exploration." *Planetary Report* 17, no. 6 (November–December 1997): 4–9.

Vertesi, Janet Amelia. "Seeing like a Rover: Images in Interaction on the Mars Exploration Rover Mission." Ph.D. diss., Cornell University, 2009.

Westwick, Peter. *Into the Black: JPL and the American Space Program, 1976–2004.* New Haven, CT: Yale University Press, 2007.

Wilford, John Noble. *Mars Beckons: The Mysteries, the Challenges, the Expectations of Our Next Great Adventure in Space.* New York: Alfred A. Knopf, 1990.

Wright, I. P., M. R. Sims, and C. T. Pillinger, "Scientific Objectives of the Beagle 2 Lander." *Acta Astronautica* 52 (2003): 219–25.

Zahnle, K., R. S. Freedman, and D. C. Catling. "Is There Methane on Mars?" *Icarus* 212, no. 2 (April 2011): 493–503. doi:10.1016/j.icarus.2010.11.027.

Zubrin, Robert. "The Significance of the Martian Frontier," *Ad Astra* (September–October 1994). www.nss.org/settlement/mars/zubrin-frontier.html.

Zubrin, Robert, David A. Baker, and Owen Gwynne. "Mars Direct: A Simple, Robust, and Cost Effective Architecture for the Space Exploration Initiative." In *The Case for Mars IV: The International Exploration of Mars*, ed. Thomas R. Meyer, 275–314. San Diego: American Astronautics Society, 1997.

Zurek, R. W., and S. Smrekar. "An Overview of the Mars Reconnaissance Orbiter Science Mission." *Journal of Geophysical Research* 112 (2007). doi:10.1029/2006E002701.

INDEX

Page numbers in *italic* indicate figures and tables.